BIOLOGY OF
DOMESTIC ANIMALS

BIOLOGY OF DOMESTIC ANIMALS

Editors

Colin G. Scanes

University of Wisconsin
Department of Biological Science
Milwaukee, WI
USA

Rodney A. Hill

School of Biomedical Sciences
Charles Sturt University
Wagga Wagga, New South Wales
Australia

CRC Press
Taylor & Francis Group
Boca Raton London New York

CRC Press is an imprint of the
Taylor & Francis Group, an **informa** business

A SCIENCE PUBLISHERS BOOK

Cover Illustrations
The cattle figure was sourced from the editor's personal figures collection. Reproduced with the permission of the editor, Prof. R.A. Hill.

The lower five panels are from:
Acharya, S. and Hill, R.A. (2014). High efficacy gold-KDEL peptide-siRNA nanoconstruct-mediated transfection in C2C12 myoblasts and myotubes. Nanomedicine: Nanotechnology, Biology and Medicine. 10:329-337. Reproduced with permission from the publisher and the author.

CRC Press
Taylor & Francis Group
6000 Broken Sound Parkway NW, Suite 300
Boca Raton, FL 33487-2742

First issued in paperback 2021

© 2018 by Taylor & Francis Group, LLC
CRC Press is an imprint of Taylor & Francis Group, an Informa business

No claim to original U.S. Government works

Version Date: 20170609

ISBN-13: 978-0-367-78201-6 (pbk)
ISBN-13: 978-1-4987-4785-1 (hbk)

Library of Congress Cataloging-in-Publication Data

Names: Scanes, C. G., editor. | Hill, Rodney A., editor.
Title: Biology of domestic animals / editors, Colin G. Scanes, University of
Wisconsin, Department of Biological Science, Milwaukee, WI, USA, Rodney A.
Hill, School of Biomedical Science, Charles Sturt University, Wagga Wagga,
New South Wales, Australia.
Description: Boca Raton, FL : CRC Press, 2017. | "A science publisher's
book." | Includes bibliographical references and index.
Identifiers: LCCN 2017000416| ISBN 9781498747851 (hardback : alk. paper) |
ISBN 9781498747875 (e-book : alk. paper)
Subjects: LCSH: Domestic animals. | Veterinary medicine.
Classification: LCC SF745 .B56 2017 | DDC 636.089--dc23
LC record available at https://lccn.loc.gov/2017000416

Visit the Taylor & Francis Web site at
http://www.taylorandfrancis.com

and the CRC Press Web site at
http://www.crcpress.com

Preface

It is January 2017. Estimates of the rate of growth of global knowledge posit a doubling in less than one year, with an exponentially increasing rate. With such dynamic changes, approaches to the sharing of knowledge are being challenged. The "cloud" now provides unprecedented access and the internet connects us to the unimaginably immense knowledge pool. Knowledge about domestic animals, is a tiny drop in this pool. And yet there is so much knowledge out there, even the greatest minds in the field are daunted in finding ways to package that knowledge into useful chunks. Our contribution to packaging a snap-shot of knowledge of domestic animals has brought together a group of contributing authors from across the globe, and together we have focused on bringing to the readers exciting new discoveries and consolidation of knowledge across our collective expertise.

Knowledge from the level of population, whole animal, systems, tissues, organs, cells, molecules, and from many disciplines from nutrition to physiology to genetics, and the impact of this knowledge on human biology in health and disease, on animals as human companions and on animals as part of our food production systems has been brought together. Our intent has been to provide stimulating reading for anyone with an interest in domestic animals, for scientists, and for students. This drop of knowledge brings perspective and cohesion that also provides the reader and indeed the contributors with opportunities to join some "dots" from the collection and to synthesize some new knowledge and to further advance the field.

Rodney A. Hill

Contents

Introduction

Rodney A. Hill

||||||||||||||||||||||||||||||||||||||||||||||||||||||||||||||||||||||||||||||||||||||||||||||||||||||||||||||||||||||||||||||||||||||||||||||||||||||||||||||||||||||||

Reader Guide to Scope

This book was conceived with the aim of providing a broad readership with information written by expert authors, with comprehensive coverage across the field of domestic animal biology. An important notion has been to provide information over reader-friendly themes, presented within sections.

The Section on domestic animals as comparative models to humans, provides some insights into the ways in which scientists better understand human health and disease through studying similar processes in animals. Horses have a long association with humans across most of the world, and have been bred as athletes, for work, as companions and in some cultures, for food. Ken McKeever provides great insights into a deep historically framed examination of what we have learned and how we have benefited from human association with horses.

One of the most distressing and impactful maladies, not only in affluent nations but also across many rapidly developing nations is obesity. The pig has been domesticated and has co-evolved and undergone intensive selection across widely diverse societies and cultures. Kola Ajuwon has brought focus on several examples of pigs that have either evolved the propensity to be obese, or have been deliberately selected as such. These lines (breeds) have provided science with an excellent tool to better understand the causes, underlying biology and to develop effective treatments for obesity and associated pathologies.

Chapters 3 and 4 explore animal models at a more micro level. In Chapter 3, Steve Harvey and Carlos Martinez-Moreno provide insights into eye development, structure and function using the chick model. Many animal models have been used to study eye development, and profound insights have been facilitated by study of the chick model; the utility of this model in developmental biology being effectively articulated by these experts in the field. Chapter 4, drills down to the sub-micro and to the molecular level in elucidating mechanisms of cellular secretion through the porosome. Bhanu Jena is a scientific pioneer who is the driving intellectual force in providing insights into this profoundly complex system. A combination of animal models, molecular data from cultured cells and molecular modeling have been effectively articulated through these studies. Atomic force microscopy (AFM) has been the tool of choice in generating many of the insights into this mechanism of secretion.

Section B brings the reader to topics addressing animal growth and metabolic efficiency. The costs associated with raising domestic animals has recently become a stronger interest for scientists, especially during and following the Global Fiscal Crisis (GFC) that began around 2007. The topics covered in this section address the science of growth and efficiency, which has rapidly advanced due to world-wide investment by governments and industries interested in reducing the cost of raising domestic animals. Larry Reynolds, Alison Ward and Joel Caton are experts in epigenetics; factors that are programmed into animals during early development in utero, and can affect gene expression over multiple generations. And Chapter 5 highlights knowledge in this field and provides insights of the profound multi-generational effects of management and genetic selection of animals from the earliest developmental stages of their existence. In Chapter 6, Claire Fitzsimons, Mark McGee, Kate Keogh, Sinead Waters and David Kenny, provide insights into the underlying molecular physiology and genetics of feed efficiency in beef cattle. Beef cattle and ruminants in general, provide a major source of high quality protein to humans world-wide. Controversy around the production of methane produced by ruminants has driven interest in their effects on total green-house gas production. An element lost in this argument is that ruminants have roamed the planes and savannahs of the world for centuries, and have provided a source of protein to humans by harvesting grasses on lands that do not have sufficient quality to grow agricultural crops. Thus they turn poor quality grasses into highly nutritious protein that has globally supported human development. Chapter 6 explores inter-animal responses in appetite regulation, digestion, host-ruminal microbiome interactions, nutrient partitioning, myocyte and adipocyte metabolism, metabolic processes, including mitochondrial function, as well as other physiological processes, and provides deep insights into the drivers of variation in feed efficiency in beef cattle. Chapter 7, contributed by the editors provides a comprehensive overview of metabolic processes that contribute to variation in growth and efficiency across a wide range of domestic animal species.

Section C focuses upon animal reproduction and we have highlighted a single chapter focusing upon poultry reproduction. In Chapter 8,

Murray Bakst, provides the reader with specific insights that integrate an accumulation of knowledge over 40 years and emphasizes the latest discoveries to the present day.

Animal welfare is one of the most important considerations in any context across domestic animal biology. In Section D, we have dedicated two chapters to this topic. Chapter 9, by Wendy Rauw and Gomez Raya, considers animal stress, in the context of the historical development of animal production from initial domestication through to high technology-based industrial-level animal production that provides the efficiency needed for a sustainable industry to support a rapidly growing world population. Chapter 10, contributed by Colin Scanes, Yvonne Vizzier-Thaxton and Karen Christensen, the authors probe deeply into a cross-species examination of the physiological drivers and molecular mechanisms that underpin stress reactions in domestic animals.

Section E is entitled Future Directions. Although any or all of the preceding chapters include cutting edge science and have elements that address the science of the future within their context, we have focused upon two topics that will rely on new and evolving technologies, and have exciting dimensions that will deeply benefit from the amazing innovations in scientific technology that are just around the corner. In Chapter 11, Eric Wong, Elizabeth Gilbert and Katarzyna Miska explore Nutrient Transporter Gene Expression in multiple species. These important mechanisms are at the core of our understanding of the drivers of nutrient uptake. The fundamental relationship between nutrient absorption and the regulation of these transporters is intimate. One of the keys to improving animal efficiency and reducing the environmental foot-print of animal production will be based in matching nutritional requirements (animal needs) and dietary intake, and manipulation of nutrient transporters will provide the fine-tuning needed for this optimization so that each animal is primed to reach its full genetic potential. Our final chapter, by Yajun Wang focuses upon science that has relied upon one of the many domestic animal genomes that have been sequenced through the first and second decades of the twenty-first century. This review of bioactive peptides, reveals that many of these newly discovered molecules mainly act as hormones to regulate many vital physiological processes, such as growth, development, metabolism, energy balance, stress, reproduction and immunity.

The Complexity of Biological Concepts, Scientific Interpretation and some Consequences

Scientists are trained to build a healthy skepticism and we often vary in interpretation of data. We should point out that for the lay reader, scientists' understanding of biological concepts is framed by their specialist training and experience. For example, a nutritionist has a very different perspective from a physiologist or a quantitative geneticist.

In order to more fully understand the bigger picture of the biology of domestic animals, it is important for scientists, students and lay-persons to appreciate the value of specialist knowledge from many different scientific disciplines that is needed and the value of exchange of their ideas amongst disciplines. This is a key element in shaping our interpretations and perspectives. Scientists who are highly trained in one discipline have to rely on those with expertise in completely different disciplines to contribute as studies are designed, the data collected and interpreted and new knowledge discovered. The apparent complexity of the concepts becomes greater as scientists with deep knowledge within their discipline probe their understanding at finer and deeper levels of biological detail. This is an established scientific approach to discovery. Scientists interpret the data and evolve their own perspectives from their discipline focus. This can sometimes lead to quite different and even opposing views. It is this complexity and variation in interpretation that drives scientific debate and is essential in the scientific discovery process. Unfortunately, an unintended consequence is that different messages from different scientists can lead to confusion for others such as commentators and even animal industry, producers and lay people.

The Role of New Technologies

The technologies that inform deepen and broaden the disciplines noted above are advancing very rapidly and the power of analysis of new genomics and molecular physiology data is impressive. Within the next few years, we will have substantial new knowledge and technical capacity that will allow us to link molecular physiology and genomics data to data generated from phenotype testing of animals, which will result in greater accuracy of estimates of genetic potential as well as improving our understanding of the genetic and physiological drivers of variation in animal performance. In fact, science is moving already in linking these elements together.

The Opportunity

Many scientists across multiple disciplines are making discoveries and finding ways forward. There is great industry awareness of the opportunities and progressive thinkers are out there implementing and adopting new technologies to improve animal welfare, to better adapt animals to roles as companions, to improve the efficiency of animal production and to improve profitability of businesses that rely upon combinations of these. This book brings together many of the aspects of the sciences that underpin our understanding of the biology of domestic animals, and provides a ready reference and source for students, scholars, technical experts, members of animal production industries, commentators and lay-persons. It is my hope that the text will also stimulate discussion in barns, coffee shops, laboratories and classrooms that lead to further insights to improve our understanding of the biology of domestic animals.

References

Ajuwon, K.M. 2017. The pig model for the study of obesity and associated metabolic diseases. (Chapter 2). *In*: C.G. Scanes and R.A. Hill (eds.). The Biology of Domestic Animals. CRC Press. Boca Raton.

Bakst, M.R. 2017. Reproduction in poultry: An overview. (Chapter 8). *In*: C.G. Scanes and R.A. Hill (eds.). The Biology of Domestic Animals. CRC Press. Boca Raton.

Fitzsimons, C., M. McGee, K. Keogh, S.M. Waters, and D.A. Kenny. 2017. Molecular Physiology of feed efficiency in beef cattle. (Chapter 6). *In*: C.G. Scanes and R.A. Hill (eds.). The Biology of Domestic Animals. CRC Press. Boca Raton.

Harvey, S., and C.G. Martinez-Moreno. 2017. Growth hormone and the chick eye (Chapter 3). *In*: C.G. Scanes and R.A. Hill (eds.). The Biology of Domestic Animals. CRC Press. Boca Raton.

Jena, B.P. 2017. Porosome enables the establishment of fusion pore at its base and the consequent kiss-and-run mechanism of secretion from cells. (Chapter 4). *In*: C.G. Scanes and R.A. Hill (eds.). The Biology of Domestic Animals. CRC Press. Boca Raton.

McKeever, K.H. 2017. Equine exercise physiology: A historical perspective (Chapter 1). *In*: C.G. Scanes and R.A. Hill (eds.). The Biology of Domestic Animals. CRC Press. Boca Raton.

Rauw, W.M., and L.G. Raya. 2017. Stress in livestock animals: From domestication to factory production. (Chapter 9). *In*: C.G. Scanes and R.A. Hill (eds.). The Biology of Domestic Animals. CRC Press. Boca Raton.

Reynolds, L.P., A.K. Ward, and J.S. Caton. 2017. Epigenetics and developmental programming in ruminants—long-term impacts on growth and development. (Chapter 5). *In*: C.G. Scanes and R.A. Hill (eds.). The Biology of Domestic Animals. CRC Press. Boca Raton.

Scanes, C.G., and R.A. Hill. 2017. Hormonal control of energy substrate utilization and energy metabolism in domestic animals. (Chapter 7). *In*: C.G. Scanes and R.A. Hill (eds.). The Biology of Domestic Animals. CRC Press. Boca Raton.

Scanes, C.G., Y. Vizzier-Thaxton, and K. Christensen. 2017. Biology of Stress in livestock and poultry (Chapter 10). *In*: C.G. Scanes and R.A. Hill (eds.). The Biology of Domestic Animals. CRC Press. Boca Raton.

Wang, Y. 2017. Novel peptides in poultry: A case study of the expanding glucagon peptide superfamily in chickens (*Gallus gallus*). (Chapter 12). *In*: C.G. Scanes and R.A. Hill (eds.). The Biology of Domestic Animals. CRC Press. Boca Raton.

Wong, E.A., E.R. Gilbert, and K.R. Miska11. 2017. Nutrient transporter gene expression in poultry, livestock and fish. (Chapter 11). *In*: C.G. Scanes and R.A. Hill (eds.). The Biology of Domestic Animals. CRC Press. Boca Raton.

Section A

Domestic Animals as Comparative Models to Humans

CHAPTER-1

Equine Exercise Physiology
A Historical Perspective

Kenneth H. McKeever

||||||||||||||||||||||||||||||||||||||||||||||||||||||||||||||||||||||||||||||||||||||||||||||||||||||||||||||||||||||||||||||||||||||||||||||||||||||||||

INTRODUCTION

One only has to look at the cave paintings in Lascaux, France to know that humans and horses have had a close relationship since thousands of years. No one knows when the first person bravely mounted the back of a horse and rode fast and furiously across the land. We do know that the horse has been the work animal that plowed fields, pulled wagons, and transported people from village to village, fought in wars and served as an inspiration to poets. The natural athletic prowess of the horse has taken it from a work animal to an athlete with a wide range of competitive endeavors. Many of those activities have their origins in the work performed by the horse. Racing, driving, pulling competitions, reining, dressage, etc. all have histories as varied as those who have owned these magnificent animals. The goal of the present chapter is to present an overview of the study of the athletic horse from ancient times to the present with a focus on the use of the horse as an animal model in classic physiological and medical experiments and later in experiments designed to elucidate what makes the horse a great athlete. All of these are historical and scientific basis for the modern field of Equine Exercise Physiology.

Understanding the anatomy and physiology of the horse no doubt evolved with the human horse relationship. A greater knowledge of feeding for work, care of the foot, treatment of ailments and injuries all have been necessary for economic survival and triumphant use of the horse

Equine Science Center, Department of Animal Science, Rutgers the State University of New Jersey, New Brunswick, NJ 08901.
Email: mckeever@aesop.rutgers.edu

agricultural work and for battle. Simon of Athens was one of the earliest to write about horse in the mid-400's BC. Latter works by Xenonphon expanded the written record on horse selection and horsemanship with his treatise "The Art of Horsemanship" which was primarily guide for cavalry commanders (McCabe, 2007). The philosopher Aristotle wrote extensively about nature, biology, and medical sciences and is credited with the first published treatise on equine locomotion (van Weeren, 2001). The Roman physician Galenus, conducted many medical experiments using animals including horses in what has been called the founding of comparative medicine (van Weeren, 2001). The science of locomotion and gait analysis finds its beginnings in the early observations, experiments, and publications published between the 13th and 16th centuries. Of prominence are the contributions of Jordanus Ruffus, Leonardo da Vinci, Carlo Ruini, Giovanni Borelli, William Cavendish, and De Solleysel (van Weeren, 2001). The dawn of the modern age of locomotion science begins with the landmark experiments of Eadweard Muybridge and Jules Marey who were the first to use technology to advance the science of movement and locomotion (Muybridge, 1887; Silverman, 1996).

The Horse as an Animal Model for Physiological Experiments

In the 1700's and 1800's, the horse was vital to everyday life and was being used for agriculture, for transportation of people and goods, and for war. Knowledge of horse selection, husbandry, foot care, veterinary care, and insight into the athletic prowess of horses was essential knowledge for success in the pre-industrial age economy. In 1761, the first veterinary school was established in France with other countries following soon with the establishment their own institutions (van Weeren 2001). The biology of the horse soon transitioned to experimental physiology with the horse at the forefront as a subject for investigations. For example, in 1733 the Reverend Stephen Hales used the horse in his studies of blood pressure and hemorrhagic shock (Hoff et al., 1965; Fregin and Thomas, 1983). Those experiments were the first to measure blood pressure using cannulation of arteries, a technique that can be done in the horse because it has large vessels and it can be trained to stand quietly while making measurements. Blood pressure is routinely measured in modern equine exercise physiology studies that focus on the control of cardiovascular function. Another early study, performed in England, was published in 1794 by Sir John Hunter. In his paper "A Treatise on the Blood, Inflammation, and Gunshot Wounds" he was the first to describe the effects of excessive exercise on the clotting system of the horse. In this classic article he noted that blood failed to clot in horses that had been run to death. Hunter did not know about the clotting cascade and its many important factors; however, his research was important for equine exercise physiologists demonstrated that intense exercise affects multiple parts of the clotting cascade, making the blood hypercoagulable during exercise to prevent bleeding. Post-exercise it was observed that fibrinolytic activity was enhanced, most likely due to an increase in the clot busting substance tissue plasminogen activating factor (tPa). Studies from other species have demonstrated that tPa increases during intense exercise, and during periods of dehydration. Recent investigations of the horse have focused on the role alterations in clotting mechanisms play in exercise-induced pulmonary hemorrhage, a condition that affects 70–90% of race horses. Of relevance to the present timeline is the fact that the observations published by Hunter marked a point in time where the horse became a key animal for many classic studies of physiology and medicine.

The 1800's saw the horse being used for a large number of important discoveries, in particular studies that advanced our understanding of the cardiovascular system. Heart sounds were studied in various equid species. Experiments that used donkeys were performed by James Hope in 1835 gave input into the timing of the cardiac cycle and heart sounds (Hope, 1835; McCrady et al., 1966; Fregin and Thomas, 1983). A number of years later Jean-Baptiste Auguste Chauveau and J. Faivre used the horse to build upon the work of Hope linking the events of the cardiac cycle with internal changes in the heart (Chauveau and Faivre, 1856; Fregin and Thomas, 1983). The horse was used as an animal model when Claude Bernard, the "Father of Modern Physiology" and his colleague Francoise Magendie performed the first cardiac catheterization (Fregin and Thomas, 1983). In 1861 Chauveau and Etienne-Jules Marey (Figure 1) went on to make the first measurement of ventricular blood pressure using a horse that had been catheterized (Geddes et al., 1965). Those of us who routinely measure central cardiovascular pressure dynamics during exercise owe them a note of thanks for this valuable technique. In more functional studies, Marey was the first to demonstrate the relationship between blood pressure and heart rate using the horse as his animal model (Fregin and Thomas, 1983). The concept, known "Marey's Law" states that there is a reflex increase heart rate when blood pressure decreases and a decrease in heart rate when blood pressure goes up. This was an important discovery and researchers have built on it to understand the integration of various neuroendocrine mechanisms in the regulation of blood pressure. These are important concepts for an understanding cardiovascular dynamics during exercise and for understanding mechanisms behind the dysregulation that leads to hypertension and other cardiovascular disease processes. It is important to note, that the Marey reflex depends on the interplay between the sympathetic and parasympathetic influences on heart rate. When one looks at various species, autonomic control of heart rate in the dog is parasympathetically driven and in the monkey the sympathetic system

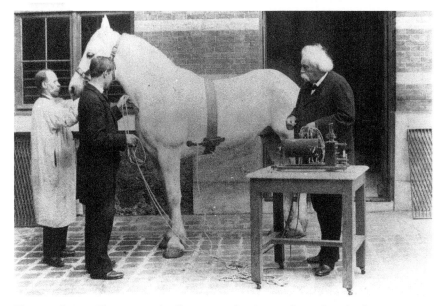

Figure 1: August Chauveau and colleagues performing cardiac catheterization to measure intracardiac pressure in the horse. Downloaded July 25, 2016. http://physiologie.envt.fr/spip/IMG/jpg/chauveau_et_ses_sondes.jpg.

predominates. Horses and humans rely on a balance and interplay between both sides of the autonomic nervous system, thus, the choice of the horse by Marey was fortuitous for those applying their discoveries from animal models to humans.

Initial Exercise Physiology Studies Utilizing the Horse as an Animal Model

It was a natural progression from these landmark studies of the cardiovascular system functions at rest to experiments designed to test hypotheses about the effects of the challenge of exercise. In 1887, Chauveau and Kaufmann measured muscle blood flow in horses while they chewed on hay (Tomás et al., 2002). There experiment not only demonstrated changes in muscle blood flow during exercise (i.e., chewing), it also demonstrated that the concentration of glucose went down in the venous blood returning flow from the masseter muscle, an observation that has been characterized by Tomas as the "first to demonstrate that glucose uptake is enhanced" when exercise is performed by a muscle (Tomás et al., 2002).

A turning point in field of exercise physiology in general took place in the laboratory of Nathan Zuntz, the German Physiologist who many refer to as the "Father of Aviation Physiology" (Gunga, 2009). Zuntz published many

Figure 2: Portable equine calorimeter used in field studies (from Derlitzki and Huxdorf, 1930).

classic papers on respiratory and cardiovascular physiology, studies that were made possible by his development and use of innovative equipment such as the Zuntz-Geppert respiratory apparatus (Gunga, 2009). This portable indirect calorimeter allowed for the measurement of gas exchange and during exercise in the field or under controlled conditions on the treadmill ("Zuntz-Lehnann'sche Laufband") he fabricated in 1889 (Gunga, 2009). His many experiments on the physiological effects of high altitude earned him the distinction of being called the "Father of Aviation Physiology" (Gunga, 2009). However, more germane to the present article were the studies that Zuntz and his colleagues conducted using the exercising horse. Zuntz and his colleague Kaufmann conducted on the effects of exercise on blood pressure (Fregin and Thomas 1983). Zuntz and Kaufmann also conducted the first experiment to demonstrate that cardiac output increases with exercise. Those classic experiments performed on a treadmill used the Fick Principle to determine cardiac output (Gunga, 2009). As aside note the horse used in the experiment was named Babylon. Zuntz appears to have had some competition in what may have been the beginning of the field of equine exercise physiology. At the end of the century Smith in England published studies that focused on the horse per se rather than using the horse as an animal model. Smith published several papers of note including his studies of respiration, sweat composition, and muscle function (Smith, 1890a,b, 1896).

The 1900's saw the continued use of the horse in physiological experiments of importance to current day exercise physiologists. In 1922, Van Slyke used the horse blood to develop his calculations regarding the oxyhemoglobin dissociation curve (Roughton 1965). Energetics and work

Figure 3: Calorimeter and treadmill used by Proctor and Brody to measure oxygen consumption in horses during exercise (with University of Missouri Agricultural Experiment Station).

physiology were important areas of research and in 1933 Werner Huxdorff used a portable calorimeter to assess the energy efficiency of draft horses (Huxdorff, 1933a,b). Huxdorff (1933a and 1933b) compared the efficiency of horses and tractors of that time and found that horses were more efficient with 28 percent of potential energy being transduce into energy for work.

Robert Proctor and Samuel Brody later studied the energetic requirements of the horse using a treadmill and calorimeter (Proctor et al., 1935). Brody later published hundreds of papers on the energetics of a wide variety

Figure 4: Field study being conducted at the Ohio State University in the late 1970's.

of species and his work established the relationship between, body mass, body surface area and the energetics (Brody, 1945).

The Pre-Treadmill Era of Research

Between 1940 and the late 1960's the papers published on equine exercise physiology were very limited. Most were field investigations and descriptive clinical papers related to the care of the athletic horse (Marlin, 2015). A key factor limiting the ability to collect meaningful data in field studies at that time was the lack of instrumentation and methods to measure physiological function during the actual exercise bout of exercise. Blood samples and measurements of physiological function could be made before and after a bout of exercise, but the ability to make measurements during the actual period perturbation of exercise was near to impossible (Figure 4).

One cannot totally discount the studies conducted in the late 1950's and 1960's. Many of those studies focused on the hematological alterations associated with performance (Abildgaard et al., 1965; Verter et al., 1966; Rotenberg and Czerniak, 1967; Soliman and Nadim, 1967; Dusek, 1967), EKG monitoring (Holmes and Alps, 1966; Bassan and Ott, 1967; Glendinning,

1969), and other indirect measures that could be used to assess the equine athlete (Cardinet et al., 1963a,b; Archer and Clabby, 1965; Carlson et al., 1965; Marsland, 1968). Still, researchers recognized the limitations of information from field trials that were fraught at that time with lack of instrumentation, control problems related to the ability to exercise intensity, running surface, influence of the environment, and other sources of experimental variation.

The High Speed Treadmill Era

The turning point in the field of Equine Exercise Physiology came with dissertation research of Dr. Sune Persson where he examined the role of blood volume plays in the aerobic and exercise capacity of the athletic horse (Persson, 1967). Horses could run at maximal speed on the high speed treadmill and measurements could be made during exercise using exacting protocols and controlled environmental conditions. Persson coupled his treadmill work with measurements made on hundreds of horses in the field to give the amazing first insight into the relationship between blood volume and in particular red cell volume, and the ability of the horse to perform exercise (Persson, 1967). His dissertation and other papers from his initial work examined the relationship between the spleen's extra 12 liters of red cell rich blood and the phenomenal cardiac outputs and rates of oxygen consumption in the athletic horse (Persson, 1967; Persson et al., 1973a,b; Persson and Lydin, 1973; Persson and Bergsten, 1975). Persson was joined in Uppsala by a team of researchers including Arne Lindhom, Birgitta Essén-Gustavsson, Peter Kallings, Stig Drevemo and others (Essén-Gustavsson et al., 2013). The group at the Swedish Agricultural University set the standard for the developing field of equine exercise physiology with papers ranging in topics from cardiopulmonary physiology, exercise biochemistry and muscle function, pharmacology and locomotion. Sune Persson is fondly known as the Father of Modern Equine Exercise Physiology and his ground breaking body of scholarly work is an inspiration for all who follow in his footsteps.

Equine exercise physiology research grew in the in the 1970s and early 1980s paralleling the boom in human exercise physiology research (Marlin, 2015). Many of these research teams started conducting research in the field, but the real revolution came about through the availability of commercially built high speed treadmills and other equipment that allowed for the development of true exercise physiology laboratories (Figure 5). In Sydney, Australia Rueben Rose, David Hodgson, and David Evans conducted classic studies on exercise biochemistry, fluid and electrolyte balance, and cardiovascular function to name a few (Marlin, 2015). David Snow at The Animal Health Trust in Newmarket, UK conducted groundbreaking studies on muscle metabolism and biochemistry with Roger Harris and others (Marlin, 2015). In Switzerland, Hans Hoppler and his colleagues established

Figure 5: Standardbred horse running on the high speed treadmill in the Equine Exercise Physiology Laboratory at the Rutgers Equine Science Center.

a very productive laboratory focused on cardiopulmonary function. Hans Meyer in Germany and Harold Hintz at Cornell University were leaders in experiments focused on the nutritional requirements of the exercising horse.

Many of these studies were presented at the first International Conference on Equine Exercise Physiology (ICEEP). If one examines the table of contents of the proceedings of the first ICEEP, one can get a sense of the groups conducting research at that time and the areas they were exploring (Snow, 1983). Most of the studies presented at the meeting held in Oxford were conducted before the availability of the commercially produced high speed treadmill. Papers presented by Gillespie (UC Davis, USA); Robinson and Derksen (Michigan State University, USA), Attenburrow (University of Exeter), Hornike (Germany), Littlejohn (South Africa) were focused on the respiratory system. Cardiovascular papers were presented by Fregin and Thomas (Cornell, USA); Manohar and Parks (University of Illinois, USA); Physick-Sheard (University of Guelph; Canada); Stewart and Rose (University of Sydney). Key papers on muscle physiology were presented by Snow (Glasgow), Lindholm and Essén-Gustavsson (Sweden), Hans Hoppler (Switzerland), Thornton (Queensland, Australia) and Hodgson (Sydney, Australia). Locomotion and gait studies were presented by Pratt (MIT); Goodship (University of Bristol) and Gunn (Scotland). Nutrition of the equine athlete was discussed by Hintz (Cornell, USA). Thermoregulation was discussed by Carlson (UC Davis, USA) and Lindner (Germany). Papers on hematology were presented by Snow (Glasgow); Persson (Sweden); Allen (Animal Health Trust, UK); and Bayly (Washington State University).

The analysis of fitness and state of training category included papers from Persson (Sweden) Thornton and Wilson (Queensland, Australia), Rose (Sydney, Australia) and Gabel (the Ohio State University). The final category of papers presented in at the first ICEEP was those related to drugs and performance. Authors included Tobin (Kentucky, USA); Gabel (OSU); and Kallings (Sweden). The first ICEEP meeting served as a meeting of the minds and a forum for those focused on the physiology of the athletic horse. Subsequent ICEEP meetings have been held every four years around the world (ICEEP2—San Diego, USA; ICEEP3—Uppsala, Sweden; ICEEP4—Brisbane, Australia; ICEEP5 Utsunomiya, Japan, ICEEP6—Lexington, USA; ICEEP collaboration of researchers.

Following the 1st ICEEP there was a boom in the establishment of Equine Exercise Physiology laboratories in the United States that paralleled what was seen being in the rest of the world (Marlin, 2015). In the early 1980s, Howard Seeherman built one of the first high speed treadmill laboratories in the United States at Tufts University. Soon after that Howard Erickson and Jerry Gillespie established the lab at Kansas State University (Erickson, Gillespie) and Jim Jones, Gary Carlson, and others established one at UC Davis in California. At Washington State University, Warwick Bayly, Phil Gollnick, David Hodgson, Barry Grant and others built a treadmill laboratory where they conducted classic studies of respiration, muscle function, exercise biochemistry, thermoregulation, and applied exercise physiology. In 1987 Ken McKeever was brought in to establish the laboratory at the Ohio State University. He was joined there by Ken Hinchcliff, a veterinarian and PhD student who would later join the faculty as a clinician and researcher. Other active research labs established in this period were built at Michigan State University by Ed Robinson, at Cal Poly Pomona, by Steve Wickler, and at the University of Pennsylvania by Larry Soma, at Rutgers University by Ken McKeever (Figure 5), at Virginia Tech by David Kronfeld, and at the University of Florida, by Pat Conahan. The late 1990's saw a large number of laboratories established in veterinary colleges around the world.

There has been a huge increase in equine exercise related publications in the scientific literature in the last 60 years and a recent analysis by Marlin (2015) has demonstrated that the rate of publications in the field of Equine Exercise Physiology has steadily grown since the 1960's from a few papers each year to over 200 in each of the ICEEP cycles with the exception of the last ICEEP cycle where the organizers decided to forgo a full refeed proceedings. One cannot do justice to the number of excellent papers published around the world over the years. But one can see have the field has grown as has the breadth of areas of exploration with a range of topics that now extends from the descriptive to experiments focused on cellular and molecular mechanisms. Topics presented in the literature have also come full circle with the horse being used as an animal model for applied physiology research. For example, at Rutgers the United States army has funded several projects using the horse to test food extracts that had been shown to reduce the inflammation in a cell culture model (Figure 6). The goal of those experiments was to use the horse with its similarities to humans physiologically and its love of exercise to conduct experiments that would test the efficacy of the extracts as non-NSAID

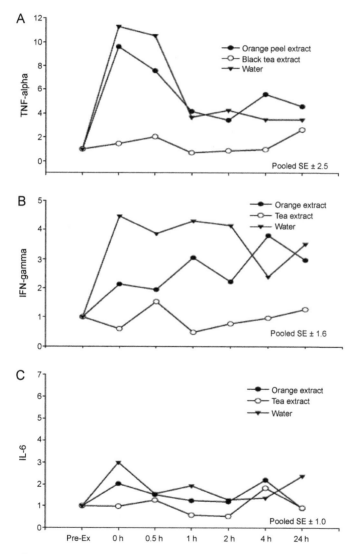

Figure 6: Effects of food extracts on selected markers of inflammation in the horse (used with permission from Streltsova et al., 2006).

compounds to reduce the inflammation caused by intense exercise (Streltsova et al., 2006; Liburt et al., 2009, 2010). Studies were also performed on the horse because one could measure concentrations of the active flavanols as well as inflammatory cytokines simultaneously in blood and muscle.

Another example of coming full circle with the use of the horse as a model for humans has been the large number of studies that focused on the effects of aging (Figure 7) and training on systems related to metabolic dysregulation and obesity (Walker et al., 2009; McKeever, 2016).

The field of equine exercise physiology has come full circle and has room for growth. The exciting advances in molecular physiology and genomics that have the potential to be coupled with traditional whole animal studies represent a phenomenal opportunity to elucidate mechanisms for many of the responses to acute exercise as well as the adaptive responses to exercise training. The challenge for future is to continue to find funding sources that value comparative, integrative, physiology that explores questions relevant both to equine health and human health.

Keywords: Horse, exercise physiology, history

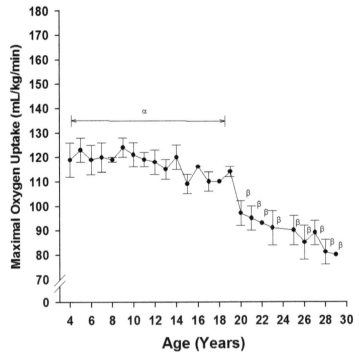

Figure 7: Effect of age on maximal aerobic capacity in the horse (used with permission from Walker et al., 2009).

References

Abildgaard, C.F., C.F. Bildgaard, and R.P. Link. 1965. Blood coagulation and hemostasis in thoroughbred horses. Proc. Soc. Exp. Biol. Med. 119: 212–215.

Archer, R.K., and J. Clabby. 1965. The effect of excitation and exertion on the circulating blood of horses. Vet. Rec. 77: 689–690.

Bassan, L., and W. Ott. 1967. Radiotelemetric studies of the heart rate in race horses at rest and in all paces (walk, trot, gallop). Arch. Exp. Veterinarmed 22: 57–75.

Brody, S. 1945. Bioenergetics and Growth. Reinhold Pub. Corp., New York.

Cardinet, G.H., M.E. Fowler, and W.S. Tyler. 1963a. Heart rates and respiratory rates for evaluating performance in horses during endurance trail ride competition. J. Am. Vet. Med. Assoc. 143: 1303–1309.

Cardinet, G.H., M.E. Fowler, and W.S. Tyler. 1963b. The effects of training, exercise, and tying-up on serum transaminase activities in the horse. Am. J. Vet. Res. 24: 980–984.

Carlson, L.A., S. Froeberg, and S. Persson. 1965. Concentration and turnover of the free fatty acids of plasma and concentration of blood glucose during exercise in horses. Acta Physiol. Scand. 63: 434–441.

Chauveau, A., C. Bertolus, and F. Laroyenne. 1860. Mémoire sur la vitesse de la circulation dans les artères du cheval (in French, Memoir on the velocity of the circulation in the horse arteries). J. Physiologie de l'Homme et Animaux 3: 695–711.

Chauveau, A., and J. Faivre. 1856. Nouvelles recherches expérimentales sur les mouvéménts et les bruits normaux du coeur envisagés au point de vue la physiologie medicale. Gaz. Med. Paris 37: 569–573.

Chauveau, A., and M. Kaufmann. 1887. Expériences pour la determination du coefficient de l'activité nutritive et respiratoire des muscles en respos et en travail. Compt. Rend. Ac. Sci. 104: 1126–1132.

Chaveau, J.B.A. and E.J. Marey. 1861. Determination graphique des rapports de la pulsation cardiaque avec les mouvements de l'ouellette et du ventricule, obtenue au moyen d'un appereil enreistreur. Gaz. Med. Paris 30: 675–678.

Derlitzki, G., and W. Huxdorf. 1930. Landarbeit wird erforscht. Die Umschau 34: 523–525.

Dusek, J. 1967. The changes in blood cell sedimentation rate in horses from the viewpoint of the work load. II. Arch. Exp. Veterinarmed. 21: 593–601.

Essén-Gustavsson, B., M. Jensen-Waern, A. Lindholm, S. Valberg, and G.P. Carlson. 2013. Curriculum vitae paper—Sune G.B. Persson (1931–2009). Comparative Exercise Physiology 9: 223–225.

Fregin, G.F., and D.P. Thomas. 1983. Cardiovascular response to exercise in the horse: A review. pp. 78–90. *In*: D.H. Snow, S.G.B. Persson, and R.J. Rose (eds.). Equine Exercise Physiology. Granata Editions, Cambridge, UK.

Geddes, L.A., J.D. McCrady, and H.E. Hoff. 1965. The contribution of the horse to knowledge of the heart and circulation. II. Cardiac catheterization and ventricular dynamics. Conn. Med. 29: 864–876.

Glendinning, S.A. 1969. The use of telemetering in the horse. Proc. Royal. Soc. Med. 62: 454.

Gunga, H.C., and K.A. Kirsch. 1995. Nathan Zuntz (1847–1920)—a German pioneer in high altitude physiology and aviation medicine, Part II: Scientific work. Aviat Space Environ. Med. 66: 172–176.

Gunga, H.C. 2009. Nathan Zuntz: His Life and Work in the Fields of High Altitude Physiology and Aviation Medicine. Academic Press, New York. 252 pp.

Hoff, H.E., and L.A. Geddes. 1975. A historical perspective on physiological monitoring: Chauveau's projecting kymograph and the projecting physiograph. Cardiovasc. Res. Cent. Bull. 14: 3–35.

Holmes, J.R., and B.J. Alps. 1966. The effect of exercise on rhythm irregularities in the horse. Vet. Rec. 78: 672–683.

Hope, J. 1835. A Treatise on the Diseases of the Heart and Great Vessels. William Kidd, London. p. 572.

Huxdorff, W. 1933a. Respirationsversuche an landwirtschaftlichen Zutieren, ein Mittel zu besseren Beurteilung und rationelleren Nutzung der tiereschen Arbeitskrafte. Zuchtungskunde 8: 6–18.

Huxdorff, W. 1933b. Die Beurteilung von Zucht- und Arbeitspferden auf Grund der Ergebnisse von Respirationsversuchen. Zuchtungskunde 8: 250–255.

Liburt, N.R., K.H. McKeever, J.M. Streltsova, W.C. Franke, M.E. Gordon, H.C. Manso, Filho, D.W. Horohov, R.T. Rosen, C.T. Ho, A.P. Singh, and N. Vorsa. 2009. Effects of cranberry and ginger on the physiological response to exercise and markers of inflammation following acute exercise in horses. Equine Comp. Exercise Physiol. 6: 157–169.

Liburt, N.R., A. Adams, A. Betancourt, D.W. Horohov, and K.H. McKeever. 2010. Exercise-induced increases in cytokine markers of inflammation in muscle and blood in horses. Equine Vet. J. 42(S38): 280–288.

Marlin, D. 2015. Has the golden age of equine exercise physiology passed and if so, have we answered all the big questions? Journal of Equine Veterinary Science 35: 354–360.

Marsland, W.P. 1968. Heart rate response to submaximal exercise in the Standardbred horse. J. Appl. Physiol. 24: 98–101.

McCabe, A.A. 2007. Byzantine Encyclopaedia of Horse Medicine: The Sources, Compilation, and transmission of the Hippiatrica. Oxford University Press, Oxford. 323 pp.

McCrady, J.D., H.E. Hoff and L.A. Geddes. 1966. The contributions of the horse to knowledge of the heart and circulation. IV. James Hope and the heart sounds. Conn. Med. 30: 126–131.

McKeever, K.H. 2016. Exercise, and rehabilitation of the older horse. *In*: C. McGowan (ed.). Veterinary Clinics of North America, Equine Practice, Geriatric Medicine 32: 317–332.

Muybridge, E. 1887. Animal Locomotion. An Electrophotographic Investigation of Consecutive Phases of Animal Movements 1872–1885, Plates Vol IX Horses. University of Pennsylvania, Philadelphia.

Persson, S.G.B. 1967. On blood volume and working capacities in horses. Acta Veterinaria Scandinavica 19: 1–189.

Persson, S.G., and G. Lydin. 1973. Circulatory effects of splenectomy in the horse. III. Effect on pulse-work relationship. Zentralblatt für Veterinarmedizin A 20: 521–530.

Persson, S.G., L. Ekman, G. Lydin, and G. Tufvesson. 1973a. Circulatory effects of splenectomy in the horse. I. Effect on red-cell distribution and variability of haematocrit in the peripheral blood. Zentralblatt für Veterinarmedizin A 20: 441–455.

Persson, S.G., L. Ekman, G. Lydin, and G. Tufvesson. 1973b. Circulatory effects of splenectomy in the horse. II. Effect on plasma volume and total and circulating red-cell volume. Zentralblatt für Veterinarmedizin A 20: 456–468.

Persson, S.G., and G. Bergsten. 1975. Circulatory effects of splenectomy in the horse. IV. Effect on blood flow and blood lactate at rest and during exercise. Zentralblatt für Veterinarmedizin A 22: 801–807.

Proctor, R.C., S. Brody, M. Jones, and D.W. Chittenden. 1935. Growth and development with special reference to domestic animals: XXXIII Efficiency of work horses of different ages and weights. Further investigations of surface areas in energy metabolism. Univ. of Missouri Res. Bull. 209.

Rotenberg, S., and W. Czerniak. 1967. Total and unesterified cholesterol in serum of stallions before and after physical exercise. Acta Physiol. Pol. 18: 47–51.

Roughton, F.J.W. 1965. The oxygen equilibrium of mammalian hemoglobin—some old and new physiochemical studies. Journal of General Physiology 49: Suppl: 105–126.

Silverman, M.E. 1996. Etienne-Jules Marey: Nineteenth century cardiovascular physiologist and inventor of cinematography. Clin. Cardiol. 19: 339–341.

Smith, F. 1890a. The chemistry of respiration in the horse during rest and work. J. Physiol. 11(1-2): 65–158.5.

Smith, F. 1890b. Note on the composition of the sweat of the horse. J. Physiol. 11: 497–503.

Smith, F. 1896. The maximum muscular effort of the horse. J. Physiol. 19: 224–226.

Snow, D.H., S.G.B. Persson, and R.J. Rose. 1983. Equine Exercise Physiology, Granata Editions, Cambridge, UK. 543 pp.

Soliman, M.K., and M.A. Nadim. 1967. Calcium, sodium and potassium level in the serum and sweat of healthy horses after strenuous exercise. Zentralbl Veterinarmed A. 14: 53–56.

Streltsova, J.M., K.H. McKeever, N.R. Liburt, H.C. Manso, M.E. Gordon, D. Horohov, R. Rosen, and W. Franke. 2006. Effect of orange peel and black tea extracts on markers of performance and cytokine markers of inflammation in horses. Equine Comp. Exercise Physiol. 3: 121–130.

Tomás, E., A. Zorzano, and N.B. Ruderman. 2002. Exercise and insulin signaling: A historical perspective. J. Appl. Physiol. 93: 765–772.

van Weeren, P.R. 2001. History of locomotor research. pp. 1–35. *In*: W. Back and H.M. Clayton (eds.). Equine Locomotion. W.B. Saunders, New York.

Verter, W., H. Mix, and J. Müller. 1966. The changes in the cell number and some biochemical data on the blood of race horses in walk, trot and gallop. Arch. Exp. Veterinarmed. 20: 417–426.

Walker, A., S.M. Arent, and K.H. McKeever. 2009. Maximal aerobic capacity (VO_{2MAX}) in horses: a retrospective study to identify the age-related decline. Equine Comp. Exercise Physiol. 6: 177–181.

CHAPTER-2

The Pig Model for the Study of Obesity and Associated Metabolic Diseases

Kolapo M. Ajuwon

INTRODUCTION

The pig has been used as a human biomedical model through the millennia. In ancient Greece, Erasistratus (304–250 B.C.) used pigs to investigate the mechanics of breathing (Rozkot et al., 2015). In Rome, Galen (130–200 A.D.) demonstrated the mechanics of blood circulation on the pig model (Brain, 1986). As a major source of animal protein for human consumption, these long-established roles for pigs show the great value they bring to human well-being. Despite these positives, on the negative ledger, pigs are difficult to herd and move for long distances. However, their positive value has led to their long-term utilization in settled farming practice. Extensive domestication and rearing of the pig for meat was made possible, perhaps because of its relative ease of breeding, large litter size, short generation interval and an omnivorous feeding habit. The modern pig descended from the Eurasian Wild Boar (*Sus scrofa*). Domestication of the pig occurred as early as 13,000 years ago at multiple centers in Europe, Eurasia and China (Larson et al., 2005). Pigs have played major roles in the evolution of modern human societies, an example was in ancient Persia, where the pig was a popular subject for statuettes.

Pigs continue to play important roles in today's world. In Papua New Guinea, a "pig culture" exists in which the pig is highly valued as a ceremonial animal, as a means of exchange, or as a form of savings (Reay,

Department of Animal Sciences, Purdue University, 915 West State Street, West Lafayette, IN 47907, USA.
Email: kajuwon@purdue.edu

1984). As a source of insulin for the treatment of type I diabetes early in the 20th century, the pig was vital for the health of many diabetics in the US and around the world (Wright et al., 1979). Although many animal models of human metabolic diseases abound, the pig is unique in that the anatomy and morphology of its digestive system and in its metabolic physiology are similar to humans. At present there are about 67 known breeds of pigs (http://www.ansi.okstate.edu/breeds/swine). This wide genetic variation in the pig allows for selection and breeding for traits of interest to scientists and producers. As a source of tissues, the pig is a source of easily accessible biopsy and post-mortem samples that can be harvested for mechanistic studies. In addition, because billions of pigs are raised annually for human consumption (Figure 1) (den Hartog, 2004), commercial swine production is a ready source of tissues and animals that can be directed to biomedical endpoints (Vodicka et al., 2005). An important biomedical use of the pig is as a model for the understanding of human obesity. Obesity is a global epidemic with a prevalence doubling between 1980 and 2008 (Stevens et al., 2012). In the United States for example, one in three adults is obese, and among African Americans and Hispanics, one in two is obese (Ogden et al., 2014). Obesity is also strongly associated with a number of debilitating diseases such as hyperlipidemia, gout, high blood pressure, insulin resistance, diabetes, coronary heart disease, infertility, arthritis, restrictive lung disease, gall bladder disease and cancer (Kwon et al., 2013;

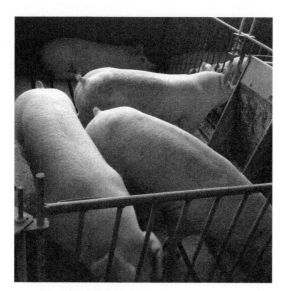

Figure 1: Pigs in a barn at Purdue University. Pigs are raised throughout the world as an important source of high quality protein. Pigs have a high efficiency of conversion of feed to meat. They are found in all climatic regions of the world, are omnivorous like humans and very adaptable to intensive production and management systems.

Lee et al., 2015; Osman and Hennessy, 2015). The pig is an excellent model for the study of human obesity because of its natural genetic propensity for depositing substantial adipose tissue, making it a suitable model for the investigation of both genetic and diet induced forms of obesity (Table 1). In this chapter we will review some of the recent work on the use of the pig as a model for the study of obesity and related diseases.

Table 1: Metabolic characteristics of the pig as a model of obesity.

Characteristic	Mouse/Rats	Humans	Pig
Genetic diversity	Several mouse and rat strains with different genetic backgrounds exist. Several knockout and transgenic models.	Varying human genetic backgrounds, for example, differences in racial and ethnic susceptibility to obesity and metabolic syndrome.	Wide genetic pool. Multiple breeds exist with different levels of susceptibility to obesity and metabolic syndrome. Availability of pigs raised for agriculture for biomedical uses is a major advantage (den Hartog, 2004). Now much easier to manipulate the pig genome that ever before.
Obesity susceptibility	There are strains of mice (e.g., C57BL6) and rats (e.g., Zucker rats) that are susceptible to diet induced obesity.	Most humans are susceptible to diet induced obesity.	Pigs are some of the fattest animals on the planet with natural ability to deposit substantial adipose tissue (Gu et al., 1992; Wojtysiak and Poltowicz, 2014).
Genetic basis of obesity	Several monogenic basis of obesity exist, e.g., the ob/ob and db/db mice.	Several quantitative trait loci (QTL) exist for monogenic and polygenic basis of obesity susceptibility (Mutch and Clement, 2006).	Just like humans, fat deposition in pigs is a polygenic trait. Over 400 QTL for fatness have been identified in the pig (Rothschild et al., 2007).
Synteny of genomic regions that affect fat deposition and energy metabolism	Different genetic arrangement of key genes compared to humans.	Similar arrangement of genes as in the pig. For example, the UCP1 is located on different chromosomal region (HSA4q31) from UCP2 and UCP3 (HSA11q13) (Cassard et al., 1990).	Like in humans UCP1 is located on SSC8q21-22, whereas UCP2 and UCP3 are located on SSC9p24.

Heterogeneity of Pig Genotypes for Obesity Research

Pigs are found in almost all climatic regions of the world, from the extreme hot climates of tropical Africa and Asia to the frigid temperate regions of North America and Europe. Significant genetic diversity exists in the pig, reflecting both natural and artificial selection pressures for survival and economic traits over the generations. The enormous genetic diversity in the pig is reflected in its physical features and body composition in terms of muscling, intramuscular fat, subcutaneous and visceral fat deposition patterns (Gu et al., 1992; Wojtysiak and Połtowicz, 2014). This variability makes the pig one of the best models for the study of polygenic basis of obesity development (Pomp, 1997). Like all complex human diseases, development of an appropriate animal model for the study of obesity is paramount in order to make rapid progress towards understanding its etiology. Although other animal models such as *C. elegans* (*Caenorhabditis elegans*), fruit fly (*Drosophila melanogaster*), zebrafish (*Danio rerio*) and rodents (especially rats and mice) are available for the study of genetic factors that lead to obesity, these animal models do not have the same level of similarity to humans obtainable with the pig. The similarity between humans and pigs occurs also at the genome level. The two genomes are very similar in terms of nucleotide base similarities, chromosomal location of genes and the overall number of genes. There are about 2.7 billion base pairs in the pig genome (Groenen et al., 2012). For example, a look at the pig chromosome 13 and human chromosome 3 reveals great synteny in gene contents and arrangement (Van Poucke et al., 1999). Humans and pigs also share similar genetic basis of fatness. Genome scanning reveals the presence of quantitative trait loci (QTL) for obesity in the human genome (Mutch and Clement, 2006) and there are over 120 recognized candidate genes for monogenic or polygenic obesity (Dahlman and Arner, 2007; Ichihara and Yamada, 2008). Likewise in the pig, more than 400 quantitative trait loci (QTL) for fatness have been identified in the pig genome (Rothschild et al., 2007). There is great interest in identifying QTLs for fat deposition in the pig because meat quality traits are affected by the level of fatness (or marbling). Therefore, it is of great interest to identify candidate genes that affect fat deposition in pigs. Because these results are directly translatable to human obesity. Recent efforts in this area have resulted in the mapping of genes encoding transcription factors involved in control of adipogenesis (Szczerbal et al., 2007a; Szczerbal and Chmurzynska, 2008) and fatty-acid-binding proteins family (Szczerbal et al., 2007b). Studies providing a low resolution picture of genome wide syntenic relationships between the pig and man have been published, and these are based upon physical or linkage maps (Rettenberger et al., 1995; Vingborg et al., 2009).

Although initial comparative sequence analysis have been performed (Jorgensen et al., 2005; Wernersson et al., 2005), more detailed comparisons are still impossible because the full annotation and assembly of the porcine genome is not yet complete. However, similar arrangement of genes that are implicated in obesity such as uncoupling proteins 1, 2 and 3 (UCP1, UCP2 and UCP3) and adrenergic receptor beta (ADRB) further illustrates the synteny of critical regions of the two genomes that affect metabolism and body composition. For example UCP1 is located on a different chromosome in the pig and humans compared to UCP2 and 3. The UCP1 is located on region SSC8q21-22 in the pig (Nowacka-Woszuk et al., 2008) and HSA4q31 in humans (Cassard et al., 1990). On the other hand UCP2 and UCP3 are located in SSC9p24 in the pig (Nowacka-Woszuk et al., 2008) and HSA11q13 in humans (Solanes et al., 1997). In the same vein, ADRB1, ADRB2 and ADRB3 are assigned to different loci, 14q28, 2q29 and 15q13-14, respectively in the pig (Nowacka-Woszuk et al., 2008). This is very similar to the organization of the ADRB proteins in the human genome that are located in HSA10q24-26, HSA5q32-34 and HSA8p, respectively (Yang-Feng et al., 1990; Bruskiewich et al., 1996). Thus multiple genomic comparisons show the similarity between the pig and humans vis-à-vis comparisons between the domestic the mouse and man (Hart et al., 2007; Rettenberger et al., 1995; Thomas et al., 2003). Therefore, the similar organization of the pig and human genomes further illustrates the closeness between the pig and humans and further justifies the use of the pig as a human model of obesity research.

A major advantage of using the other non-porcine models of obesity research is the relative ease with which their genomes can be manipulated for loss or gain of function studies compared to the pig. However, recent advances in the sequencing and annotation of the pig genome and arrival of cutting edge genome editing technologies such as RNAi and CRISPR/Cas9 are allowing the editing of the porcine genome in ways not conceivable just a decade ago (Yang et al., 2015). Additional progress on the annotation of the pig genome, and extensive characterization of porcine proteomes and transcriptomes will further enhance the use of the new technologies for manipulating the pig genome to create models for the study of adipose tissue and obesity development.

Hormonal and Developmental Basis of Adipose Tissue Expansion in the Pig Model

Obesity is a disease of excessive adipose tissue expansion. This expansion is in the form of preadipocyte hyperplasia and differentiation. Thus to limit obesity, it is imperative to understand the mechanisms of adipose expansion and the identities of cells within adipose tissue, and their relative contribution to the expansion process. There are multiple cells types within adipose tissue, notable among which are endothelial cells, fibroblasts, stem cells, macrophages, preadipocytes and adipocytes (Esteve, 2014). However, preadipocytes and adipocytes typically constitute the majority of cells in adipose tissue. These cells are highly responsive to physiological cues for adipose tissue expansion, typically as a result of chronic positive energy balance. Hormones play a major role in the regulation of adipocyte proliferation and differentiation, and several animal models have been used to understand the mechanism of hormone action on adipocyte differentiation. Earlier work in rats showed that thyroid hormones and glucocorticoids enhanced adipocyte development (Yukimura and Bray, 1978; Picon and Levacher, 1979; Levacher et al., 1984; Freedman et al., 1986; Bray et al., 1992). Results from pig studies also provide very detailed information on both prenatal and postnatal regulation of adipocyte differentiation. Developmentally, term fetal pigs are similar in characteristics to 25–30 week human fetuses, and both have been shown to have similar rates of fat accretion during earlier periods of gestation (Wilkerson and Gortner, 1932; Gortner, 1945; Widdowson, 1950, 1968).

The development of subcutaneous adipose tissue in the fetal pig has been used as a model to evaluate hormonal regulation of adipogenesis (Hausman et al., 1993; Hausman and Hausman, 1993). For example, fetal hypophysectomy and gestational diabetes have significant impact on adipose tissue development in the fetal pig due to severe impacts on serum levels of critical hormones implicated in the regulation of adipocyte differentiation (thyroid hormones, growth hormone and glucocorticoids) (Hausman et al., 1986, 1990). In addition, in utero hypophysectomy of fetal pigs around day 70 with surgical delivery at day 110 resulted in decreases in fat cell number and increases in fat cell size, as well as increases in lipoprotein lipase activity (LPL) and lipogenesis (Hausman and Hausman, 1993). This is perhaps linked to a reduction in the levels of growth hormone (GH) and insulin-like growth factor (IGF-1) after hypophysectomy (Jewell et al., 1989). Studies in the pig model ultimately led to the identification of a critical period of adipose tissue development from day 70 of fetal life (Hausman et al., 1986; Hausman and Hausman, 1993). Several *in vitro* experiments in the pig preadipocyte model also showed the utility of the pig model of adipocyte differentiation as a tool for understanding the regulation of adipocyte differentiation. For example the use of *in vitro* cultures led to the understanding that glucocorticoids (or analog dexamethasone), enhanced preadipocyte recruitment and differentiation (Kras et al., 1999). Hausman and Yu (1998) showed that glucocorticoid treatment of hypophysectomized pig fetuses led to enhancement of preadipocyte recruitment of stroma vascular (SV) cells. However, the importance of porcine fetal age in the responsiveness of preadipocytes to glucocorticoid is reflected in the fact that responsiveness to glucocorticoids increased as a fetus advances in age, perhaps due to elevated expression of preadipocyte glucocorticoid receptor in older fetuses (Chen et al., 1995). The use of the hypophysectomized fetal pig model also led to a deeper understanding of the role of thyroxine (T4) in adipose tissue development. Thyroxine treatment increases adipocyte cell number, lipid deposition, *de novo* lipogenesis and hormone-induced lipolysis in hypophysectomized, but not intact fetuses (Hausman, 1992). These are invaluable contributions to current understanding of developmental regulation of adipose development.

Adipose Tissue Role in the Regulation of Inflammation and Immune Response

Obesity is connected to the presence of systemic and adipose tissue chronic inflammation and the discovery that adipose tissue expresses tumor necrosis factor alpha (TNFα) (Hotamisligil et al., 1993) opened a new chapter in the understanding of adipose tissue role beyond lipid storage. This work generated interests that led to the identification of immune cells, especially macrophages (Xu et al., 2003; Lumeng et al., 2007) within adipose tissue. This breakthrough further deepened our understanding of the role of adipose tissue in the regulation of immune response. We have known for some time of the presence of numerous lymph nodes in adipose tissue, especially within the mesenteric fat pads (Pond, 1996). We are also aware that secretions from the lymph nodes may act in a paracrine manner to influence surrounding adipose tissue. It has been established that adipose tissue surrounding lymph nodes shows a higher expression of type I tumor necrosis factor (TNF) receptors than those that are remotely located (MacQueen and Pond, 1998). These findings substantiate a role for adipose tissue in the regulation of immune function (Cousin et al., 1999). The discovery of TLR4 expression on adipocytes is another piece of key evidence of adipocyte role in immune response (Shi et al., 2006; Bès-Houtmann et al., 2007), solidifying evidence that adipocytes are able to recognize pathogen associated molecular patterns (PAMP). Our work in pig adipocytes has confirmed that pig adipocytes also express TLR4 and are able to respond to bacterial lipopolysaccharide (LPS), leading to activation of NFkB and induction of inflammatory genes such as TNFα and IL-6 (Ajuwon et al., 2004). Thus the use of the pig model has contributed to the understanding of the role of adipose tissue in immune or inflammatory response (Table 2).

Pig Adipose Tissue Adipokines and Evidence for their Role in Regulation of Metabolism

The discovery of leptin, the ob gene, in 1994 (Zhang et al., 1994), confirmed the long held view among scientists of the presence of an adipose derived lipostat that serves as an indicator of body energy reserve. Adiponectin,

Table 2: Porcine adipocytes and evidence of involvement in inflammatory and immune response.

Feature	Evidence in the Pig
Expression of toll like receptors	Pig adipose tissue and adipocytes express functional toll like receptors (Gabler et al., 2008)
Expression of inflammatory cytokine	Pig adipocytes respond to bacteria component, lipopolysaccharide with induction of inflammatory cytokines such as TNFα and IL-6 (Ajuwon et al., 2004)
Adipokine role in inflammation	Porcine adiponectin induces PPARγ and downregulates inflammatory cytokine expression in adipocytes (Ajuwon et al., 2004) and macrophages (Wulster-Radcliffe et al., 2004)
Obesity-induced inflammation	Obesity results in increased expression of inflammatory cytokines such as TNFα and increased macrophage infiltration (Pawar et al., 2015)

another adipokine with far reaching effects on peripheral metabolism was discovered soon thereafter (Scherer et al., 1995; Tomas et al., 2002; Yamauchi et al., 2002). Both leptin (Margetic et al., 2002) and adiponectin (Yamauchi et al., 2002) have been shown to regulate metabolism through autocrine, paracrine and endocrine mechanisms.

Since their initial discovery in mice, substantial work has been done in other animal models, including the pig, on the understanding of the role of these adipokines in the regulation of metabolism. Polymorphisms of both leptin, adiponectin and their receptors are associated with growth and body composition differences in the pig. The leptin receptor is located in a QTL region for backfat thickness (BF), fat area ratios, and serum leptin concentration, and it is associated with polymorphisms that are highly correlated to these variables (Uemoto et al., 2012). Furthermore, two novel polymorphisms between adiponectin promoters in the pig are significantly correlated to loin measurements in the Polish Landrace pigs (Cieslak et al., 2013). Polymorphisms within the coding region of adiponectin are also significantly correlated to serum cholesterol, low-density lipoprotein (LDL) triglyceride levels (Castelló et al., 2014). Receptors for both hormones, leptin receptor (ObR-b) and adiponectin receptor (ADIPOR1 and ADIPOR2), have been identified in multiple porcine tissues, including the hypothalamus, myoblasts, endometrial glands and oocytes, an indication of their involvement in the regulation of several aspects of peripheral metabolism (Will et al., 2012; Moreira et al., 2014; Kaminski et al., 2014; Smolinska et al., 2014). Adiponectin regulates porcine adipocyte proliferation and differentiation potential because silencing of adiponectin with siRNA leads to significant reduction in preadipocyte proliferation and expression of mature adipocyte markers such as PPARγ and AP2 (Gao et al., 2013). Porcine leptin also inhibits protein breakdown in the C2C12 myoblast model and increases fatty acid oxidation (Ramsay, 2003), suggesting a direct effect of leptin in the regulation of muscle growth and metabolism. Adiponectin attenuates proliferation of porcine myoblasts, but leptin does not (Will et al., 2012), suggesting that the two adipokines have non-overlapping functions.

There is evidence that leptin is involved in regulation of postnatal growth. Leptin directly influences GH production in the pituitary, perhaps through a mechanism that includes induction of nitric oxide synthesis (Baratta et al., 2002). Chronic exogenous leptin administration resulted in a reduction in feed intake in the pig (Ajuwon et al., 2003a). This indicates an intact feedback system for appetite regulation in the pig involving leptin action. Administration of leptin to piglets with intra-uterine growth restriction (IUGR) causes an increase in body weight and the relative weights of the liver, spleen, pancreas, kidneys, and small intestine (Attig et al., 2013). Both acute and chronic leptin treatment of porcine adipocytes result in significant induction of basal lipolysis and attenuation of the suppression of isoproterenol-stimulated lipolysis by insulin, suggesting that leptin may be involved in the partitioning of energy away from lipid accretion within porcine adipose tissue by promoting lipolysis directly, and indirectly by reducing insulin-mediated inhibition of lipolysis (Ramsay, 2001; Ajuwon et al., 2003b). Comparative studies in mouse and porcine hepatocytes also reveal that, whereas leptin suppresses gluconeogenesis in in mouse hepatocytes, it has no effect in porcine hepatocytes, pointing to possible species differences in the regulation of gluconeogenesis by leptin (Raman et al., 2004). Nevertheless, there is strong evidence that leptin may increase IGF-1 expression in the liver, independent of GH effect (Ajuwon et al., 2003a), suggesting that leptin may modulate postnatal growth through IGF-1 induction. Nutritionally, adiponectin and leptin may also mediate the effects of soy isoflavones because administration of soy isoflavones regulate plasma glucose, leptin and adiponectin concentration in pigs with a concomitant regulation of leptin and adiponectin expression (Yang et al., 2013).

Pig Model of Diabetes Research

Diabetes mellitus (DM), a complex disease characterized by high blood glucose levels, polydipsia, polyuria, weight loss as well as diabetic ketoacidosis and non-ketotic hyperosmolar syndrome, is a major human disease that currently afflicts about 346 million people worldwide (World Health Organization, 2012). The International Diabetes Federation has predicted that a total of 552 million persons will suffer from diabetes by 2030 (International Diabetes Federation, 2012). Efforts at limiting the incidence of diabetes involve both preventive and therapeutic strategies. Although most diabetics are classified as either insulin dependent (type I) or non-insulin dependent (type II), the majority of diabetes sufferers (90–95%) have type II diabetes which is caused by insulin resistance and an inadequate compensatory insulin secretion. Studies in the pig have provided some of the fundamental knowledge on diabetes cause, prevention and therapy (Table 3). Although rodents have traditionally being used to study diabetes, rodents have limitations for translational research in this area.

The similar anatomy of the digestive tracts and physiology of digestion of the pig and humans allows the use of humanized diets for their effects on both obesity and diabetes development in the pig (Miller and Ullrey, 1987). In addition, there is similarity in the shape, size, blood supply and location of the porcine exocrine and endocrine pancreas (Murakami et al., 1997). Metabolically, serum glucose levels are in the same range (75–115

Table 3: Porcine models of obesity and metabolic syndrome and their characteristics.

Feature	Characteristics of the pig model
Diet induced obesity	Pigs get obese and moderately diabetic on humanized diets (Miller and Ullrey, 1987; Xi et al., 2004; Yin et al., 2004).
Pancreatic structure and glucose homeostasis and diabetes development	Similarity in the shape, size, blood supply and location of the porcine exocrine and endocrine pancreas (Murakami et al., 1997). Serum glucose levels are in the same range (75–115 mg/dl) in pigs and humans (Kraft and Dürr, 2005; Kerner and Brückel, 2012). In pigs, islet architecture is less defined. Endocrine cells are clustered in the islets of Langerhans or spread around as individual cells or in miniature clusters as opposed to rodents (Steiner et al., 2010), and their islets have a similar composition of endocrine cell types. The mass of beta cells in both pigs and humans corresponds to beta-cell function (Butler et al., 2003; Renner et al., 2010). Type 2 diabetes occurs in both pigs and humans with advancing age (Larsen et al., 2001).
Streptozotocin induced diabetes and pancreatic manipulation	Pigs are a good model for streptozotocin (STZ) induced pancreatic beta cell toxin and diabetes (Lee et al., 2010; Manell et al., 2014). Pigs are a good model for pathophysiological impact of total pancreatectomy (Kobayashi et al., 2004), and simplified techniques of pancreas transplantation (Fonouni et al., 2015).
Cardiovascular disease research and susceptibility	The minipig (Yucatan and Ossabaw) pig models are especially valuable models for the study of cardiovascular disease. Easy to reproduce the neointimal formation and thrombosis like in humans (Touchard and Schwartz, 2006). Develop metabolic syndrome and cardiovascular disease when fed a high-calorie atherogenic diet (Wang et al., 2009). Ossabaw minipigs pigs have high LDL, hypertriglyceridemia, hypertension, and early coronary atherosclerosis when on high cholesterol high fat diet (Dyson et al., 2006). In adult Göttingen minipig model of chronic heart failure after myocardial infarction, heart failure is reproducible, mimicking the pathophysiology in patients who have experienced myocardial infarction (Schuleri et al., 2008).

mg/dl) in pigs and humans (Kraft and Dürr, 2005; Kerner and Brückel, 2012) (Table 3). Furthermore, islet architecture in pigs and humans is less defined, because in both species endocrine cells are clustered in the islets of Langerhans or spread around as individual cells or in miniature clusters as opposed to rodents (Steiner et al., 2010), and their islets have a similar composition of endocrine cell types. In addition, the mass of beta cells in both pigs and humans corresponds to beta-cell function (Butler et al., 2003; Renner et al., 2010). Type 2 diabetes occurs in both pigs and humans with advancing age (Larsen et al., 2001). Therefore, due to these similarities between humans and pigs, numerous studies have been conducted in the pig for a better understanding of diabetes in humans. As in humans, diet-induced diabetes occurs in some minipigs, albeit with only a moderate rise in blood glucose (Xi et al., 2004; Yin et al., 2004). Although there is experimental evidence that visceral adiposity is associated with glucose intolerance or prediabetes in the pig (McKnight et al., 2012), generally it is very hard to induce diabetes in the pig due to robust islet function even in the face of a harmful metabolic environment.

Several studies conducted in the pig to investigate beta-cell function have used surgical or pharmacological approaches because of the difficulty of inducing diabetes in the pig using milder experimental approaches like diet induced obesity. Therefore, the minipig has been successfully used in numerous studies showing the effectiveness of the pancreatic beta cell toxin streptozotocin (STZ) (Lee et al., 2010; Manell et al., 2014). In addition, the pig model has been used to establish the pathophysiological impact of total pancreatectomy (Kobayashi et al., 2004). The pig has also been used to investigate simplified techniques of pancreas transplantation (Fonouni et al., 2015). A porcine prediabetic model was used to study expression of GLUT-4 translocation related genes (Kristensen et al., 2015). The pig is also an excellent candidate for xenotransplantation studies. Shin et al. (2015) reported the successful use of pig islets as an alternative source for islet transplantation to treat type 1 diabetes (T1D) using immunosuppressed nonhuman primates (NHP) as recipients. Renner et al. (2013) reported the development of the INS(C94Y) transgenic pig model of type I diabetes with a reduced secretion of normal insulin. These animals exhibit elevated blood glucose, 41% reduced body weight at 4.5 months of age, 72% decreased β-cell mass, and 60% lower fasting insulin levels compared with littermate controls. The animals also developed diabetic complications such as development of cataracts. An additional transgenic pig model expressed mutant HNF1A and these animals have defective insulin secretion and deranged glucose homeostasis (Nyunt et al., 2009). Thus the advancement in genome manipulation technologies has made it possible to manipulate the pig genome in ways not possible before, leading to greater understanding of pathophysiology of diseases such as diabetes and its complications.

Fetal and Postnatal Programming of Obesity in the Pig

The role of nutrition during the fetal and early postnatal period on postnatal potential for development of metabolic diseases such as obesity, cardiovascular diseases is exemplified in the Dutch winter famine that occurred between November 1944 and April 1945 (Stein et al., 1995). Offspring from pregnant mothers who experienced poor nutrition in the first trimester of pregnancy during this period developed obesity and a more truncal and abdominal fat distribution in adulthood (Ravelli et al., 1999). This study formed the basis of the "Barker" hypothesis of fetal programming, which states that fetal undernutrition leads to development of a thrifty genotype in the fetus as the fetus attempts to adapt to an adverse condition encountered *in utero* (Hales and Barker, 2001). Notably, as in the Dutch famine, prenatal undernutrition followed by postnatal over nutrition is very harmful, leading to adverse risks for development of metabolic syndrome phenotype such as abdominal obesity, insulin resistance or type 2 diabetes, hypertension, and cardiovascular diseases. Because of its importance for development of critical tissues during fetal life, glucocorticoids are thought to be a major link between poor fetal nutrition and development of adult metabolic diseases (Seckl et al., 2000).

The work of Ovilo et al. (2014) demonstrates that maternal gestational undernutrition in Iberian pigs results in heavier and fatter offspring with higher concentrations of cortisol, lower hypothalamic expression of anorexigenic peptides, leptin receptor and proopiomelanocortin (POMC), than the controls. Under normal fetal nutrition, maternal glucocorticoids (cortisol/corticosterone) are oxidized to an inactive form by the enzyme 11β-hydroxysteroid dehydrogenase type 2 (11β-HSD2) in the placenta. This is to prevent these hormones from entering the placenta where they can affect fetal growth (Seckl, 2004). However, during maternal undernutrition in pregnancy, placental 11β-HSD2 levels are significantly attenuated, allowing glucocorticoids to cross into the placenta where they come in contact with the fetus (Langley-Evans et al., 1996; Bertram et al., 2001; Lesage et al., 2001). Fetal contact with high level of glucocorticoids causes impaired fetal growth and increased risk for development of hypertension and other metabolic diseases (Nyirenda et al., 1998; Seckl, 2004). Increased fetal exposure to glucocorticoids causes altered rate of maturation of critical fetal organs and organ systems such as the hypothalamic-pituitary-adrenal (HPA) axis and dopaminergic motor systems (Seckl, 2004). Elevated fetal exposure to glucocorticoids may also result in reduced levels of trophic hormones such as insulin-like growth factors (IGFs) (Oue et al., 1999).

The effects of elevated glucocorticoid exposure and other factors known to have lasting effects on offspring at risk for chronic disease are probably linked to epigenetic effects. Epigenetics is an important emerging field that provides an explanation at the molecular level for the altered gene expression patterns during fetal programming (Gluckman and Hanson, 2008; Heerwagen et al., 2010). We recently used the pig model to investigate the impact of maternal overconsumption of calories on postnatal metabolic characteristics and growth in the offspring (Arentson-Lantz et al., 2014). Although maternal overconsumption of calories did not result in differences in offspring birth weights, offspring from mothers that overconsumed calories during gestation who were themselves fed a high calorie post-natal diet had elevated blood glucose, insulin, but lower concentrations of non-esterified fatty acids than offspring from the same mothers fed a normal calorie post-natal diet. This demonstrates the interaction between prenatal maternal calorie consumption and postnatal offspring calorie consumption on metabolic phenotype in the pig. At present, the mechanism of adipose tissue programming in the offspring of obese mothers with excessive gestational calorie overconsumption is poorly understood. The difficulty of conducting such a study in humans is obvious due to ethical considerations, and in rodents, because rodent pups lack substantial adipose tissue at birth. Our study in the pig reveals increased adipose expression of genes such as steroid receptor coactivator 1 (SRC1), soluble frizzle related receptors SFRP2, set domain-containing protein 8 (SETD8), glucocorticoid receptor (GCR) and downregulation of nuclear receptor corepressor 1 (NCOR1), a suppressor of peroxisome proliferator activated receptor γ (PPARγ) activity, in subcutaneous adipose tissue in piglets from mothers which consumed a high calorie gestational diet, reflecting a unique adipose tissue effect of maternal high calorie diet (Ajuwon et al., 2016). The induction of SETD8 indicates that maternal diet may induce epigenetic changes in the genome of the offspring that may have a consequence on postnatal adiposity. This is because there is a positive feedback loop between SETD8 and PPARγ during adipogenesis, and the suppression of SETD8 suppresses adipogenesis (Wakabayashi et al., 2009). Activation of SETD8 also results in increased PPARγ H4K20 monomethylation and an enhanced transcriptional activity and adipogenesis (Wakabayashi et al., 2009). The upregulation of SFRP2 in the adipose tissue of piglets from mothers on high energy diet also supports programming for increased adipogenic potential in those offspring. It is well known that soluble frizzle related receptors (SFRPs) are negative regulators of Wnt signaling (Surana et al., 2014). The SFRP proteins are natural wnt antagonists, preventing wnt proteins from inhibiting adipogenesis (Park et al., 2008). As shown previously, SFRP 1-4 are adipokines that are elevated in human obesity (Ehrlund et al., 2013).

We also observed that genes involved in adipocyte differentiation (PPARγ, CCAAT Enhancer binding proteinα, CEBPα and fatty acid binding protein 4, FABP4) are elevated in the adipose tissue of piglets from mothers that consumed excessive calories during gestation (Ajuwon et al., 2016). Furthermore, the induction of GCR in offspring from mothers that consumed a high amount of calories during gestation is consistent with the established effect of glucocorticoids in increasing adipogenesis (Ringold et al., 1986). However, it is quite interesting that at a later stage of life (3 months of age), most of the programming effects seen in the adipose tissue within the immediate post-natal period of life (within 48 hours and at 3 weeks) as a result of maternal consumption of high calorie diets are no longer apparent, but postnatal high calorie consumption still resulted in higher SFRP5 expression in adipose tissue (Ajuwon et al., 2016). This suggests that a significant programming of adiposity by maternal diet occurs in the immediate postnatal period in the pig, whereas beyond this period, effects of postnatal dietary energy intake is more important. Thus the pig may represent a good animal model for determining effects of maternal nutrition on adipose tissue programming in the immediate postnatal period.

Cardiovascular Disease Research in the Pig Model

Cardiovascular diseases, such as atherosclerotic coronary artery disease (CAD), are a major cause of mortality in humans around the world. This disease is increased at least 2-fold in human patients who have metabolic syndrome (Grundy, 2007). Coronary artery disease causes severe microvascular dysfunction that impedes coronary blood flow (Camici and Crea, 2007). Metabolic syndrome affects about 27% of the American population (Wilson et al., 2005). The pig has been used for basic research on CAD because it can be used to precisely replicate metabolic syndrome and the accompanying CAD (Table 4). The minipig (Yucatan and Ossabaw) pig models are especially valuable models for the study of cardiovascular disease because it is very easy to reproduce the neointimal formation and thrombosis that occurs in humans (Touchard and Schwartz, 2006). The Ossabaw miniature swine (Martin et al., 1973) (Figure 2) is an exceptionally good model for the study of cardiovascular diseases because these animals develop metabolic syndrome and cardiovascular disease when fed a high-calorie atherogenic diet (Wang et al., 2009). However, there may be a gender bias in the development of cardiovascular disease in the Ossabaw swine because only females show severe metabolic syndrome and obesity: doubling of body fat, showed insulin resistance, impaired glucose tolerance, high LDL, hypertriglyceridemia, hypertension, and early coronary atherosclerosis when fed high cholesterol high fat diet to induce atherosclerosis (Dyson et al., 2006).

Although rodents (mice and rats) have been used to investigate the link between atherosclerotic coronary artery disease and metabolic syndrome (Bellinger et al., 2006; Christoffersen et al., 2007), they do not reproduce well enough the combined symptoms of metabolic syndrome and CAD as in the pig. Recently Phillips-Eakley et al. (2015) used the Ossabaw minipig

Table 4: Porcine models of pharmacological and surgical approaches for obesity prevention and control.

Feature	Evidence in the Pig
Pig models of weight loss	Göttingen pigs can double their weight from obesity and have similar meal patterns during the daylight hours as humans (Raun et al., 2007). Drugs such as liraglutide, exert similar effects in humans and obese Göttingen minipigs causing weight loss and appetite suppression (Astrup et al., 2009).
Surgical manipulation and weight loss	Significant reduction in growth rate in pigs subjected to gastric fundus invagination compared to sham (Darido et al., 2012).
Surgical manipulation and glycemic control	The RYGB procedure is effective in correcting type 2 diabetes (T2D) and weight loss (Buchwald et al., 2009; Birck et al., 2013), reduction in food intake (Jackness et al., 2013; Sham et al., 2014); changes to the circulating levels of gut hormones (e.g., glucagon-like peptide 1, GLP-1); through effects on the islets (Schauer et al., 2003); direct effects on the gut microbiota (Sweeney and Morton, 2013); effects on intestinal glucose sensing (Breen et al., 2012) and bile acids (Kohli et al., 2013). Data on RYGB effect on glucose metabolism obtained from rodents cannot be extrapolated to humans, because of pronounced differences in pancreatic anatomy and physiology between the humans and rodents (Seyfried et al., 2011). Gastric bypass increased postprandial insulin and GLP-1 concentrations in non-obese minipigs (Verhaeghe et al. 2014). Compared to sham-operated pigs, RYGB pigs, displayed improved glycemic control, increased in β-cell mass, islet number, and number of extra islet β-cells, elevated pancreatic expression of insulin and glucagon, and had increased number of glucagon-like peptide 1 receptor expressing cells (Lindqvist et al., 2014).
Surgical manipulation and nutrient absorption	RYGB procedure led to significant reduction in calcium, fat, and ash digestibility, compared to SGIT or IT procedures in pigs (Gandarillas et al., 2015).

to assess the impact of high calcium intake on coronary artery calcification using an innovative calcium tracer kinetic modeling in Ossabaw swine with diet-induced metabolic syndrome. The adult Göttingen is another minipig model that is being used as a model for the study of chronic heart failure after myocardial infarction. Schuleri et al. (2008), used magnetic resonance imaging, angiography and Multidetector Computed Tomography of the heart and showed that the Göttingen minipig is a useful model for evaluating cardiac anatomy and physiology prior to myocardial infarction and during follow-up. In this animal model, heart failure was found to be reproducible, mimicking the pathophysiology in patients who have experienced myocardial infarction.

The Yucatan miniature pigs, especially males, may not readily develop a compromised metabolic phenotype by feeding atherogenic diets like the Ossabaw or Göttingen minipig (Witczak et al., 2005), but are being used for

Figure 2: The Ossabaw miniature pig in an experimental pen at Purdue. Originally from the Ossabaw Island off the coast of the state of Georgia in the US, Ossabaw pigs can reach a mature weight of approximately 150 kg. They are highly prone to diet-induced obesity and cardiovascular disease.

cardiovascular disease investigations. Davis et al. (2014) generated Yucatan miniature pigs with targeted disruptions of the low-density lipoprotein receptor (LDLR) gene as an improved large animal model of familial hypercholesterolemia and atherosclerosis. Homozygote animals with a total deletion of the LDLR gene had elevated total and LDL cholesterol with atherosclerotic lesions in the coronary arteries and abdominal aorta that mimics human atherosclerosis under low or high fat diets. This animal model of cardiovascular disease could be an important resource for investigating development and testing of novel detection and treatment strategies for coronary and aortic atherosclerosis and its complications. Researchers at Purdue University have also used the pig to study impact of exercise on cardiovascular and muscle physiology with results that show parallels between the pig and humans on the impact of exercise on critical hemodynamic, cardiovascular and muscle function (Taheripour et al., 2014, Figure 3).

Figure 3: A pig exercising on a treadmill. Pigs can be used to study the impact of exercise on muscle development, cardiovascular disease and obesity. Picture courtesy S.C. Newcomer, Purdue University. (Present address, California State University, San Marcos.)

Gut Microbiome and Obesity Research in the Pig

The pig has also recently been used to investigate the link between the microbiome and obesity susceptibility and the interaction between diet, gut metabolite profile and peripheral metabolism. Recent discoveries that the gut microbiota plays a major role in determining susceptibility to obesity (Gibson et al., 2004; Topping and Clifton, 2001) has fueled studies aimed at investigating the mechanistic links between the two. These discoveries also hint at a possibility of directly altering the gut microbiome to achieve obesity prevention or treatment (Ley et al., 2006). Notably, because obesity is a metabolic disorder that involves chronic positive energy balance, studies on how diet or lifestyle choices affect the microbiome and the implication on energy balance and whole-body metabolism are critically needed (Gibson et al., 2004; de Lange et al., 2010). One of the earliest changes in the gut microbiome is the reduction in its diversity in response to obesogenic diets (Duncan et al., 2008). Recent analysis of the human hindgut microbiome suggests a strong association between changes in the microbiome composition and obesity susceptibility (Gibson et al., 2004; Ley et al., 2006). The close correlation between obesity susceptibility and the composition of the microbiome suggests that alteration of the gut microbial community could be used as an approach for obesity prevention and treatment (Ley et al., 2006). In this respect, studies are urgently needed that go beyond characterization of the microbiome in response to diets, to those aimed at addressing the functionality of the microbiome as it relates to obesity susceptibility with respect to its integrated effects on the host. One of the recent findings of obesity effects on the gut and the microbiota relates to the increased leakage of LPS from the gut into systemic circulation during obesity, causing low-grade inflammation (Cani et al., 2007).

However, studies of the gut microbiota in human subjects are limited by profound individual variation in microbial community composition, and sometimes ethical considerations related to invasive approaches for obtaining human biological samples. To get around this problem, germ-free mice, often highly in-bred, are used as the animal model of choice, but there are large differences between mice and humans in their physiology and gut microbial communities, primarily due to the significant differences in gut architecture and dietary requirements between the two species. The pig is an ideal animal for investigating the effect of dietary components on bacterial communities and metabolic changes because of the similarities in its dietary requirements, and the anatomy and physiology of its digestive tract and that of humans (Pang et al., 2007). Obesity prone minipig models are especially useful for such studies due to their compact size and the genetic disposition for obesity. We (Yan et al., 2013) have investigated the changes in the microbiome in the Ossabaw minipig in response to consumption of two types of dietary fiber (cellulose and inulin), and determined metabolic markers in the liver, muscle, adipose and intestinal tissues. One of our key findings was that feeding inulin resulted in increases in observed concentrations of volatile fatty acids in the cecum. Feeding inulin also causes lower body weight gain and adiposity. Mechanistically consumption of fermentable fiber such as inulin could lead to stimulation of peroxisomal β-oxidation of fatty acid (Lazarow and Deduve, 1976). In addition, inulin feeding resulted in lower expression of sterol regulatory element binding protein -1c (SREBP-1c) a transcription factor that regulates expression of multiple lipogenic genes (Horton et al., 2002). Consumption of a high fat also causes reduction in the diversity of the microbiome, and feeding inulin prevented this change (Yan et al., 2013).

The work by Pedersen et al. (2013) in which the microbiome of lean and obese Göttingen and Ossabaw minipigs were compared indicate clear differences in the microbiome in these two models depending on their obesity status. Using 16S sequencing of cecal content, the investigators found that lean Göttingen minipigs had a higher abundance of Firmicutes relative to Bacteroidetes in the cecum. However, higher ratio of Firmicutes to Bacteroidetes was found in obese Ossabaw minipigs in the terminal ileum and colon. Thus, although both minipig models are useful for the study of diet-induced obesity, genetic and dietary differences may predominate to determine the composition of their microbiome. This is similar to several human studies where diet and genetics have been identified as key determinants of microbial composition (Mar Rodríguez et al., 2015;

Davenport et al., 2015), and further show the usefulness of the pig as a model to investigate the importance of these factors in determining the relationship between the microbiome and obesity susceptibility.

The use of the Pig Model for Surgical Approach in Obesity Therapy

The rise in cases of morbid obesity across the globe and the attendant increase in the incidence of type 2 diabetes and cardiovascular disease complications has led to the use of drastic surgical approaches, often as last resort, to induce rapid weight loss and mitigate the side effects of morbid obesity (Driscoll et al., 2016). Gastric fundus invagination, Roux-en-Y gastric bypass (RYGB) and vertical sleeve gastrectomy are some of the common surgical approaches for weight loss and have varying degrees of effectiveness. Minipig models are currently being used to evaluate surgical approaches for obesity therapy (Table 4). The obese Göttingen minipig is judged to be superior for the study of severe obesity compared to rodents (Raun et al., 2007). Unlike rodents, the obese minipig body composition is very similar to that reported for severely obese people. Göttingen pigs can double their weight from obesity and have similar meal patterns during the daylight hours as humans (Raun et al., 2007). In addition, drugs such as liraglutide, a human glucagon-like peptide-1 analog, exert similar effects in humans and obese Göttingen minipigs through effects in causing weight loss and appetite suppression (Astrup et al., 2009).

Darido et al. (2012) evaluated the effects of laparotomy, stomach manipulation, short gastric vessel ligation and gastric fundus invagination on juvenile pig growth and found significant reduction in growth rate in pigs subjected to gastric fundus invagination compared to sham. The RYGB procedure is also highly effective in correcting type 2 diabetes (T2D) in just a few days after surgery (Buchwald et al., 2009). The surgical procedures are thought to work through effecting reduction in food intake (Jackness et al., 2013) and causing changes to the circulating levels of gut hormones such as incretins (glucagon-like peptide 1, GLP-1), through effects on the islets (Schauer and Buchwald, 2003), direct effects on the gut microbiota (Sweeney and Morton, 2013) or by acting on intestinal glucose sensing (Breen et al., 2012) and bile acids (Kohli et al., 2013). Indeed, Verhaeghe et al. (2014) found that gastric bypass increased postprandial insulin and GLP-1 concentrations in non-obese minipigs. Birck et al. (2013) used morbidly obese Göttingen minipigs to evaluate the effect of the RYGB procedure in correcting obesity and found that the surgery led to weight loss and reduction in food intake.

The pig model has also been used to elucidate the effects of RYGB on effect on β-cell mass (Lindqvist et al., 2014). Unlike pigs, data on RYGB effect on glucose metabolism obtained from rodents cannot be extrapolated to humans, because of pronounced differences in pancreatic anatomy and physiology between the humans and rodents (Seyfried et al., 2011). The work by Lindqvist et al. (2014) showed that, compared to sham-operated pigs, RYGB pigs, despite a lack of weight loss, displayed improved glycemic control, increased in β-cell mass, islet number, and number of extra islet β-cells. They also show elevated pancreatic expression of insulin and glucagon, and had increased number of glucagon-like peptide 1 receptor expressing cells. Thus the pig model of RYGB is helping to provide information on the importance of improved β-cell function and β-cell mass for the improved glucose tolerance after RYGB.

Using the Ossabaw minipig model, Sham et al. (2014) also studied three surgical procedures in the Ossabaw pig, RYGB, gastrojejunostomy (GJ), gastrojejunostomy with duodenal exclusion (GJD), and showed that RYGB promoted weight loss, correction of insulin resistance, and increased AUCinsulin/AUCglucose, compared to the smaller changes with GJ and GJD. Their results pointed to a combination of upper and lower gut mechanisms in improving glucose homeostasis. Because the effect of surgical procedures on nutrient utilization is unclear, Gandarillas et al. (2015) evaluated the effects of three surgical procedures, ileal transposition (IT), sleeve gastrectomy with ileal transposition (SGIT) and RYGB on protein, lipid, fiber, energy, calcium, and phosphorous digestibility in a swine model. Their results show that digestibility values for dry matter, fiber, phosphorus, and energy showed were not different among surgical types. However, they found significant differences of surgical procedure on fat, protein, ash, and calcium digestibilities. In addition, they concluded that the RYGB procedure led to significant reduction in calcium, fat, and ash digestibilities, compared to SGIT or IT procedures. Thus the use of the pig model is facilitating a greater understanding of effects of surgical interventions in obesity therapy on nutrient metabolism in ways not possible with human patients or rodent models.

Summary and Conclusions

The pig has proven over and over again as one of man's most useful animal species. From its initial relevance as a source of high quality animal protein, the pig is now one of the most valuable animal models of disease. As a model for comparative medicine, the pig has unique resemblance to humans in the etiology of multiple human metabolic diseases. The reliance of porcine insulin for many decades to treat diabetics exemplify how the pig has proven to be an indispensable animal for human civilization. Indeed, the increasing incidence of metabolic diseases such as obesity, diabetes and cardiovascular diseases point to a bright future for the use of the pig as a biomedical model for the investigation of cause, prevention and cure for these diseases. The advancement of genome editing technologies will further enhance global efforts to use the pig as a large animal model of metabolic diseases. The pig is not a "large mouse", but an animal that more closely resembles humans than rodents in its genetics, anatomy, morphology of its organ systems and the physiology of its metabolic processes. Thus, for the pig, better days are still ahead.

Acknowledgement

The author acknowledges the contribution of multiple graduate students to his research program at Purdue University, especially Hang Lu and Hui Yan, who made useful suggestions in the preparation of this chapter.

Keywords: obesity, pig, adipose, minipig, cardiovascular, disease model, diabetes, nutrition, microbiome, gastric surgery

References

Ajuwon, K.M., J.L. Kuske, D. Ragland, O. Adeola, D.L. Hancock, D.B. Anderson, and M.E. Spurlock. 2003a. The regulation of IGF-1 by leptin in the pig is tissue specific and independent of changes in growth hormone. J. Nutr. Biochem. 14: 522–530.

Ajuwon, K.M., J.L. Kuske, D.B. Anderson, D.L. Hancock, K.L. Houseknecht, O. Adeola, and M.E. Spurlock. 2003b. Chronic leptin administration increases serum NEFA in the pig and differentially regulates PPAR expression in adipose tissue. J. Nutr. Biochem. 14: 576–583.

Ajuwon, K.M., S.K. Jacobi, J.L. Kuske, and M.E. Spurlock. 2004. Interleukin-6 and interleukin-15 are selectively regulated by lipopolysaccharide and interferon-gamma in primary pig adipocytes. Am. J. Physiol-Reg. I 286: R547–R553.

Ajuwon, K.M., E.J. Arentson-Lantz, and S.S. Donkin. 2016. Excessive gestational calorie intake in sows regulates early postnatal adipose tissue development in the offspring. BMC Nutrition. 2:29 DOI 10.1186/s40795-016-0069-3.

Arentson-Lantz, E.J., K.K. Buhman, K. Ajuwon, and S.S. Donkin. 2014. Excess pregnancy weight gain leads to early indications of metabolic syndrome in a swine model of fetal programming. Nutr. Res. 34: 241–249.

Astrup, A., S. Rossner, L. Van Gaal, A. Rissanen, L. Niskanen, M. Al Hakim, J. Madsen, M.F. Rasmussen, and M.E. Lean. 2009. Effects of liraglutide in the treatment of obesity: A randomised, double-blind, placebo-controlled study. Lancet 374: 1606–1616.

Attig, L., D. Brisard, T. Larcher, M. Mickiewicz, P. Guilloteau, S. Boukthir, C.N. Niamba, A. Gertler, J. Djiane, D. Monniaux, and L. Abdennebi-Najar. 2013. Postnatal leptin promotes organ maturation and development in IUGR piglets. PLoS One 8: e64616.

Baratta, M., R. Saleri, G.L. Mainardi, D. Valle, A. Giustina, and C. Tamanini. 2002. Leptin regulates GH gene expression and secretion and nitric oxide production in pig pituitary cells. Endocrinology 143: 551–557.

Bellinger, D.A., E.P. Merricks, and T.C. Nichols. 2006. Swine models of type 2 diabetes mellitus: Insulin resistance, glucose tolerance, and cardiovascular complications. Ilar. J. 47: 243–258.

Bertram, C., A.R. Trowern, N. Copin, A.A. Jackson, and C.B. Whorwood. 2001. The maternal diet during pregnancy programs altered expression of the glucocorticoid receptor and type 2 11 beta-hydroxysteroid dehydrogenase: Potential molecular mechanisms underlying the programming of hypertension *in utero*. Endocrinology 142: 2841–2853.

Bes-Houtmann, S., R. Roche, L. Hoareau, M.P. Gonthier, F. Festy, H. Caillens, P. Gasque, C. Lefebvre d'Hellencourt, and M. Cesari. 2007. Presence of functional TLR2 and TLR4 on human adipocytes. Histochem. Cell Biol. 127: 131–137.

Birck, M.M., A. Vegge, M. Stockel, I. Gogenur, T. Thymann, K.P. Hammelev, P.T. Sangild, A.K. Hansen, K. Raun, P. von Voss, and T. Eriksen. 2013. Laparoscopic Roux-en-Y gastric bypass in super obese Gottingen minipigs. Am. J. Transl. Res. 5: 643–653.

Bray, G.A., J.S. Stern, and T.W. Castonguay. 1992. Effect of adrenalectomy and high-fat diet on the fatty Zucker rat. Am. J. Physiol. 262: E32–E39.

Breen, D.M., B.A. Rasmussen, A. Kokorovic, R.N.A. Wang, G.W.C. Cheung, and T.K.T. Lam. 2012. Jejunal nutrient sensing is required for duodenal-jejunal bypass surgery to rapidly lower glucose concentrations in uncontrolled diabetes. Nat. Med. 18: 950–955.

Brian, P. 1986. Galen on Bloodletting: A Study of the Origins, Development and Validity of his Opinions, with a Translation of the Three Works. Cambridge University Press, London, UK.

Bruskiewich, R., T. Everson, L. Ma, L. Chan, M. Schertzer, J.P. Giacobino, P. Muzzin, and S. Wood. 1996. Analysis of CA repeat polymorphisms places three human gene loci on the 8p linkage map. Cytogenet. Cell Genet. 73: 331–333.

Buchwald, H., R. Estok, K. Fahrbach, D. Banel, M.D. Jensen, W.J. Pories, J.P. Bantle, and I. Sledge. 2009. Weight and type 2 diabetes after bariatric surgery: systematic review and meta-analysis. Am. J. Med. 122: 248–281.

Butler, A.E., J. Janson, S. Bonner-Weir, R. Ritzel, R.A. Rizza, and P.C. Butler. 2003. Beta-cell deficit and increased beta-cell apoptosis in humans with type 2 diabetes. Diabetes 52: 102–110.

Camici, P.G., and F. Crea. 2007. Coronary microvascular dysfunction—Reply. New. Engl. J. Med. 356: 2325–2325.

Cani, P.D., J. Amar, M.A. Iglesias, M. Poggi, C. Knauf, D. Bastelica, A.M. Neyrinck, F. Fava, K.M. Tuohy, C. Chabo, A. Waget, E. Delmée, B. Cousin, T. Sulpice, B. Chamontin, and J. Ferrières. 2007. Metabolic endotoxemia initiates obesity and insulin resistance. Diabetes 56: 1761–1772.

Cassard, A.M., F. Bouillaud, M.G. Mattei, E. Hentz, S. Raimbault, M. Thomas, and D. Ricquier. 1990. Human uncoupling protein gene—structure, comparison with rat gene, and assignment to the long arm of chromosome-4. J. Cell Biochem. 43: 255–264.

Castelló, A., R. Quintanilla, C. Melo, D. Gallardo, A. Zidi, A. Manunza, J.L. Noguera, J. Tibau, J. Jordana, R.N. Pena, and M. Amills. 2014. Associations between pig adiponectin (ADIPOQ) genotype and serum lipid levels are modulated by age-specific modifiers. J. Anim. Sci. 92: 5367–5373.

Chen, N.X., B.D. White, and G.J. Hausman. 1995. Glucocorticoid receptor-binding in porcine preadipocytes during development. J. Anim. Sci. 73: 722–727.

Christoffersen, B.O., N. Grand, V. Golozoubova, O. Svendsen, and K. Raun. 2007. Gender-associated differences in metabolic syndrome-related parameters in Göttingen minipigs. Comp. Med. 57: 493–504.

Cieslak, J., T. Flisikowska, A. Schnieke, A. Kind, M. Szydlowski, M. Switonski, and K. Flisikowski. 2013. Polymorphisms in the promoter region of the adiponectin (ADIPOQ) gene are presumably associated with transcription level and carcass traits in pigs. Anim. Genet. 44: 340–343.

Cousin, B., O. Munoz, M. Andre, A.M. Fontanilles, C. Dani, J.L. Cousin, P. Laharrague, L. Casteilla, and L. Pénicaud. 1999. A role for preadipocytes as macrophage-like cells. FASEB J. 13: 305–312.

Dahlman, I., and P. Arner. 2007. Obesity and polymorphisms in genes regulating human adipose tissue. Int. J. Obesity 31: 1629–1641.

Darido, E., D.W. Overby, K.A. Brownley, and T.M. Farrell. 2012. Evaluation of gastric fundus invagination for weight loss in a porcine model. Obes. Surg. 22: 1293–1297.

Davenport, E.R., D.A. Cusanovich, K. Michelini, L.B. Barreiro, C. Ober, and Y. Gilad. 2015. Genome-wide association studies of the human gut microbiota. PLoS One 10: e014031.

Davis, B.T., X.J. Wang, J.A. Rohret, J.T. Struzynski, E.P. Merricks, D.A. Bellinger et al. 2014. Targeted disruption of LDLR causes hypercholesterolemia and atherosclerosis in yucatan miniature pigs. PLoS One 9: e93457.

de Lange, C.F.M., J. Pluske, J. Gong, and C.M. Nyachoti. 2010. Strategic use of feed ingredients and feed additives to stimulate gut health and development in young pigs. Livest. Sci. 134: 124–134.

den Hartog, L. 2004. Developments in global pig production. Adv. Pork Prod. 15: 17–24.

Driscoll, S., D.M. Gregory, J.M. Fardy, and L.K. Twells. 2016. Long-term health-related quality of life in bariatric surgery patients: A systematic review and meta-analysis. Obesity 24: 60–70.

Duncan, S.H., G.E. Lobley, G. Holtrop, J. Ince, A.M. Johnstone, P. Louis, and H.J. Flint. 2008. Human colonic microbiota associated with diet, obesity and weight loss. Int. J. Obesity 32: 1720–1724.

Dyson, M.C., M. Alloosh, J.P. Vuchetich, E.A. Mokelke, and M. Sturek. 2006. Components of metabolic syndrome and coronary artery disease in female Ossabaw swine fed excess atherogenic diet. Comp. Med. 56: 35–45.

Ehrlund, A., N. Mejhert, S. Lorente-Cebrian, G. Astrom, I. Dahlman, J. Laurencikiene, and M. Rydén. 2013. Characterization of the wnt inhibitors secreted frizzled-related proteins (SFRPs) in human adipose tissue. J. Clin. Endocr. Metab. 98: E503–E508.

Esteve, R.M. 2014. Adipose tissue: cell heterogeneity and functional diversity. Endocrinol. Nutr. 61: 100–112.

Fonouni, H., M.T. Rad, M. Esmaeilzadeh, M. Golriz, A. Majlesara, and A. Mehrabi. 2015. A simplified technique of pancreas transplantation in a porcine model. Eur. Surg. Res. 54: 24–33.

Freedman, M.R., B.A. Horwitz, and J.S. Stern. 1986. Effect of adrenalectomy and glucocorticoid replacement on development of obesity. Am. J. Physiol. 250: R595–R607.

Gabler, N.K., J.D. Spencer, D.M. Webel, and M.E. Spurlock. 2008. n-3 PUFA attenuate lipopolysaccharide-induced down-regulation of toll-like receptor 4 expression in porcine adipose tissue but does not alter the expression of other immune modulators. J. Nutr. Biochem. 19: 8–15.

Gandarillas, M., S. Hodgkinson, J. Riveros, and F. Bas. 2015. Effect of three different bariatric obesity surgery procedures on nutrient and energy digestibility using a swine experimental model. Exp. Biol. Med. 240: 1158–1164.

Gao, Y., F.J. Li, Y.H. Zhang, L.S. Dai, H. Jiang, H.Y. Liu, S. Zhang, C. Chen, and J. Zhang. 2013. Silencing of ADIPOQ efficiently suppresses preadipocyte differentiation in porcine. Cell Physiol. Biochem. 31: 452–461.

Gibson, G.R., H.M. Probert, J. Van Loo, R.A. Rastall, and M.B. Roberfroid. 2004. Dietary modulation of the human colonic microbiota: updating the concept of prebiotics. Nutr. Res. Rev. 17: 259–275.

Gluckman, P.D., and M.A. Hanson. 2008. Developmental and epigenetic pathways to obesity: an evolutionary-developmental perspective. Int. J. Obesity 32: S62–S71.

Gortner, W.A. 1945. The Lipids of the pig during embryonic development. J. Biol. Chem. 159: 135–143.

Groenen, M.A.M., A.L. Archibald, H. Uenishi, C.K. Tuggle, Y. Takeuchi, M.F. Rothschild, C. Rogel-Gaillard, C. Park, D. Milan, H.J. Megens, S. Li, D.M. Larkin, H. Kim, L.A. Frantz, M. Caccamo, H. Ahn, B.L. Aken, A. Anselmo, C. Anthon, L. Auvil, B. Badaoui, C.W. Beattie, C. Bendixen, D. Berman, F. Blecha, J. Blomberg, L. Bolund, M. Bosse, S. Botti, Z. Bujie, M. Bystrom, B. Capitanu, D. Carvalho-Silva, P. Chardon, C. Chen, R. Cheng, S.H. Choi, W. Chow, R.C. Clark, C. Clee, R.P. Crooijmans, H.D. Dawson, P. Dehais, F. De Sapio, B. Dibbits, N. Drou, Z.Q. Du, K. Eversole, J. Fadista, S. Fairley, T. Faraut, G.J. Faulkner, K.E. Fowler, M. Fredholm, E. Fritz, J.G. Gilbert, E. Giuffra , J. Gorodkin, D.K. Griffin, J.L. Harrow, A. Hayward, K. Howe, Z.L. Hu, S.J. Humphray, T. Hunt, H. Hornshøj, J.T. Jeon, P. Jern, M. Jones, J. Jurka, H. Kanamori, R. Kapetanovic, J. Kim, J.H. Kim, K.W. Kim, T.H. Kim, G. Larson, K. Lee, K.T. Lee, R. Leggett, H.A. Lewin, Y. Li, W. Liu, J.E. Loveland, Y. Lu, J.K. Lunney, J. Ma, O. Madsen, K. Mann, L. Matthews, S. McLaren, T. Morozumi, M. P. Murtaugh, J. Narayan, D.T. Nguyen, P. Ni, S.J. Oh, S. Onteru, F. Panitz, E.W. Park, H.S. Park, G. Pascal, Y. Paudel, M. Perez-Enciso, R. Ramirez-Gonzalez, J.M. Reecy, S. Rodriguez-Zas, G.A. Rohrer, L. Rund, Y. Sang, K. Schachtschneider, J.G. Schraiber, J.

Schwartz, L. Scobie, C. Scott, S. Searle, B. Servin, B.R. Southey, G. Sperber, P. Stadler, J. V. Sweedler, H. Tafer, B. Thomsen, R. Wali, J. Wang, J. Wang, S. White, X. Xu, M. Yerle, G. Zhang, J. Zhang, J. Zhang, S. Zhao, J. Rogers, C. Churcher, and L.B. Schook. 2012. Analyses of pig genomes provide insight into porcine demography and evolution. Nature 491: 393–398.

Grundy, S.M. 2007. Controversy in clinical endocrinology—metabolic syndrome: A multiplex cardiovascular risk factor. J. Clin. Endocr. Metab. 92: 399–404.

Gu, Y., A.P. Schinckel, and T.G. Martin. 1992. Growth, development, and carcass composition in 5 genotypes of swine. J. Anim. Sci. 70: 1719–1729.

Hales, C.N., and D.J.P. Barker. 2001. The thrifty phenotype hypothesis. Br. Med. Bull. 60: 5–20.

Hart, E.A., M. Caccamo, J.L. Harrow, S.J. Humphray, J.G. Gilbert, S. Trevanion, T. Hubbard, J. Rogers, and M.F. Rothschild. 2007. Lessons learned from the initial sequencing of the pig genome: Comparative analysis of an 8 Mb region of pig chromosome 17. Genome Biol. 8: R168.

Hausman, G.J., R.J. Martin, and D.R. Campion. 1986. Regulation of adipose tissue development in the fetus: The fetal pig model. pp. 997–1006. In: M.E. Tumbleson (ed.). Swine in Biomedical Research. Plenum Press, New York, USA.

Hausman, G.J., J.T. Wright, D.E. Jewell, and T.G. Ramsay. 1990. Fetal adipose-tissue development. Int. J. Obesity 14: 177–185.

Hausman, G.J. 1992. The influence of thyroxine on the differentiation of adipose-tissue and skin during fetal development. Pediatr. Res. 32: 204–211.

Hausman, G.J., and D.B. Hausman. 1993. Endocrine regulation of porcine adipose-tissue development—cellular and metabolic aspects. pp. 49–73. In: G.R. Hollis (ed.). Growth of the Pig. CAB International, London, UK.

Hausman, G.J., J.T. Wright, R. Dean, and R.L. Richardson. 1993. Cellular and molecular aspects of the regulation of adipogenesis. J. Anim. Sci. 71: 33–55.

Hausman, G.J., and Z.K. Yu. 1998. Influence of thyroxine and hydrocortisone in vivo on porcine preadipocte recruitment, development and expression of C/EBP binding proteins in fetal stromal vascular (S-V) cell cultures. Grow. Dev. Ag. 62: 107–118.

Heerwagen, M.J.R., M.R. Miller, L.A. Barbour, and J.E. Friedman. 2010. Maternal obesity and fetal metabolic programming: A fertile epigenetic soil. Am. J. Physiol-Reg. I 299: R711–R722.

Horton, J.D., J.L. Goldstein, and M.S. Brown. 2002. SREBPs: activators of the complete program of cholesterol and fatty acid synthesis in the liver. J. Clin. Invest. 109: 1125–1131.

Hotamisligil, G.S., N.S. Shargill, and B.M. Spiegelman. 1993. Adipose expression of tumor-necrosis-factor-alpha—direct role in obesity-linked insulin resistance. Science 259: 87–91.

Ichihara, S., and Y. Yamada. 2008. Genetic factors for human obesity. Cell Mol. Life Sci. 65: 1086–1098.

International Diabetes Federation. 2012. IDF Diabetes Atlas. 5th ed.

Jackness, C., W. Karmally, G. Febres, I.M. Conwell, L. Ahmed, M. Bessler, D.J. McMahon, and J. Korner. 2013. Very low-calorie diet mimics the early beneficial effect of Roux-en-Y gastric bypass on insulin sensitivity and beta-cell function in type 2 diabetic patients. Diabetes 62: 3027–3032.

Jewell, D.E., G.J. Hausman, and D.R. Campion. 1989. Fetal hypophysectomy causes a decrease in preadipocyte growth and insulin-like growth factor-I in Pigs. Domest. Anim. Endocrin. 6: 243–252.

Jorgensen, F.G., A. Hobolth, H. Hornshoj, C. Bendixen, M. Fredholm, and M.H. Schierup. 2005. Comparative analysis of protein coding sequences from human, mouse and the domesticated pig. BMC Biol. 3: 2.

Kaminski, T., N. Smolinska, A. Maleszka, M. Kiezun, K. Dobrzyn, J. Czerwinska, K. Szeszko, and A. Nitkiewicz. 2014. Expression of adiponectin and its receptors in the porcine hypothalamus during the oestrous cycle. Reprod. Domest. Anim. 49: 378–386.

Kerner, W., and J. Brückel. 2012. Definition, diagnosis and classification of diabetes mellitus. Diabetol. Stoffwechs 7: S84–S87.

Kobayashi, K., N. Kobayashi, T. Okitsu, C. Yong, T. Fukazawa, H. Ikeda, Y. Kosaka, M. Narushima, T. Arata, and N. Tanaka. 2004. Development of a porcine model of type 1 diabetes by total pancreatectomy and establishment of a glucose tolerance evaluation method. Artif. Organs 28: 1035–1042.

Kohli, R., K.D.R. Setchell, M. Kirby, A. Myronovych, K.K. Ryan, S.H. Ibrahim, J. Berger, K. Smith, M. Toure, S.C. Woods, and R.J. Seeley. 2013. A surgical model in male obese rats uncovers protective effects of bile acids post-bariatric surgery. Endocrinology 154: 2341–2351.

Kraft, W., and U.M. Dürr. 2005. Klinische Labordiagnostik in der Tiermedizin. Schattauer, Stuttgart. GE.

Kras, K.M., D.B. Hausman, G.J. Hausman, and R.J. Martin. 1999. Adipocyte development is dependent upon stem cell recruitment and proliferation of preadipocytes. Obes. Res. 7: 491–497.

Kristensen, T., M. Fredholm, and S. Cirera. 2015. Expression study of GLUT4 translocation-related genes in a porcine pre-diabetic model. Mamm. Genome. 26: 650–657.

Kwon, B.J., D.W. Kim, S.H. Her, D.B. Kim, S.W. Jang, E.J. Cho, S.H. Ihm, H.Y. Kim, H.J. Youn, K.B. Seung, J.H. Kim, and T.H. Rho. 2013. Metabolically obese status with normal weight is associated with both the prevalence and severity of angiographic coronary artery disease. Metabolism 62: 952–960.

Langley-Evans, S.C., G.J. Phillips, R. Benediktsson, D.S. Gardner, C.R. Edwards, A.A. Jackson, and J.R. Seckl. 1996. Protein intake in pregnancy, placental glucocorticoid metabolism and the programming of hypertension in the rat. Placenta 17: 169–72.

Larsen, M.O., B. Rolin, M. Wilken, R.D. Carr, O. Svendsen, and P. Bollen. 2001. Parameters of glucose and lipid metabolism in the male Göttingen minipig: Influence of age, body weight, and breeding family. Comp. Med. 51: 436–442.

Larson, G., K. Dobney, U. Albarella, M.Y. Fang, E. Matisoo-Smith, J. Robins, S. Lowden, H. Finlayson, T. Brand, E. Willerslev, P. Rowley-Conwy, L. Andersson, and A. Cooper. 2005. Worldwide phylogeography of wild boar reveals multiple centers of pig domestication. Science 307: 1618–1621.

Lazarow, P.B., and C. Deduve. 1976. Fatty acyl-Coa oxidizing system in rat-liver peroxisomes— enhancement by clofibrate, a hypolipidemic drug. Proc. Natl. Acad. Sci. USA 73: 2043–2046.

Lee, J., J.Y. Lee, J.H. Lee, S.M. Jung, Y.S. Suh, J.H. Koh, S.K. Kwok, J.H. Ju, K.S. Park, and S.H. Park. 2015. Visceral fat obesity is highly associated with primary gout in a metabolically obese but normal weighted population: A case control study. Arthritis Res. Ther. 17: 79.

Lee, P.Y., S.G. Park, E.Y. Kim, M.S. Lee, S.J. Chung, S.C. Lee, D.Y. Yu, and K.H. Bae. 2010. Proteomic analysis of pancreata from mini-pigs treated with streptozotocin as type I diabetes models. J. Microbiol. Biotechnol. 20: 817–820.

Lesage, J., B. Blondeau, M. Grino, B. Breant, and J.P. Dupouy. 2001. Maternal undernutrition during late gestation induces fetal overexposure to glucocorticoids and intrauterine growth retardation, and disturbs the hypothalamopituitary adrenal axis in the newborn rat. Endocrinology 142: 1692–1702.

Levacher, C., C. Sztalryd, M.F. Kinebanyan, and L. Picon. 1984. Effects of thyroid-hormones on adipose-tissue development in Sherman and Zucker rats. Am. J. Physiol. 246: C50–C56.

Ley, R.E., P.J. Turnbaugh, S. Klein, and J.I. Gordon. 2006. Microbial ecology—human gut microbes associated with obesity. Nature 444: 1022–1023.

Lindqvist, A., P. Spegel, M. Ekelund, E.G. Vaz, S. Pierzynowski, M.F. Gomez, H. Mulder, J. Hedenbro, L. Groop, and N. Wierup. 2014. Gastric bypass improves b-Cell function and increases b-Cell mass in a porcine model. Diabetes 63: 1665–1671.

Lumeng, C.N., S.M. DeYoung, J.L. Bodzin, and A.R. Saltiel. 2007. Increased inflammatory properties of adipose tissue macrophages recruited during diet-induced obesity. Diabetes 56: 16–23.

MacQueen, H.A., and C.M. Pond. 1998. Immunofluorescent localisation of tumour necrosis factor-alpha receptors on the popliteal lymph node and the surrounding adipose tissue following a simulated immune challenge. J. Anat. 192: 223–231.

Manell, E.A.K., A. Ryden, P. Hedenqvist, M. Jacobson, and M. Jensen-Waern. 2014. Insulin treatment of streptozotocin-induced diabetes re-establishes the patterns in carbohydrate, fat and amino acid metabolisms in growing pigs. Lab. Anim. 48: 261–269.

Margetic, S., C. Gazzola, G.G. Pegg, and R.A. Hill. 2002. Leptin: A review of its peripheral actions and interactions. Int. J. Obes. Relat. Metab. Disord. 26: 1407–1433.

Mar Rodríguez, M., D. Pérez, F. Javier Chaves, E. Esteve, P. Marin-Garcia, G. Xifra, J. Vendrell, M. Jové, R. Pamplona, W. Ricart, M. Portero-Otin, M.R. Chacón, and J.M. Fernández Real. 2015. Obesity changes the human gut mycobiome. Sci. Rep. 5: 14600.

Martin, R.J., J.L. Gobble, T.H. Hartsock, H.B. Graves, and J.H. Ziegler. 1973. Characterization of an obese syndrome in pig. Proc. Soc. Exp. Biol. Med. 143: 198–203.

McKnight, L.L., S.B. Myrie, D.S. MacKay, J.A. Brunton, and R.F. Bertolo. 2012. Glucose tolerance is affected by visceral adiposity and sex, but not birth weight, in Yucatan miniature pigs. Appl. Physiol. Nutr. Metab. 37: 106–114.

Miller, E.R., and D.E. Ullrey. 1987. The pig as a model for human-nutrition. Annu. Rev. Nutr. 7: 361–382.

Moreira, F., S.M.M. Gheller, R.G. Mondadori, A.S. Varela, C.D. Corcini, and T. Lucia. 2014. Presence of leptin and its receptor in the hypothalamus, uterus and ovaries of swine females culled with distinct ovarian statuses and parities. Reprod. Domest. Anim. 49: 1074–1078.

Murakami, T., S. Hitomi, A. Ohtsuka, T. Taguchi, and T. Fujita. 1997. Pancreatic insulo-acinar portal systems in humans, rats, and some other mammals: scanning electron microscopy of vascular casts. Microsc. Res. Tech. 37: 478–488.

Mutch, D.M., and K. Clement. 2006. Unraveling the genetics of human obesity. Plos Genet. 2: 1956–1963.

Nowacka-Woszuk, J., I. Szczerbal, H. Fijak-Nowak, and M. Switonski. 2008. Chromosomal localization of 13 candidate genes for human obesity in the pig genome. J. Appl. Genet. 49: 373–7.

Nyirenda, M.J., R.S. Lindsay, C.J. Kenyon, A. Burchell, and J.R. Seckl. 1998. Glucocorticoid exposure in late gestation permanently programs rat hepatic phosphoenolpyruvate carboxykinase and glucocorticoid receptor expression and causes glucose intolerance in adult offspring. J. Clin. Invest. 101: 2174–2181.

Nyunt, O., J.Y. Wu, I.N. McGown, M. Harris, T. Huynh, G.M. Leong, D.M. Cowley, and A. M. Cotterill. 2009. Investigating maturity onset diabetes of the young. Clin. Biochem. Rev. 30: 67–74.

Ogden, C.L., M.D. Carroll, B.K. Kit, and K.M. Flegal. 2014. Prevalence of childhood and adult obesity in the United States, 2011–2012. Jama-J. Am. Med. Assoc. 311: 806–814.

Osman, M., and B. Hennessy. 2015. Obesity correlation with metastases development and response to first line metastatic chemotherapy in breast cancer. Eur. J. Cancer 51: S273–S273.

Oue, T., Y. Taira, H. Shima, E. Miyazaki, and P. Puri. 1999. Effect of antenatal glucocorticoid administration on insulin-like growth factor I and II levels in hypoplastic lung in nitrofen-induced congenital diaphragmatic hernia in rats. Pediatr. Surg. Int. 15: 175–179.

Ovilo, C., A. Gonzalez-Bulnes, R. Benitez, M. Ayuso, A. Barbero, M.L. Perez-Solana et al. 2014. Prenatal programming in an obese swine model: sex-related effects of maternal energy restriction on morphology, metabolism and hypothalamic gene expression. Br. J. Nutr. 111: 735–746.

Pang, X.Y., X.G. Hua, Q. Yang, D.H. Ding, C.Y. Che, L. Cui, W. Jia, P. Bucheli, and L. Zhao. 2007. Inter-species transplantation of gut microbiota from human to pigs. Isme. J. 1: 156–162.

Park, J.R., J.W. Jung, Y.S. Lee, and K.S. Kang. 2008. The roles of Wnt antagonists Dkk1 and sFRP4 during adipogenesis of human adipose tissue-derived mesenchymal stem cells. Cell Prolif. 41: 859–874.

Pawar, A.S., X.Y. Zhu, A. Eirin, H. Tang, K.L. Jordan, J.R. Woollard, A. Lerman, and L.O. Lerman. 2015. Adipose tissue remodeling in a novel domestic porcine model of diet-induced obesity. Obesity 23: 399–407.

Pedersen, R., H.C. Ingerslev, M. Sturek, M. Alloosh, S. Cirera, B.O. Christoffersen, S.G. Moesgaard, N. Larsen, and M. Boye. 2013. Characterisation of gut microbiota in Ossabaw and Göttingen minipigs as models of obesity and metabolic syndrome. PLoS One 8: e56612.

Phillips-Eakley, A.K., M.L. McKenney-Drake, M. Bahls, S.C. Newcomer, J.S. Radcliffe, M.E. Wastney, W.G. Van Alstine, G. Jackson, M. Alloosh, B.R. Martin, M. Sturek, and C.M. Weaver. 2015. Effect of high-calcium diet on coronary artery disease in Ossabaw miniature swine with metabolic syndrome. J. Am. Heart Assoc. 4: e001620.

Picon, L., and C. Levacher. 1979. Thyroid-hormones and adipose-tissue development. J. Physiol. 75: 539–543.

Pomp, D. 1997. Genetic dissection of obesity in polygenic animal models. Behav. Genet. 27: 285–306.

Pond, C.M. 1996. Functional interpretation of the organization of mammalian adipose tissue: Its relationship to the immune system. Biochem. Soc. Trans. 24: 393–400.

Raman, P., S.S. Donkin, and M.E. Spurlock. 2004. Regulation of hepatic glucose metabolism by leptin in pig and rat primary hepatocyte cultures. Am. J. Physiol.-Reg. I 286: R206–R216.

Ramsay, T.G. 2001. Porcine leptin alters insulin inhibition of lipolysis in porcine adipocytes *in vitro*. J. Anim. Sci. 79: 653–657.

Ramsay, T.G. 2003. Porcine leptin inhibits protein breakdown and stimulates fatty acid oxidation in C2C12 myotubes. J. Anim. Sci. 81: 3046–3051.

Raun, K., P. von Voss, and L.B. Knudsen. 2007. Liraglutide, a once-daily human glucagon-like peptide-1 analog, minimizes food intake in severely obese minipigs. Obesity 15: 1710–1716.

Ravelli, A.C.J., J.H.P. van der Meulen, C. Osmond, D.J.P. Barker, and O.P. Bleker. 1999. Obesity at the age of 50 y in men and women exposed to famine prenatally. Am. J. Clin. Nutr. 70: 811–816.

Reay, M. 1984. A high pig culture of the new guinea highlands. Canberra Anthropology 7: 71–77.

Renner, S., C. Fehlings, N. Herbach, A. Hofmann, D.C. von Waldthausen, B. Kessler, K. Ulrichs, I. Chodnevskaja, V. Moskalenko, W. Amselgruber, B. Göke, A. Pfeifer, R. Wanke, and E. Wolf. 2010. Glucose intolerance and reduced proliferation of pancreatic beta-cells in transgenic pigs with impaired glucose-dependent insulinotropic polypeptide function. Diabetes 59: 1228–1238.

Renner, S., C. Braun-Reichhart, A. Blutke, N. Herbach, D. Emrich, E. Streckel, A. Wünsch, B. Kessler, M. Kurome, A. Bähr, N. Klymiuk, S. Krebs, O. Puk, and H. Nagashima. 2013. Permanent neonatal diabetes in INSC94Y transgenic pigs. Diabetes 62: 1505–1511.

Rettenberger, G., C. Klett, U. Zechner, J. Kunz, W. Vogel, and H. Hameister. 1995. Visualization of the conservation of synteny between humans and pigs by heterologous chromosomal painting. Genomics 26: 372–378.

Ringold, G.M., A.B. Chapman, and D.M. Knight. 1986. Glucocorticoid control of developmentally regulated adipose genes. J. Steroid Biochem. 24: 69–75.

Rothschild, M.F., Z.L. Hu, and Z.H. Jiang. 2007. Advances in QTL mapping in pigs. Int. J. Biol. Sci. 3: 192–197.

Rozkot M., E. Václavková, and J. Bělková. 2015. Minipigs as laboratory animals—review. Res. Pig. Breeding 9: 2.

Schauer, P.R., and H. Buchwald. 2003. Effect of laparoscopic Roux-En Y gastric bypass on type 2 diabetes mellitus—discussion. Ann. Surg. 238: 484–485.

Schauer, P.R., B. Burguera, S. Ikramuddin, D. Cottam, W. Gourash, G. Hamad, G.M. Eid, and S. Mattar. 2003. Effect of laparoscopic Roux-En Y gastric bypass on type 2 diabetes mellitus. Ann. Surg. 238: 467–483.

Scherer, P.E., S. Williams, M. Fogliano, G. Baldini, and H.F. Lodish. 1995. A novel serum-protein similar to C1q, produced exclusively in adipocytes. J. Biol. Chem. 270: 26746–26749.

Schuleri, K.H., A.J. Boyle, M. Centola, L.C. Amado, R. Evers, J.M. Zimmet, K.S. Evers, K.M. Ostbye, D.G. Scorpio, J.M. Hare, and A.C. Lardo. 2008. The adult Göttingen minipig as a model for chronic heart failure after myocardial infarction: Focus on cardiovascular imaging and regenerative therapies. Comp. Med. 58: 568–579.

Seckl, J.R., M. Cleasby, and M.J. Nyirenda. 2000. Glucocorticoids, 11 beta-hydroxysteroid dehydrogenase, and fetal programming. Kidney Int. 57: 1412–1417.

Seckl, J.R. 2004. Prenatal glucocorticoids and long-term programming. Eur. J. Endocrinol. 151: U49–U62.

Seyfried, F., C.W. le Roux, and M. Bueter. 2011. Lessons learned from gastric bypass operations in rats. Obes. Facts 4: 3–12.

Sham, J.G., V.V. Simianu, A.S. Wright, S.D. Stewart, M. Alloosh, M. Sturek, D.E. Cummings, and D.R. Flum. 2014. Evaluating the mechanisms of improved glucose homeostasis after bariatric surgery in Ossabaw miniature swine. J. Diabetes Res. 2014: 526972.

Shi, H., M.V. Kokoeva, K. Inouye, I. Tzameli, H. Yin, and J.S. Flier. 2006. TLR4 links innate immunity and fatty acid-induced insulin resistance. J. Clin. Invest. 116: 3015–3025.

Shin, J.S., J.M. Kim, J.S. Kim, B.H. Min, Y.H. Kim, H.J. Kim, J.Y. Jang, I.H. Yoon, H.J. Kang, J. Kim, E.S. Hwang, D.G. Lim, W.W. Lee, J. Ha, K.C. Jung, S.H. Park, S.J. Kim, and C.G. Park. 2015. Long-term control of diabetes in immunosuppressed nonhuman primates (NHP) by the transplantation of adult porcine islets. Am. J. Transplant 15: 2837–2850.

Smolinska, N., K. Dobrzyn, A. Maleszka, M. Kiezun, K. Szeszko, and T. Kaminski. 2014. Expression of adiponectin and adiponectin receptors 1 (AdipoR1) and 2 (AdipoR2) in the porcine uterus during the oestrous cycle. Anim. Reprod. Sci. 146: 42–54.

Solanes, G., A. Vidal-Puig, D. Grujic, J.S. Flier, and B.B. Lowell. 1997. The human uncoupling protein-3 gene—Genomic structure, chromosomal localization, and genetic basis for short and long form transcripts. J. Biol. Chem. 272: 25433–25436.

Stein, A.D., A.C.J. Ravelli, and L.H. Lumey. 1995. Famine, 3rd-trimester pregnancy weight-gain, and intrauterine growth—the Dutch famine birth cohort study. Hum. Biol. 67: 135–150.

Steiner, D.J., A. Kim, K. Miller, and M. Hara. 2010. Pancreatic islet plasticity interspecies comparison of islet architecture and composition. Islets 2: 135–145.

Stevens, G.A., G.M. Singh, Y. Lu, G. Danaei, J.K. Lin, M.M. Finucane, A.N. Bahalim, R.K. McIntire, H.R. Gutierrez, M. Cowan, C.J. Paciorek, F. Farzadfar, L. Riley, and M. Ezzati. 2012. National, regional, and global trends in adult overweight and obesity prevalences. Popul. Health. Metr. 10: 22.

Surana, R., S. Sikka, W. Cai, E.M. Shin, S.R. Warner, H.J.G. Tan, F. Arfuso, S.A. Fox, A.M. Dharmarajan, and A.P. Kumar. 2014. Secreted frizzled related proteins: Implications in cancers. Biochim. Biophys. Acta. 1845: 53–65.

Sweeney, T.E., and J.M. Morton. 2013. The human gut microbiome: A review of the effect of obesity and surgically induced weight loss. JAMA Surg. 148: 563–569.

Szczerbal, I., L. Lin, M. Stachowiak, A. Chmurzynska, M. Mackowski, A. Winter, K. Flisikowski, R. Fries, and M. Switonski. 2007a. Cytogenetic mapping of DGAT1, PPARA, ADIPOR1 and CREB genes in the pig. J. Appl. Genet. 48: 73–76.

Szczerbal, I., A. Chmurzynska, and M. Switonski. 2007b. Cytogenetic mapping of eight genes encoding fatty acid binding proteins (FABPs) in the pig genome. Cytogenet. Genome. Res. 118: 63–66.

Szczerbal, I., and A. Chmurzynska. 2008. Chromosomal localization of nine porcine genes encoding transcription factors involved in adipogenesis. Cytogenet. Genome Res. 121: 50–54.

Taheripour, P., M.A. DeFord, E.J. Arentson-Lantz, S.S. Donkin, K.M. Ajuwon, and S.C. Newcomer. 2014. Impact of excess gestational and post-weaning energy intake on vascular function of swine offspring. BMC Pregnancy Childbirth. 14: 405.

Thomas, J.W., J.W. Touchman, R.W. Blakesley, G.G. Bouffard, S.M. Beckstrom-Sternberg, E.H. Margulies et al. 2003. Comparative analyses of multi-species sequences from targeted genomic regions. Nature 424: 788–793.

Tomas, E., T.S. Tsao, A.K. Saha, H.E. Murrey, C.C. Zhang, S.I. Itani, H.F. Lodish, and N.B. Ruderman. 2002. Enhanced muscle fat oxidation and glucose transport by ACRP30 globular domain: Acetyl-CoA carboxylase inhibition and AMP-activated protein kinase activation. Proc. Natl. Acad. Sci. USA 99: 16309–16313.

Topping, D.L., and P.M. Clifton. 2001. Short-chain fatty acids and human colonic function: Roles of resistant starch and nonstarch polysaccharides. Physiol. Rev. 81: 1031–1064.

Touchard, A.G., and R.S. Schwartz. 2006. Preclinical restenosis models: Challenges and successes. Toxicol. Pathol. 34: 11–18.

Uemoto, Y., Y. Soma, S. Sato, T. Shibata, H. Kadowaki, K. Katoh, E. Kobayashi, and K. Suzuki. 2012. Mapping QTL for fat area ratios and serum leptin concentrations in a Duroc purebred population. Anim. Sci. J. 83: 187–193.

Van Poucke, M., A. Tornsten, M. Mattheeuws, A. Van Zeveren, L.J. Peelman, and B.P. Chowdhary. 1999. Comparative mapping between human chromosome 3 and porcine chromosome 13. Cytogenet. Cell Genet. 85: 279–284.

Verhaeghe, R., C. Zerrweck, T. Hubert, B. Trechot, V. Gmyr, M. D'Herbomez, P. Pigny, F. Pattou, and R. Caiazzo. 2014. Gastric bypass increases postprandial insulin and GLP-1 in nonobese minipigs. Eur. Surg. Res. 52: 41–49.

Vingborg, R.K.K., V.R. Gregersen, B.J. Zhan, F. Panitz, A. Hoj, K.K. Sorensen, L.B. Madsen, K. Larsen, H. Hornshøj, X. Wang, and C. Bendixen. 2009. A robust linkage map of the porcine autosomes based on gene-associated SNPs. BMC Genomics 10: 134.

Vodicka, P., K. Smetana, B. Dvorankova, T. Emerick, Y.Z. Xu, J. Ourednik, V. Ourednik, and J. Motlík. 2005. The miniature pig as an animal model in biomedical research. Stem Cell Biol. Dev. Plast 1049: 161–171.

Wakabayashi, K., M. Okamura, S. Tsutsumi, N.S. Nishikawa, T. Tanaka, I. Sakakibara, J. Kitakami, S. Ihara, Y. Hashimoto, T. Hamakubo, T. Kodama, H. Aburatani, and J. Sakai. 2009. The peroxisome proliferator-activated receptor gamma/retinoid X receptor alpha heterodimer targets the histone modification enzyme PR-Set7/Setd8 gene and regulates adipogenesis through a positive feedback loop. Mol. Cell Biol. 29: 3544–3555.

Wang, H.W., I.M. Langohr, M. Sturek, and J.X. Cheng. 2009. Imaging and quantitative analysis of atherosclerotic lesions by CARS-based multimodal nonlinear optical microscopy. Arterioscl. Throm. Vas. 29: 1342–1348.

Wernersson, R., M.H. Schierup, F.G. Jorgensen, J. Gorodkin, F. Panitz, H.H. Staerfeldt, O.F. Christensen, T. Mailund, H. Hornshøj, A. Klein, J. Wang, B. Liu, S. Hu, W. Dong, W. Li, G.K. Wong, J. Yu, J. Wang, C. Bendixen, M. Fredholm, S. Brunak, H. Yang, and L. Bolund. 2005. Pigs in sequence space: A 0.66X coverage pig genome survey based on shotgun sequencing. BMC Genomics 6: 70.

Widdowson, E.M. 1950. Chemical composition of newly born mammals. Nature 166: 626–628.

Widdowson, E.M. 1968. Growth and composition of the fetus and newborn. pp. 324–346. *In*: M. Asalim (ed.). The Biology of Gestation, Vol. 2. Academic Press, New York, USA.

Wilkerson, V.A., and R.A. Gortner. 1932. The chemistry of embryonic growth III. A biochemical study of the embryonic growth of the pig with special reference to nitrogenous compounds. Am. J. Physiol. 102: 153–166.

Will, K., C. Kalbe, J. Kuzinski, D. Losel, T. Viergutz, M.F. Palin, and C. Rehfeldt. 2012. Effects of leptin and adiponectin on proliferation and protein metabolism of porcine myoblasts. Histochem. Cell Biol. 138: 271–287.

Wilson, P.W.F., R.B. D'Agostino, H. Parise, L. Sullivan, and J.B. Meigs. 2005. Metabolic syndrome as a precursor of cardiovascular disease and type 2 diabetes mellitus. Circulation 112: 3066–3072.

Witczak, C.A., E.A. Mokelke, R. Boullion, J. Wenzel, D.H. Keisler, and M. Sturek. 2005. Noninvasive measures of body fat percentage in male Yucatan swine. Comp. Med. 55: 445–451.

Wojtysiak, D., and K. Poltowicz. 2014. Carcass quality, physico-chemical parameters, muscle fibre traits and myosin heavy chain composition of m. longissimus lumborum from Pulawska and Polish Large White pigs. Meat Sci. 97: 395–403.

World Health Organization. Diabetes—Factsheet. 2012. http://www.who.int/mediacentre/factsheets/fs312/en/index.html.

Wulster-Radcliffe, M.C., K.M. Ajuwon, J. Wang, J.A. Christian, and M.E. Spurlock. 2004. Adiponectin differentially regulates cytokines in porcine macrophages. Biochem Biophys. Res. Commun. 316: 924–929.

Wright, A.D., C.H. Walsh, M.G. Fitzgerald, and J.M. Malins. 1979. Very pure porcine insulin in clinical practice. Br. Med. J. 1: 25–27.

Xi, S.M., W.D. Yin, Z.B. Wang, M. Kusunoki, X. Lian, T. Koike, J. Fan, and Q. Zhang. 2004. A minipig model of high-fat/high-sucrose diet-induced diabetes and atherosclerosis. Int. J. Exp. Pathol. 85: 223–231.

Xu, H.Y., G.T. Barnes, Q. Yang, Q. Tan, D.S. Yang, C.J. Chou, J. Sole, A. Nichols, J.S. Ross, L.A. Tartaglia, and H. Chen. 2003. Chronic inflammation in fat plays a crucial role in the development of obesity-related insulin resistance. J. Clin. Invest. 112: 1821–1830.

Yamauchi, T., J. Kamon, Y. Minokoshi, Y. Ito, H. Waki, S. Uchida, S. Yamashita, M. Noda, S. Kita, K. Ueki, K. Eto, Y. Akanuma, P. Froguel, F. Foufelle, P. Ferre, D. Carling, S. Kimura, R. Nagai, B.B. Kahn, and T. Kadowaki. 2002. Adiponectin stimulates glucose utilization and fatty-acid oxidation by activating AMP-activated protein kinase. Nat. Med. 8: 1288–1295.

Yan, H., R. Potu, H. Lu, V.V. de Almeida, T. Stewart, D. Ragland, A. Armstrong, O. Adeola, C.H. Nakatsu, and K.M. Ajuwon. 2013. Dietary fat content and fiber type modulate hind gut microbial community and metabolic markers in the pig. PLoS One 8: e59581.

Yang, H.S., F.N. Li, X. Xiong, X.F. Kong, B. Zhang, X.X. Yuan, J. Fan, Y. Duan, M. Geng, L. Li, and Y. Yin. 2013. Soy isoflavones modulate adipokines and myokines to regulate lipid metabolism in adipose tissue, skeletal muscle and liver of male Huanjiang mini-pigs. Mol. Cell Endocrinol. 365: 44–51.

Yang, L.H., M. Guell, D. Niu, H. George, E. Lesha, D. Grishin, J. Aach, E. Shrock, W. Xu, J. Poci, R. Cortazio, R.A. Wilkinson, J.A. Fishman, and G. Church. 2015. Genome-wide inactivation of porcine endogenous retroviruses (PERVs). Science 350: 1101–1104.

Yang-Feng, T.L., F.Y. Xue, W.W. Zhong, S. Cotecchia, T. Frielle, M.G. Caron, R.J. Lefkowitz, and U. Francke. 1990. Chromosomal organization of adrenergic-receptor genes. Proc. Natl. Acad. Sci. USA 87: 1516–1520.

Yin, W., D. Liao, M. Kusunoki, S. Xi, K. Tsutsumi, Z. Wang, X. Lian, T. Koike, J. Fan, Y. Yang, and C. Tang. 2004. NO-1886 decreases ectopic lipid deposition and protects pancreatic beta cells in diet-induced diabetic swine. J. Endocrinol. 180: 399–408.

Yukimura, Y., and G.A. Bray. 1978. Effects of adrenalectomy on body-weight and the size and number of fat-cells in the Zucker (Fatty) rat. Endocr. Res. Commun. 5: 189–198.

Zhang, Y.Y., R. Proenca, M. Maffei, M. Barone, L. Leopold, and J.M. Friedman. 1994. Positional cloning of the mouse obese gene and its human homolog. Nature 372: 425–432.

CHAPTER-3

Growth Hormone and the Chick Eye

Steve Harvey[1],* and *Carlos G. Martinez-Moreno*[2]

INTRODUCTION

The chick embryo is a classical model to study eye development (Adler and Canto-Soler, 2007; Belecky-Adams et al., 2008; Goodall et al., 2009; Vergara and Canto-Soler, 2012). As pituitary growth hormone (GH) is an established growth factor (Harvey, 2013), it was tacitly assumed that it was causally involved in embryogenesis and eye growth. This belief was, however, re-evaluated with the realization that embryogenesis was a *growth-without GH syndrome* (Geffner, 1996) and that pituitary somatotrophs only arise ontogenetically during the last trimester of incubation (Harvey et al., 1998), after the completion of approximately 40 of the 46 Hamilton and Hamburger stages of development (Figure 1) (Hamburger and Hamilton, 1951). Nevertheless, with the demonstration that GH gene expression is not confined to pituitary somatotrophs and occurs widely in extrapituitary tissues (Figure 2) (Harvey, 2010), it is likely that GH is involved in the embryogenesis and eye growth as a local autocrine or paracrine factor (Harvey and Baudet, 2014).

Ocular Growth Hormone

It is now well established that GH is present in the eyes of all vertebrate groups (Harvey et al., 2007a,b). In avian species GH immunoreactivity is abundantly present in the neural retina by ED (embryonic day) 5 of the

[1] Department of Physiology, University of Alberta, Edmonton, T6G 2H7, Canada.
[2] Departamento de Neurobiología Celular y Molecular, Instituto de Neurobiología, Campus Juriquilla, Universidad Nacional Autónoma de México, Querétaro, Qro., 76230, México.
* Corresponding author: steve.harvey@ualberta.ca

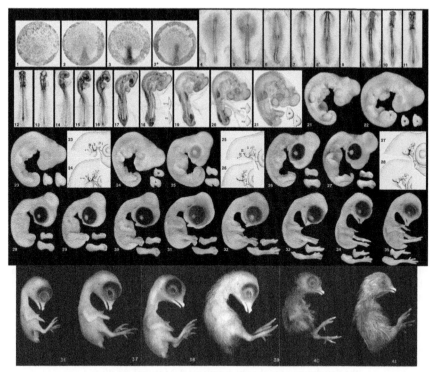

Figure 1: A series of normal stages in the development of the chick embryo from Hamburger, V., and H.L. Hamilton. 1951. J. Morph. 88: 49–92. Reprinted in Developmental Dynamics 195, 231–72 (1992). Poster by Drew M. Noden, Cornell University; sponsored by the American Association of Anatomists; production by Wiley-Liss, Inc.

21 day incubation period (Harvey et al., 1998). This immunoreactivity is widespread through most layers of the neural retina (Harvey et al., 2001, 2003; Baudet et al., 2003) but is especially abundant in the ganglion cell layer (GCL), in which it is located within cytoplasmic and nuclear compartments (Baudet et al., 2003; Sanders et al., 2005). This immunoreactivity appears to reflect a secretory form, since submonomer GH proteins of 15- and 16-kDa are released into incubation media following the culture of ED6-9 retinal explants (Baudet et al., 2003; Sanders et al., 2005). This immunoreactivity appears to be secreted from the retinal ganglion cells (RGCs) of the neural retina, as it is co-localized with islet-1, a nuclear antigen specific for RGCs (Baudet et al., 2007a). GH immunoreactivity is thus found in the vitreous fluid of chicken embryo eyes and the eyes of neonatal chicks (Baudet et al., 2003). Within the vitreous, the secreted GH becomes bound to opticin, a unique proteoglycan binding protein produced in the neural retina (Sanders et al., 2003).

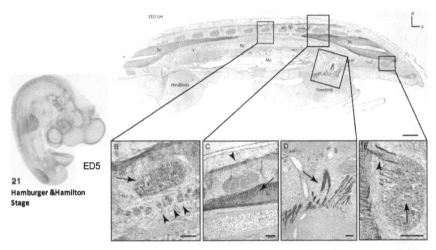

Figure 2: Growth hormone (GH) immuno-reactivity within a cross section of an ED5 chick embryo body. (A) Low magnification photograph of GH immuno-reactivity visualized in an ED5 chick embryo body. (B–E) High magnification images of: a vertebral condensation (arrow) and dorsal root ganglia (arrowhead) (B); spinal nerves (arrow) and bone collar (arrowhead) (C); spinal nerves innervating the limb (arrow) (D); spinal nerves (arrow) innervating a vertebral condensation (E). Abbreviations: a, anterior; d, dorsal; Me, Mesonephros; Nc, notochord; Sc, spinal cord; Ve, ventricle. Scale bars, (A) 1 mm; (B–E) 100 μm. (From Harvey, S., and M.L. Baudet. 2014. Extrapituitary growth hormone and growth? Gen. Comp. Endocrinol. 205: 55–61 and ED5 embryo from Hamburger, V., and H.L. Hamilton. 1951. A series of normal stages in the development of the chick embryo. J. Morphol. 88: 49–92).

Although retinal GH is present in the cytoplasm of the RGCs (Figure 3), it is also present in the axons derived from these cells and it accumulates in the fasciles of the optic fiber layer (OFL) (Baudet et al., 2003, 2007a). The OFL coalesce at the optic nerve head (ONH) by ED5 (Figure 4A) and the GH-staining optic nerve (ON) exit the eye by ED7 (Figure 4D). GH is also present in the optic chiasm (OC) by ED7-8, when the axons of the optic nerve decussate (Figure 4E). The retinofugal optic nerve then enters into the *striatum opticum* (SO) and *striatum griseum et fibrosum superficiale* (SGFS) of the optic tectum (OT), where the RGC axons terminate (Figure 4I–K) (Baudet et al., 2007a). Within the OFL, ONH, ON, OC, SO and OT, the GH staining of the retinofugal fibres are colocalized with neurofilament associated proteins, reflecting their neural origin. Interestingly, GH is detected in the developing OT at ED4, before the first RGC axons reach this tissue (at ED6). Its presence in the OT is therefore not due to sequestration from the retinofugal fibers. This was confirmed, by the finding of GH mRNA in the cytoplasm of cells in the ventricular zone of OT at this early developmental stage (Baudet et al., 2007a).

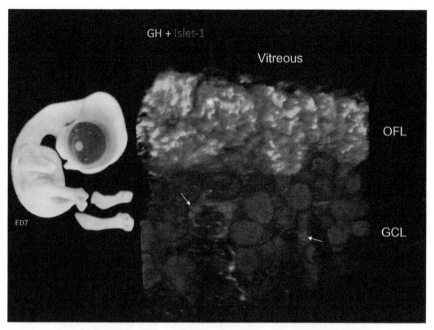

Figure 3: GH immuno-reactivity in the neural retina of E7 chick embryos. GH immuno-reactivity shown in green in fascicles (asterisk) of the optic fiber layer (OFL) and in the cytoplasm/perinuclear area (arrow) of large rounded cells in the retinal GCL. Immuno-reactivity for islet-1, a nuclear marker of RGCs, is also shown (in red). Arrows represent illustrative RGCs (in red) containing cytoplasmic GH staining (in green). (unpublished Z-stack of GH immuno-reactivity in the ED 7 chick neural retina by Marie-Laure Baudet and Steve Harvey) (ED7 chick embryo from Hamburger, V., and H.L. Hamilton. 1951. A series of normal stages in the development of the chick embryo. J. Morphol. 88: 49–92.)

The retinofugal fibres synapse with the visual centres in the brain by ED10 and ED12 (Baudet et al., 2007a), suggesting the presence of GH in the OFL might result in its anterograde transport in RCG axons, it is therefore possible that GH in the neural retina acts as a neurotrophic factor involved in axonal growth or guidance prior to the synapsing of those fibres to the visual centres in the brain. This possibility is supported by the finding that there is a loss of GH from the OFL between ED14 and ED18, at the time when GH might not be required for synaptogenesis (Baudet et al., 2007a; Harvey et al., 2007a). This possibility is also supported by the fact that proteins involved in neurite development are not expressed in the neural retina of GHR-knockout mice, as in normal mice (Baudet et al., 2008).

While the GH gene expressed in the neural retina of chick embryos codes for the full-length monomer protein (Figure 5A), GH immunoreactivity is largely associated with proteins of 15-16 kDa. These submonomer proteins are also present in pituitary extracts, in which the 22-kDa monomer (or 25-kDa in reducing conditions) is the most abundant moiety (Figure 5B)

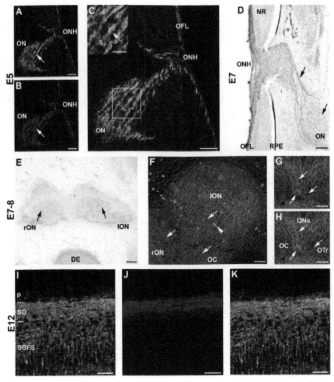

Figure 4: GH immuno-reactivity in the RGC axons (retinofugal tract) of chick embryos. Confocal microscopy of GH immune-reactivity (green) (A) and neurofilament-associated protein immuno-reactivity (red) (B) at the back of the eye of E5 chick embryos. (C) Image overlay of A and B showing co-localization (yellow–orange coloration) in the OFL, optic nerve head (ONH) and optic nerve (ON). A higher magnification of GH in the fibers of the ONH is shown in the inset. The arrow points to an illustrative fascicle of the ON where GH (A) and neurofilament-associated protein (B) are co-localized. (D) Brightfield microscopy of GH immuno-reactivity (brown) in the OFL of the neural retina (NR), ONH and ON in the eyes of E7 chick embryos. Note the presence of GH in fascicles of the ON (arrows) (RPE, retinal pigmented epithelium). (E) Brightfield microscopy of GH immuno-reactivity (brown) in the left (lON) and right (rON) optic nerves, just prior to decussation in the optic chiasm (OC) at E7 (DE, diencephalon). (F) Co-localization (yellow–orange coloration) of GH immuno-reactivity (green) in fibers immuno-reactive for neurofilament-associated protein (red) in lON, rON and OC (arrows) of E7 chick embryos. Nuclei are stained in blue with DAPI. Note the presence of GH in cells of the ON (arrowhead). (G) GH-immuno-reactivity (green) in both ONs, OC and optic tracts (OTr) (arrows) of E8 embryonic chicks. (H) Co-localization (yellow–orange coloration) of GH immuno-reactivity (green) and neurofilament-associated protein (red) in ONs, OC and OTr. (I) GH immuno-reactivity in the *stratum opticum* (SO) of the optic tectum (OT) in the brain of E12 chick embryos and in the *stratum griseum et fibrosum superficiale* (SGFS) where RGC axons terminate (p, pia). (J) Neurofilament-associated immuno-reactivity in the SO of E12 chick embryo OT. (K) Merged overlay of I and J showing colocalization (yellow–orange coloration) in the SO, but not in the SGFS. Scale bars = 20 μm (A–C, F, I–K), 10 μm (C, inset), 50 μm (D, E, G, H). (From Baudet, M.L., D. Rattray, and S. Harvey. 2007a. Growth hormone and its receptor in projection neurons of the chick visual system: retinofugal and tectobulbar tracts. Neuroscience 148: 151–163.)

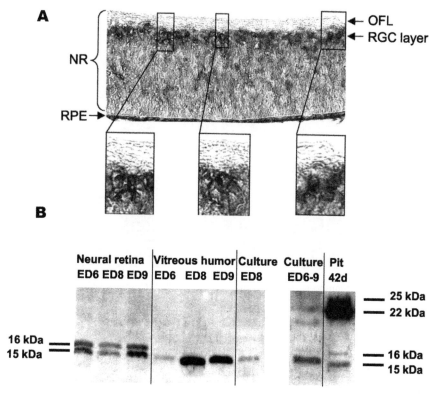

Figure 5: GH mRNA and immuno-reactivity in embryonic retinal ganglion cells (RGCs). (A) *In situ* hybridization of GH mRNA in the neural retina of ED7 chick embryos. (A) Specific hybridization with a 690-bp DIG-labeled HindIII antisense probe for GH mRNA is shown throughout the neural retina (NR), particularly in a layer of large cells in the RGC layer below the optic fiber layer (OFL). (B) Western blotting of GH immuno-reactive proteins in the neural retina and vitreous humor of chick embryos (pooled tissues from at least three embryos) at ED6, ED8, and ED9, in comparison with proteins in the pituitary (Pit) glands of slaughterhouse (42-d-old) chickens and proteins in culture media after the incubation of retinal tissues. (From Baudet, M.L., E.J. Sanders, and S. Harvey. 2003. Retinal growth hormone in the chick embryo. Endocrinology 144: 5459–5468.)

(Baudet et al., 2003). In addition to the neural retina, the same 15-kDa GH protein is also found in the OT of ED7-8 embryos (Baudet et al., 2007a). It is thus highly likely that after translation, the full-length monomer is rapidly degraded within the neural retina, especially as exogenous recombinant chicken GH is similarly found to be degraded into 15-16 kDa moieties when incubated with embryonic retinal extracts (Harvey et al., 2007a).

In addition to the GH gene that codes for the full-length monomer protein, a second transcript which codes for a small (16.5-kDa) protein (s-cGH, small-chicken GH) was also discovered in the eyes of ED17 chicken embryos (Takeuchi et al., 2001). This novel GH mRNA is transcribed from

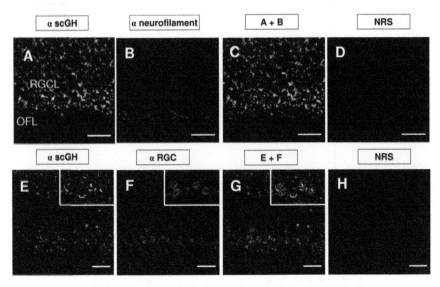

Figure 6: scGH immuno-reactivity in the NR of ED7 chick embryos. (A), scGH immuno-reactivity (arrows) is seen in the cytoplasm of cells in the retinal ganglion cell layer (RGCL). No scGH immuno-reactivity is present in the OFL. Staining for neurofilament immuno-reactivity is shown (B) and is confined to the OFL. The image overlay (C) clearly shows that scGH immuno-reactivity and neurofilament immuno-reactivity are discrete. The specificity of detection is shown in a control, NRS section (D). The immuno-reactivity for scGH in the NR (E) is also compared with immuno-reactivity for a RGC marker (F). The yellow orange coloration in the image overlay (G) demonstrates the presence of scGH in RGC cytoplasm and nuclei. The specificity of detection is shown in a control normal rabbit serum (NRS)-treated section (H). Bars, 20 µm. (From Baudet, M.L., B. Martin, Z. Hassanali, E. Parker, E.J. Sanders, and S. Harvey. 2007. Expression, translation, and localization of a novel, small growth hormone variant. Endocrinology 148: 103–115.)

the middle of intron 3 of the GH gene. The deduced protein is a cytosolic protein of 16.5-kDa with 140 amino acid residues (Figure 6). s-cGH lacks the signal peptide and the N-terminal amino acid residues of 22-kDa GH, replacing them with 20 aberrant amino acid residues. Immunoreactivity for this protein is found in the retinal pigmented epithelium (RPE) of the retina from ED10, increasing in abundance until it reaches a peak at ED17. By hatching, s-cGH immunoreactivity rapidly decreases and it is not detectable after hatching. Using a specific antibody against the unique N-terminus of s-cGH, immunoreactivity was found to be associated with neural retinal proteins of approximately 16-kDa, comparable with its predicted size (Baudet et al., 2007b). Most of the s-cGH immunoreactivity detected is, however, associated with a 31-kDa moiety, suggesting s-cGH is normally present in a dimerized form. Neither proteins were, however, present in the media of human epithelial kidney (HEK) cells that had been transfected with s-cGH DNA after its insertion into an expression plasmid. This suggests s-cGH is not a secretory product, consistent with its lack of signal peptide sequence. Similar s-cGH moieties of 16- and 31-kDa were found in proteins extracted from other ocular tissues (the neural retina, RPE, cornea and choroid) of chicken embryos, although they are not consistently present in the vitreous humor, consistent with its lack of secretion within the eye (Baudet et al., 2007b). Specific s-cGH immunoreactivity was also detected in chicken ocular tissues by immunohistochemistry, but it was not detected in axons of the OFL or ONH, which were both immunoreactive for full-length chicken GH. Thus, although s-cGH is expressed and translated in chick ocular tissues, its localization in the neural retina and ONH is distinct from the full-length protein.

In addition to RGCs and their axons, GH immunoreactivity is widely distributed in the neural retina and staining is also seen in the inner and outer nuclear layers, and within the photoreceptor layer (Harvey et al., 2016). GH immunoreactivity is also present in the choroid layer, the sclera and the cornea (Harvey et al., 2003). It is also abundant in the lens, in which the epithelial lens fiber cells are intensely stained, particularly in the nuclei (Harvey et al., 2001).

Ocular GH receptors

Ocular tissues are not just sites of GH production, as they express the GH receptor (GHR) gene and are thus sites of GH actions. Ocular GHRs were first demonstrated in the chicken eye by Tanaka et al. (1996), by their demonstration of GHR mRNA in the eyes of ED16 chicks embryos. The presence of GHR proteins in ocular tissues was first shown by the finding of GHR immunoreactivity in the optic vesicle of ED3 embryos (Harvey et al., 2001) and subsequently by RT-PCR and the detection of GHR cDNA for regions coding for the extracellular and intracellular domains of the GHR in extracts of whole eye, neural retina and RPE of ED7 chicks (Harvey et al., 2003). GHR expression was also shown, by *in situ* hybridization, to be present in large rounded cells in the ganglion cell layer of the neural retina. As these are presumptive GH-secreting RGCs, GH is likely to act locally within the neural retina in autocrine or paracrine ways. GHR immunoreactivity was similarly found in the OFL and mirrored the localization of GH (Figure 7) (Harvey et al., 2007a). The expression of the GHR gene in the OT of ED7-8 embryos was also demonstrated by RT-PCR (Baudet et al., 2007a). Interestingly, GHRG (GH responsive-gene)-1 mRNA was also expressed in the OT of ED7-8 embryos (Baudet et al., 2007a). As GHRG-1 is a specific marker of GH action (Agarwal et al., 1995; Radecki et al., 1997) in chick embryo brain (Harvey et al., 2002), this suggests GHR mediated GH action within the visual tract prior to the ontogeny of pituitary somatotrophs (at ED15) (Porter et al., 1995) and the endocrine actions of circulating GH (at ED17; Harvey et al., 1979).

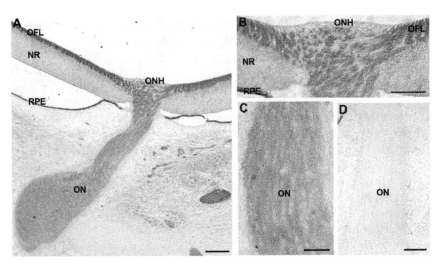

Figure 7: GHR immuno-reactivity in the neural retina of E7-8 chick embryos. (A) Brightfield microscopy showing GHR immuno-reactivity (brown) in the back of the eye of E7 embryonic chicks, in the OFL, in the optic nerve head (ONH) and in the optic nerve (ON) (RPE, retinal pigmented epithelium). (B) Higher magnification of ONH shown in A. (C) Higher magnification of ON shown in A. Negative (normal mouse serum) control of the ON (D), ONH (E), optic chiasm (OC) and optic tract (OTr). (From Baudet, M.L., D. Rattray, and S. Harvey. 2007a. Growth hormone and its receptor in projection neurons of the chick visual system: retinofugal and tectobulbar tracts. Neuroscience 148: 151–163.)

In addition to the full-length GHR, other receptors may be present in the chick eye, as s-cGH immunoneutralization inhibits RGC survival in chick embryo cultures (Baudet and Harvey, 2007). This suggests it has functional activity, yet structural analysis of the s-cGH protein shows that it cannot bind to the classical GHR (Baudet et al., 2007b). The 15 kDa GH variant expressed in the chick neuroretina may similarly not act via the classical GHR and other non-classical signaling pathways may be present (Harvey et al., 2014).

Ocular GH actions

(1) Neuroprotection

Within the chick eye, the first demonstrated action of exogenous GH was its stimulation of a 5-fold induction of IGF (insulin-like growth factor)-1 mRNA after a 48 h culture of ED8 (Baudet et al., 2003). This treatment was accompanied by a decrease in the number of TUNEL-labelled cells in the retinal explants (Sanders et al., 2005) and a decrease in the content of caspase 3 mRNA and a decrease in the content of apoptosis inducing factor (AIF)-1 mRNA (Harvey et al., 2006). This result suggested a role for retinal GH in

RGC cell survival during developmental waves of apoptosis. Within the neural retina, two developmental waves of apoptosis are known to exist, one which peaks at ED7 and a second that peaks at ED12 (Sanders et al., 2005). Retinal GH was thus thought to be neuroprotective against both of these developmental waves of apoptosis. This role in early embryogenesis was supported by the fact that the GH treatment of immuno-panned ED8 RGCs similarly reduced the number of apoptotic cells (Sanders et al., 2006). This neuroprotective action was due to a local autocrine or paracrine actions of retinal GH, as the immuno-neutralization of endogenous GH in immuno-panned ED8 RGCs tripled the incidence of apoptosis (Sanders et al., 2005). Moreover, when the same antibodies were microinjected into the optic cup of ED2 embryos, the incidence of apoptosis was disrupted in comparison with ED2 embryos microinjected with control (normal rabbit) serum (Sanders et al., 2005). Not surprisingly, GH immuno-neutralization of cultured immuno-panned RGCs was found to inhibit Akt phosphorylation and to increase the accumulation of caspase 3 and the cleavage of PARP-1, through the activation of PARP-1. The GH treatment of immuno-panned RGCs was also shown to reduce the cleavage of caspase 9 and to activate cytosolic tyrosine kinases (Trks) and extracellular-signal-related kinases (Erks), which together converge in the activation of cAMP response element binding protein (CREB) and initiate the transcription of pro- and anti-apoptotic genes (Sanders et al., 2008). These signalling pathways are common to other neurotrophins (e.g., brain-derived growth factor and transforming growth factor-1) and GH can thus be considered to be an authentic growth and differentiation factor in the development of the embryonic retina. IGF-1 has been shown to similarly promote cell survival in the neural retina through similar signaling pathways and as GH increases IGF-1 expression in ED8 retinal explants (Baudet et al., 2003), the possibility that the neuroprotective action of GH might be IGF-dependent was investigated. The simultaneous immuno-neutralization of both GH and IGF-1 did not increase the level of apoptosis in immuno-panned RGC cultures above that achieved by immuno-neutralization of GH alone, suggesting the neuroprotective action of GH is mediated in large part through the action of IGF-1 (Sanders et al., 2009a).

The immuno-neutralization of endogenous retinal GH provides evidence that it is of physiological importance as a paracrine or autocrine in cell survival. This role has also been shown by GH gene silencing in the embryonic chick retina. For instance, using siRNA to silence the local synthesis of GH (and IGF-1) in QNR/D cells there is an increase in the appearance of cells with apoptotic fragmented nucleus morphology (Sanders et al., 2010, 2011). Moreover, when the siRNA is microinjected into the eye cup of ED4 chick embryos it was found to significantly increase the number of apoptotic cells in flatmounts of the ED5, embryos in comparison

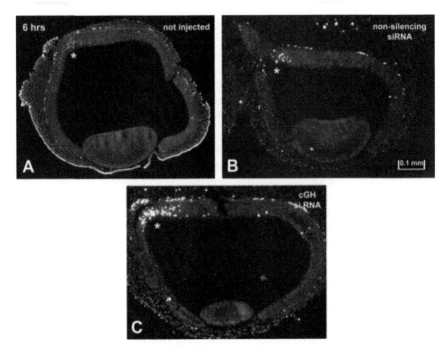

Figure 8: Sections of ED5 eyes 6 h after intra-vitreal injection. (A) A non-injected eye. (B) An eye injected with non-silencing siRNA. Note that in both (A) and (B) there are concentrations of apoptotic cells in the region of the optic fissure (asterisks). (C) An eye injected with cGH siRNA. Note the high concentration of apoptotic cells in the region of the optic fissure in comparison with (A) and (B) (asterisk). (From Sanders, E.J., W.Y. Lin, E. Parker, and S. Harvey. 2011. Growth hormone promotes the survival of retinal cells *in vivo*. Gen. Comp. Endocrinol. 172: 140–150.)

with embryos microinjected with the non-silencing siRNA (Figure 8) (Sanders et al., 2011). Within the neural retina, the apoptotic retinal cells were mostly found close to the optic fissure, which is a transient embryological structure in which the cells are more prone to apoptosis. GH expression in the neural retina therefore promotes retinal cell survival by autocrine or paracrine GH actions.

Cell survival in the neural retina was also found to be correlated with the amount of GH expressed. This was shown by the reduction on cell survival when endogenous GHRH (GH-releasing hormone) was reduced by GHRH immuno-neutralization (Martinez-Moreno et al., 2014a) thereby removing the stimulatory effect of GHRH on retinal GH synthesis and retinal GH secretion (Harvey et al., 2012; Martinez-Moreno et al., 2014a). The autocrine or paracrine control of retinal GH release therefore contributes to the autocrine or paracrine control of GH in terms of retinal cell survival.

In addition to increasing RGC survival during developmental waves of apoptosis, a similar signaling mechanism might protect RGCs, against the cell death induced by glutamate-induced excitotoxicity (Ientile et al.,

2001). The possibility that this might occur is supported by the presence of glutamate receptors in the embryonic chick neuroretina and OT and their presence in immortalized QNR/D cells, which provide an experimental model of RGC function (Martinez-Moreno et al., 2014b). The presence of these receptors was first shown by RT-PCR using oligonucleotide primers against the GluR2 and GluR3 receptor subunits of the AMPA receptor and primers for the NMDA receptor (NR1 subunit) (Martinez-Moreno et al., 2016). The presence of the metabotropic (GRM6+7 subunits) and ionotropic (GluR2+3 subunits) AMPA receptor proteins was also shown by immunohistochemistry. The toxicity of glutamate was show by its ability to induce cell death, especially in the presence of buthionine sulfoxamide (BSO), as shown by TUNEL-labeling and the release of lactate dehydrogenase (LDH) (Martinez-Moreno et al., 2016). When exogenous recombinant chicken GH was added to QNR/D cultures it was effective in a dose-related way, in reducing glutamate-induced cell death and LDH release (Martinez-Moreno et al., 2016). The GH treatment of QNR/D cells was also found to increase the abundance of pSTAT5 immunoreactivity and immunoreactivity for Bcl-2 (Martinez-Moreno et al., 2016). Bcl-2, abundance was also increased in abundance after explants of ED8 chick neuroretina were incubated in 100 nM recombinant chicken GH (Martinez-Moreno et al., 2016). The finding that GH is able to protect against excitotoxicity in the glutamate cell death assay implies its prevention of both apoptosis and necrosis. These results may also have clinical relevance, since excitotoxicity is an underlying damage mechanism involved in many neurodegenerative processes including neuroretinal diseases. Indeed, RGC death is a cause of glaucoma and while the absence of GH in human RGCs is correlated with a 100% incidence of RGC apoptosis, the presence of GH in human RGCs is associated with a 100% incidence of RGC survival (Sanders et al., 2009b).

(2) Neurite growth

Within the nervous system, a few studies have suggested that GH promotes cell proliferation and differentiation (Turnley et al., 2002; Ajo et al., 2003; McLenachan et al., 2009; Aberg et al., 2009; Aberg, 2010). In the chick embryo, exogenous GH was also found to increase the length of RGC axons sprouting from cultured immuno-panned RGCs after a 3d cultured in 10^{-6} or 10^{-9} M recombinant chicken GH (Figure 9A and C) (Baudet et al., 2009). Exogenous GH also increased the number of cells that had sprouting neurites (Figure 9C) (Baudet et al., 2009). This action reflected a neurotrophic role of endogenous, retinal GH, as the siRNA knockdown of retinal GH in immuno-panned RGCs, reduced the number of cells with sprouting neurites and reduced neurite length (Figure 9D) (Baudet et al., 2009). Retinal GH, thus acts as an autocrine or paracrine growth factor to promote axon growth in the chick neural retina.

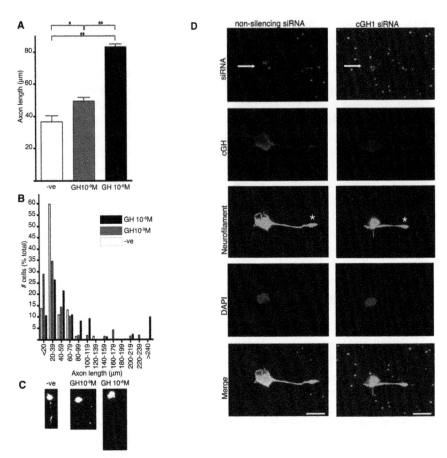

Figure 9: Growth hormone promotes axon growth in chicken (ED7) retinal ganglion cells. (A) Effect of 10^{-9} and 10^{-6} m GH treatment on axon elongation of immuno-panned RGC after 3 d in culture. Values are means ± SEM (n = 3 dishes). *, P < 0.05; **, P < 0.001 (ANOVA followed by Tukey's multiple comparison test). (B), Proportion of cells elongating axons of specific length in absence or presence of treatment. (C), Illustrative RGCs in absence or presence of cGH treatment. –ve, Negative. (D), Illustrative siRNA-transfected immuno-panned RGCs. Note the decrease in GH immuno-reactivity (red) after transfection with cGH1-labeled siRNA (green, arrow), compared with that in cells transfected with nonsilencing siRNA. Neurofilament-associated protein marker (light blue) was used to detect RGC neurites. Nuclei are stained with DAPI (dark blue). Note the difference in axon length (star) between nonsilencing and cGH1 siRNA transfected RGCs. Scale bar, 10 μm. (Baudet, M.L., D. Rattray, B.T. Martin, and S. Harvey. 2009. Growth hormone promotes axon growth in the developing nervous system. Endocrinology 150: 2758–2766.)

(3) Synaptogenesis

While actions of GH in neural differentiation are now well established (Waters and Blackmore, 2011), there is a paucity of information of the possible involvement GH in synaptogenesis. However, SNAP-25 (synaptomalassociated protein 25) immuno-reactivity in ED8-10 neuoretinal explants was shown to be increased in response to exogenous GH treatment (Fleming et al., 2016). This is just prior to the onset of synaptogenesis of amacrine cells and RGCs (Catsicas et al., 1991) and it coincides with the peak presence of GH in the OFL of the neural retina (Baudet et al., 2009). The possibility that GH is involved in synaptogenesis is also supported by the finding that the intravitreal injection of GH increases SNAP-25 immuno-reactivity in the eye of ED10 embryos (Fleming et al., 2016). This possibility is also supported by the finding that GH treatment also increased GAP-43 (growth associated protein 43) immuno-reactivity in the chick neural retina and in QNR/D cells (Fleming et al., 2016). GAP-43 is known to be critical for the development of the nervous system and for synaptogenesis (Latchney et al., 2014). Moreover, as secretoneurin is also thought to be involved in synaptogenesis (Marksteiner et al., 2002), the co-localization of secretoneurin and GH in QNR/D cells (Martinez-Moreno et al., 2015) further suggests the involvement of GH in synapse formation within the chick neural retina.

Ocular GH secretion

The expression of GH mRNA in the neural retina of chick embryos is likely to be stimulated, as in the pituitary gland, by the action of GHRH (GH-releasing hormone). GHRH immuno-reactivity was shown to be present and co-localized within the RGCs of the ganglion cell layer in ED7 chick embryos (Harvey et al., 2012). It was similarly co-localized with GH in quail-derived QNR/D cells (Martinez-Moreno et al., 2014b). When incubated with 1 μM GHRH for 24 h it significantly increased (by 2 fold) the expression of GH mRNA in comparison with controls. Exogenous GHRH also increased the amount of GH released from QNR/D cells, causing a depletion in the GH content acutely (within 15 min) but increasing the GH content after long-term (48 h) culture (Martinez-Moreno et al., 2014a).

In addition to GHRH, TRH (thyrotrophin-releasing hormone) was also shown to be immunologically present in QNR/D cells (Harvey et al., 2012) and to stimulate GH synthesis and release from QNR/D cells when incubated with TRH *in vitro* (Martinez-Moreno et al., 2014a). GH release from QNR/D cells is thus similar to that from the pituitary gland, although the actions of SRIF (somatostatin or GH-release inhibitory hormone, GHRIH) and IGF-1 are currently unknown. Both of these factors, which inhibit pituitary GH release, are however, immuno-cytochemically present in QNR/D cells (Harvey et al., 2012). It is thus possible that GH release from RGCs in the chick embryo may be regulated by a number of factors that interact in autocrine, paracrine or intracrine ways.

The release of GH from chick RGCs is likely to be in secretory granules that were identified by electron microscopy in QNR/D cells (Martinez-

Moreno et al., 2015). Within the granules inside the QNR/D cells, GH was co-localized with secretoneurin, a neuropeptide derived from secretogranins, which are similarly found in pituitary somatotrophs, in which GH is also stored and co-secreted with secretoneurin. The release of GH from QNR/D cells is also likely to involve SNAP-25 (synaptosomal-associated protein 25), with which it was also co-localized. This suggests that GH release occurs following the specific fusion of vesicles with the plasma membrane, similar to the release of GH from the anterior pituitary cells (Rotondo et al., 2008).

Pituitary Growth Hormone in Ontogenic Chick Eye Development

As pituitary GH-secreting somatotrophs are not present until the last trimester of incubation, most eye development occurs in the absence of pituitary GH action, and eye development largely reflects autocrine or paracrine actions of ocular GH. However, as pituitary somatotrophs arise ontogenetically by ED14 of incubation (Porter et al., 1995), the presence of GH in plasma, by ED17 (Harvey et al., 1979), suggests pituitary GH may contribute to development of the chick eye in the last week of incubation. This possibility is supported by the finding that Cy3-labelled GH was translocated from peripheral plasma into the neural retina of ED15 embryos, where it was internalized into RGCs (Fleming et al., 2016). Exogenous (pituitary derived) GH was similarly internalized into QNR/D cells after its addition to incubation media. The uptake of exogenous GH was by a receptor-mediated mechanism and maximal after 30–60 min and was accompanied by STAT5 phosphorylation and increased GAP43 and SNAP25 immuno-reactivity. This suggest exogenous (endocrine) and local (autocrine/paracrine) GH are both involved in retinal function in late embryogenesis, either directly or through the induction of IGF-I or other growth factors (Diaz-Casares et al., 2005).

Summary

In summary, GH is produced and acts within the chick eye during embryogenesis, in which it has functional actions as an autocrine or paracrine regulator. Pituitary GH secretion occurs in the last trimester of incubation and as it can be internalized into the RGCs of the neural retina, it may also contribute to chick eye development in late embryogenesis. Eye development in the chick thus depends upon GH-mediated signaling pathways, that may reflect autocrine, paracrine or endocrine actions.

Keywords: Ocular growth hormone, synaptogenesis, neurite growth, pituitary growth hormone, ontogenic, chick, eye development

References

Aberg, D. 2010. Role of the growth hormone/insulin-like growth factor 1 axis in neurogenesis. Endocr. Dev. 17: 63–76.

Aberg, N.D., I. Johansson, M.A. Aberg, J. Lind, U.E. Johansson, C.M. Cooper-Kuhn, H.G. Kuhn, and J. Isgaard. 2009. Peripheral administration of GH induces cell proliferation in the brain of adult J. hypophysectomized rats. J. Endocrinol. 201: 141–150.

Adler, R., and M.V. Canto-Soler. 2007. Molecular mechanisms of optic vesicle development: Complexities, ambiguities and controversies. Dev. Biol. 305: 1–13.

Agarwal, S.K., L.A. Cogburn, and J. Burnside. 1995. Comparison of gene expression in normal and growth hormone receptor-deficient dwarf chickens reveals a novel growth hormone regulated gene. Biochem. Biophys. Res. Commun. 206: 153–160.

Ajo, R., L. Cacicedo, C. Navarro, and F. Sanchez-Franco. 2003. Growth hormone action on proliferation and differentiation of cerebral cortical cells from fetal rat. Endocrinology 144: 1086–1097.

Baudet, M.L., E.J. Sanders, and S. Harvey. 2003. Retinal growth hormone in the chick embryo. Endocrinology 144: 5459–5468.

Baudet, M.L., and S. Harvey. 2007. Small chicken growth hormone (scGH) variant in the neural retina. J. Mol. Neurosci. 31: 261–271.

Baudet, M.L., D. Rattray, and S. Harvey. 2007a. Growth hormone and its receptor in projection neurons of the chick visual system: Retinofugal and tectobulbar tracts. Neuroscience 148: 151–163.

Baudet, M.L., B. Martin, Z. Hassanali, E. Parker, E.J. Sanders, and S. Harvey. 2007b. Expression, translation, and localization of a novel, small growth hormone variant. Endocrinology 148: 103–115.

Baudet, M.L., Z. Hassanali, G. Sawicki, E.O. List, J.J. Kopchick, and S. Harvey. 2008. Growth hormone action in the developing neural retina: A proteomic analysis. Proteomics 8: 389–401.

Baudet, M.L., D. Rattray, B.T. Martin, and S. Harvey. 2009. Growth hormone promotes axon growth in the developing nervous system. Endocrinology 150: 2758–2766.

Belecky-Adams, T.L., J.M. Wilson, and K. Del Rio-Tsonis. 2008. The chick as a model for retinal development and regeneration. pp. 102–119. In: P.A. Tsonis (ed.). Animal Models in Eye Research. Academic Press. An imprint of Elsevier, 525 B Street, Suite 1900, San Diego, CA 92101-4495, USA.

Catsicas, S., D. Larhammar, A. Blomqvist, P.P. Sanna, R.J. Milner, and M.C. Wilson. 1991. Expression of a conserved cell-type-specific protein in nerve terminals coincides with synaptogenesis. Proc. Natl. Acad. Sci. U.S.A. 88: 785–789.

Chapman, L.P., M.J. Epton, J.C. Buckingham, J.F. Morris, and H.C. Christian. 2003. Evidence for a role of the adenosine 5'-triphosphate-binding cassette transporter A1 in the externalization of annexin I from pituitary folliculo-stellate cells. Endocrinology 144: 1062–1073.

Diaz-Casares, A., Y. Leon, E.J. de la Rosa, and I. Varela-Nieto. 2005. Regulation of vertebrate sensory organ development: A scenario for growth hormone and insulin-like growth factors action. Adv. Exp. Med. Biol. 567: 221–242.

Fleming, T., C.G. Martinez-Moreno, J. Mora, M. Aizouki, M. Luna, C. Aramburo, and S. Harvey. 2016. Internalization and synaptogenic effect of GH in retinal ganglion cells (RGCs). Gen. Comp. Endocrinol. (in press).

Geffner, M.E. 1996. The growth without growth hormone syndrome. Endocrinol. Metab. Clin. North Am. 25: 649–663.

Goodall, N., L. Kisiswa, A. Prashar, S. Faulkner, P. Tokarczuk, K. Singh, J.T. Erichsen, J. Guggenheim, W. Halfter, and M.A. Wride. 2009. 3-Dimensional modelling of chick embryo eye development and growth using high resolution magnetic resonance imaging. Exp. Eye Res. 89: 511–521.

Hamburger, V., and H.L. Hamilton. 1951. A series of normal stages in the development of the chick embryo. J. Morphol. 88: 49–92.

Harvey, S., T.F. Davison, and A. Chadwick. 1979. Ontogeny of growth hormone and prolactin secretion in the domestic fowl (*Gallus domesticus*). Gen. Comp. Endocrinol. 39: 270–273.

Harvey, S., C.D. Johnson, P. Sharma, E.J. Sanders, and K.L. Hull. 1998. Growth hormone: A paracrine growth factor in embryonic development? Comp. Biochem. Physiol. C. Pharmacol. Toxicol. Endocrinol. 119: 305–315.

Harvey, S., C.D. Johnson, and E.J. Sanders. 2001. Growth hormone in neural tissues of the chick embryo. J. Endocrinol. 169: 487–498.

Harvey, S., I. Lavelin, and M. Pines. 2002. Growth hormone (GH) action in the brain: Neural expression of a GH-response gene. J. Mol. Neurosci. 18: 89–95.

Harvey, S., M. Kakebeeke, A.E. Murphy, and E.J. Sanders. 2003. Growth hormone in the nervous system: Autocrine or paracrine roles in retinal function? Can. J. Physiol. Pharmacol. 81: 371–384.

Harvey, S., M.L. Baudet, and E.J. Sanders. 2006. Growth hormone and cell survival in the neural retina: Caspase dependence and independence. NeuroReport 17: 1715–1718.

Harvey, S., B.T. Martin, M.L. Baudet, P. Davis, Y. Sauve, and E.J. Sanders. 2007a. Growth hormone in the visual system: Comparative endocrinology. Gen. Comp. Endocrinol. 153: 124–131.

Harvey, S., M.L. Baudet, and E.J. Sanders. 2007b. Growth hormone and developmental ocular function: Clinical and basic studies. Pediatr. Endocrinol. Rev. 5: 510–515.

Harvey, S. 2010. Extrapituitary growth hormone. Endocrine 38: 335–359.

Harvey, S., W. Lin, D. Giterman, N. El-Abry, W. Qiang, and E.J. Sanders. 2012. Release of retinal growth hormone in the chick embryo: Local regulation? Gen. Comp. Endocrinol. 176: 361–366.

Harvey, S. 2013. Growth hormone and growth? Gen. Comp. Endocrinol. 190: 3–9.

Harvey, S., and M.L. Baudet. 2014. Extrapituitary growth hormone and growth? Gen. Comp. Endocrinol. 205: 55–61.

Harvey, S., M.L. Baudet, M. Luna, and C. Aramburo. 2014. Non-classical signalling of growth hormone in the chick neural retina? Avian Biol. Res. 10: 48–57.

Harvey, S., C.G. Martinez-Moreno, J. Avila-Mendoza, M. Luna, and C. Aramburo. 2016. Growth hormone in the eye: A comparative update. Gen. Comp. Endocrinol. (in press).

Ientile, R., V. Macaione, M. Teletta, S. Pedale, V. Torre, and S. Macaione. 2001. Apoptosis and necrosis occurring in excitotoxic cell death in isolated chick embryo retina. J. Neurochem. 79: 71–78.

Latchney, S.E., I. Masiulis, K.J. Zaccaria, D.C. Lagace, C.M. Powell, J.S. McCasland, and A.J. Eisch. 2014. Developmental and adult GAP-43 deficiency in mice dynamically alters hippocampal neurogenesis and mossy fiber volume. Dev. Neurosci. 36: 44–63.

Marksteiner, J., W.A. Kaufmann, P. Gurka, and C. Humpel. 2002. Synaptic proteins in Alzheimer's disease. J. Mol. Neurosci. 18: 53–63.

Martinez-Moreno, C.G., D. Giterman, D. Henderson, and S. Harvey. 2014a. Secretagogue induction of GH release in QNR/D cells: Prevention of cell death. Gen. Comp. Endocrinol. 203: 274–280.

Martinez-Moreno, C., A. Andres, D. Giterman, E. Karpinski, and S. Harvey. 2014b. Growth hormone and retinal ganglion cell function: QNR/D cells as an experimental model. Gen. Comp. Endocrinol. 195: 183–189.

Martinez-Moreno, C.G., V.L. Trudeau, and S. Harvey. 2015. Co-storage and secretion of growth hormone and secretoneurin in retinal ganglion cells. Gen. Comp. Endocrinol. 220: 124–132.

Martinez-Moreno, C.G., J. Avila-Mendoza, Y. Wu, E.C. Arellanes-Licea, M. Louie, M. Luna, C. Aramburo, and S. Harvey. 2016. Neuroprotection by GH against excitotoxic-induced cell death in retinal ganglion cells. Gen. Comp. Endocrinol. (in press).

McLenachan, S., M.G. Lum, M.J. Waters, and A.M. Turnley. 2009. Growth hormone promotes proliferation of adult neurosphere cultures. Growth Horm. IGF Res. 19: 212–218.

Porter, T.E., G.S. Couger, C.E. Dean, and B.M. Hargis. 1995. Ontogeny of growth hormone (GH)-secreting cells during chicken embryonic development: Initial somatotrophs are responsive to GH-releasing hormone. Endocrinology 136: 1850–1856.

Radecki, S.V., L. McCann-Levorse, S.K. Agarwal, J. Burnside, J.A. Proudman, and C.G. Scanes. 1997. Chronic administration of growth hormone (GH) to adult chickens exerts marked effects on circulating concentrations of insulin-like growth factor-I (IGF-I), IGF binding proteins, hepatic GH regulated gene I, and hepatic GH receptor mRNA. Endocrine 6: 117–124.

Rotondo, F., K. Kovacs, B.W. Scheithauer, E. Horvath, C.D. Bell, R.V. Lloyd, and M. Cusimano. 2008. Immunohistochemical expression of SNAP-25 protein in adenomas of the human pituitary. Appl. Immunohistochem. Mol. Morphol. 16: 477–481.

Sanders, E.J., M.A. Walter, E. Parker, C. Arámburo, and S. Harvey. 2003. Opticin binds retinal growth hormone in the embryonic vitreous. Invest. Ophthalmol. Vis. Sci. 44: 5404–5409.

Sanders, E.J., E. Parker, C. Arámburo, and S. Harvey. 2005. Retinal growth hormone is an anti-apoptotic factor in embryonic retinal ganglion cell differentiation. Exp. Eye Res. 81: 551–560.

Sanders, E.J., E. Parker, and S. Harvey. 2006. Retinal ganglion cell survival in development: Mechanisms of retinal growth hormone action. Exp. Eye Res. 83: 1205–1214.

Sanders, E.J., E. Parker, and S. Harvey. 2008. Growth hormone-mediated survival of embryonic retinal ganglion cells: Signaling mechanisms. Gen. Comp. Endocrinol. 156: 613–621.

Sanders, E.J., M.L. Baudet, E. Parker, and S. Harvey. 2009a. Signaling mechanisms mediating local GH action in the neural retina of the chick embryo. Gen. Comp. Endocrinol. 163: 63–69.

Sanders, E.J., E. Parker, and S. Harvey. 2009b. Endogenous growth hormone in human retinal ganglion cells correlates with cell survival. Mol. Vis. 15: 920–926.

Sanders, E.J., W.Y. Lin, E. Parker, and S. Harvey. 2010. Growth hormone expression and neuroprotective activity in a quail neural retina cell line. Gen. Comp. Endocrinol. 165: 111–119.

Sanders, E.J., W.Y. Lin, E. Parker, and S. Harvey. 2011. Growth hormone promotes the survival of retinal cells *in vivo*. Gen. Comp. Endocrinol. 172: 140–150.

Takeuchi, S., M. Haneda, K. Teshigawara, and S. Takahashi. 2001. Identification of a novel GH isoform: A possible link between GH and melanocortin systems in the developing chicken eye. Endocrinology 142: 5158–5166.

Tanaka, M., Y. Hayashida, K. Sakaguchi, T. Ohkubo, M. Wakita, S. Hoshino, and K. Nakashima. 1996. Growth hormone-independent expression of insulin-like growth factor I messenger ribonucleic acid in extrahepatic tissues of the chicken. Endocrinology 137: 30–34.

Turnley, A.M., C.H. Faux, R.L. Rietze, J.R. Coonan, and P.F. Bartlett. 2002. Suppressor of cytokine signaling 2 regulates neuronal differentiation by inhibiting growth hormone signaling. Nat. Neurosci. 5: 1155–1162.

Vergara, M.N., and M.V. Canto-Soler. 2012. Rediscovering the chick embryo as a model to study retinal development. Neural Dev. 7: 22.

Waters, M.J., and D.G. Blackmore. 2011. Growth hormone (GH), brain development and neural stem cells. Pediatr. Endocrinol. Rev. 9: 549–553.

CHAPTER-4

Porosome Enables the Establishment of *Fusion Pore* at its base and the Consequent Kiss-and-Run Mechanism of Secretion from Cells

||||||||||||||||||||||||||||||||||||||||||||||||||||||||||||||||||||||||||||||||||||||||||||||||||||||||||||||||||||||||||||||||||||||||||||

INTRODUCTION

In the 1970's, Bruno Ceccarelli recognized the presence of a 'transient' mechanism of secretory vesicle fusion at the cell plasma membrane (Ceccarelli et al., 1973) enabling fractional release of intravesicular contents, and coined the term '*kiss-and-run*'. In 1990, Wolfhard Almers hypothesized, based on his own and existing studies, that the fusion pore is the continuity established between the vesicle membrane and the cell plasma membrane, and results from a "preassembled ion channel-like structure that could open and close" (Almers and Tse, 1990; Monck and Fernandez, 1992). In 1993, Erwin Neher reasoned that: "It seems terribly wasteful that, during the release of hormones and neurotransmitters from a cell, the membrane of a vesicle should merge with the plasma membrane to be retrieved for recycling only seconds or minutes later" (Neher, 1993). Our current knowledge of the presence of over 100,000 lipid species in cells, and their precise distribution with subcellular organelles, makes even more sense, the statement made by Prof. Neher. A membrane-associated portal—*the porosome* discovered in the mid 1990's, has been demonstrated to enable

Wayne State University School of Medicine, Department of Physiology, Detroit, MI, USA.
Email: bjena@med.wayne.edu

secretory vesicles to transiently establish continuity with the cell plasma membrane without collapsing, expel a portion of the vesicular contents and disengage, while remaining partially filled as demonstrated in numerous cells, including in the rat peritoneal mast cells [Figure 1]. Similarly, in acinar cells of the exocrine pancreas, partially filled Zymogen Granules (ZG), the secretory vesicles in these cells, are generated following a secretory episode as observed in electron micrographs (EM). EM morphometry of intracellularly located ZG in pancreatic acinar cells demonstrate that although the total number of ZG in cells remains unchanged following secretion, there is an increase in the number of empty and partially empty vesicles following a secretory episode, suggesting a transient kiss-and-run mechanism of intravesicular content release during cell secretion. Further confirmation of the transient kiss-and-run mechanism of cell secretion is demonstrated by the direct observation using atomic force microscopy (AFM) docked ZG at the apical plasma membrane in live pancreatic acinar cells (Figure 2). Physiological stimulation of cell secretion using the

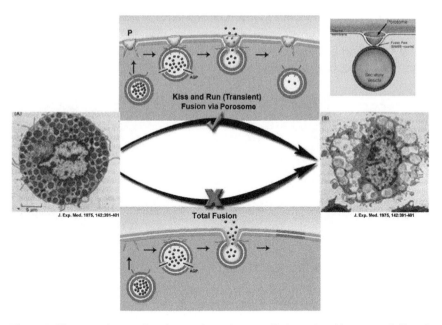

Figure 1: Electron micrographs of rat peritoneal mast cells in resting (A, extreme left) and following secretion (B, extreme right). Note the fractional release of intravesicular contents following secretion (B) (Electron micrographs obtained from J. Exp. Med. 142: 391–401, 1975). This fractional release of intravesicular contents could only be possible via the porosome (P)-mediated transient fusion mechanism shown (√). ©Bhanu Jena. The schematic drawing on the top right, illustrating the establishment of the fusion pore at the porosome base when t-SNAREs at the porosome base interact with v-SNAREs at the secretory vesicle membrane in a ring or rosette pattern, to establish continuity in the presence of calcium ions (https:// en.wikipedia.org/wiki/Porosome).

cholecystokinin analogue carbamylcholine, results first in ZG swelling followed by intravesicular content release and a consequent decrease in ZG size. ZG remain long after the completion of the secretory episode (Figure 2), demonstrating the transient or kiss-and-run mechanism of cell secretion.

The supramolecular structure at the cell plasma membrane called 'porosome' that enables the kiss-and-run mechanism of secretion in cells was discovered in the mid 1990's and initially misnamed 'fusion pore'. Porosomes are supramolecular lipoprotein structures at the cell plasma membrane ubiquitously present in all cells examined, and hence have been recognized as the universal secretory portal in cells (Schneider et al., 1997; Cho et al., 2002a,b,c; Jena et al., 2003; Jeremic et al., 2003; Cho et al., 2004). During cell secretion, secretory vesicles transiently dock and fuse at the porosome base via SNAREs to establish such a fusion pore or continuity between the secretory vesicle membrane and the porosome, to enable measured release of intravesicular contents. Immediately prior to vesicle fusion at the porosome, secretory vesicles swell via regulated active transport of water and ions, and the consequent intravesicular pressure generated, drives intravesicular contents to the outside through the fusion pore and through the porosome opening to the outside (Figure 1). As a result, the integrity of either the vesicle membrane or the cell plasma membrane is uncompromised (Schneider et al., 1997; Cho et al., 2002a,b,c;

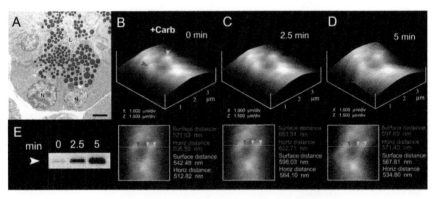

Figure 2: The volume dynamics of zymogen granules (ZG) in live pancreatic acinar cells demonstrating fractional release of ZG contents during secretion. (A) Electron micrograph of pancreatic acinar cells showing the basolaterally located nucleus (N) and the apically located electron-dense vesicles, the ZGs. The apical end of the cell faces the acinar lumen (L). Bar = 2.5 µ. (B-D) Apical ends of live pancreatic acinar cells in physiological buffer imaged by AFM, showing ZGs (red and green arrowheads) lying just below the apical plasma membrane. Exposure of the cell to a secretory stimulus (1 µ carbamylcholine), results in ZG swelling within 2.5 min, followed by a decrease in ZG size after 5 min. The decrease in size of ZGs after 5 min is due to the release of secretory products such as ⟨-amylase, as demonstrated by the immunoblot assay (E). If ZG's had fused at the plasma membrane and fully merged, it would not be visible, demonstrating transient fusion and fractional discharge on intravesicular contents during secretion in pancreatic acinar cells [28]. ©Bhanu Jena.

Jena et al., 2003; Jeremic et al., 2003; Cho et al., 2004; Lee et al., 2012; Kovari et al., 2014). In the past two decades, in further confirmation, hundreds of studies from scores of laboratories from around the world, provide evidence on the kiss-and-run mechanism of cell secretion and fractional release of intravesicular contents from cells. Studies demonstrated that "secretory granules are recaptured largely intact following stimulated exocytosis in cultured endocrine cells" (Taraska et al., 2003); "single synaptic vesicles fuse transiently and successively without loss of identity" (Aravanis et al., 2003); and "zymogen granule exocytosis is characterized by long fusion pore openings and preservation of vesicle lipid identity" (Thorn et al., 2004). Utilizing the porosome-mediated "kiss-and-run" mechanism of secretion in cells, secretory vesicles are capable of reuse for subsequent rounds of exo-endocytosis, until completely empty of contents. However, in a fast secretory cells such as the neuron, synaptic vesicles have the advantage of rapidly refilling, utilizing the neurotransmitter transporters present at the synaptic vesicle membrane.

Elucidation of the porosome structure, its chemical composition, and functional reconstitution into artificial lipid membrane (Schneider et al., 1997; Cho et al., 2002a,b,c; Jena et al., 2003; Jeremic et al., 2003; Cho et al., 2004; Lee et al., 2012; Kovari et al., 2014), and the molecular assembly of membrane-associated t-SNARE and v-SNARE proteins in a ring or rosette complex [17-25], resulting in the establishment of membrane continuity between the membrane of the porosome base and the secretory vesicle membrane to establish a *fusion pore*, has been demonstrated in great detail (Cho et al., 2002d; Jeremic et al., 2004a,b; Cho et al., 2004; Jeremic et al., 2006; Cook et al., 2008; Shin et al., 2010; Issa et al., 2010; Cho et al., 2011). Similarly, an understanding of the molecular mechanism of secretory vesicle swelling, and its requirement for intravesicular content release during cell secretion has further progressed (Jena et al., 1997; Cho et al., 2002e; Kelly et al., 2004; Jeremic et al., 2005; Lee et al., 2010; Chen et al., 2011). Collectively, these studies provide a molecular understanding of porosome-mediated kiss-and-run mechanism of fractional release of intravesicular contents from cells during secretion, resulting in a paradigm-shift in our understanding of the secretory process.

Porosome: Discovery, Isolation, Composition, and Reconstitution

In the mid 1990's using AFM, our group was the first to image the morphology and dynamics of new 100–180 nm pores in cellular structures at the apical plasma membrane of live pancreatic acinar cells (Figure 3), and demonstrated their involvement in cell secretion (Schneider et al., 1997). During secretion, the pores grew larger, and returned to their resting size following completion

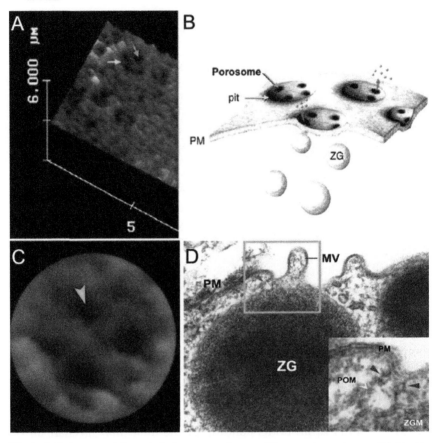

Figure 3: Porosomes at the apical plasma membrane of the exocrine pancreas. (A) Atomic Force Microscopy (AFM) micrograph depicting 'pits' (light arrow) and 'porosomes' within (upperdark arrow), at the apical plasma membrane in a live pancreatic acinar cell. (B) To the right is a schematic drawing depicting porosomes at the cell plasma membrane (PM), where membrane-bound secretory vesicles called zymogen granules (ZG), dock and fuse to release intravesicular contents. (C) A high resolution AFM image shows a single pit with four 100–180 nm porosomes within. (D) An electron micrograph depicting a porosome (arrow head within box) close to a microvilli (MV) at the apical plasma membrane (PM) of a pancreatic acinar cell. Note the association of the porosome membrane (light arrow head), and the zymogen granule membrane (ZGM) (lower right arrow head) of a docked ZG (inset). Cross section of a circular complex at the mouth of the porosome is observed (upper dark arrow heads). In the presence of the actin depolymerizing agent cytocholasin, porosome openings collapse, and there is a concomitant loss of secretion (data not shown) (Schneider et al., 1993; Jena et al., 2003). ©Bhanu Jena.

of cell secretion. In the following five years, our results demonstrated these structures to be secretory portals or porosomes, where secretory vesicles transiently dock and fuse to expel intravesicular contents to the outside during cell secretion (Cho et al., 2002a; Jena et al., 2003). Following stimulation of cell secretion, gold-conjugated antibody against the secretory starch-digesting enzyme amylase, accumulates at these pore structures, establishing them to be secretory portals in cells (Cho et al., 2002a; Jena et al., 2003). Using immuno-AFM, the presence of t-SNAREs at the porosome base facing the cytosol was determined, where ZG transiently dock and fuse during secretion (Jena et al., 2003). Next, porosomes were found to be present at the plasma membrane of growth hormone (GH) secreting cells of the pig pituitary gland (Cho et al., 2002b), and in rat chromaffin cells (Cho et al., 2002c). The porosome structure in GH cells, its dynamics, and the accumulation of GH-immuno-gold at its opening following stimulation of secretion, was further determined (Cho et al., 2002b). In 2003, the porosome from the rat exocrine pancreas was isolated, its composition determined, and it was both structurally and functionally reconstituted into lipid membrane (Jeremic et al., 2003). In the same study (Jeremic et al., 2003), morphological details of the porosome complex associated with docked secretory vesicle with established fusion pore, was observed at ultrahigh resolution using electron microscopy (EM) (Jeremic et al., 2003) (Figure 3).

In 2004, the neuronal porosome complex was discovered, isolated, and functionally reconstituted into artificial lipid membrane (Cho et al., 2004). In 2012, the proteome of the neuronal porosome complex (Lee et al., 2012), and in 2014, its lipidome were finally determined (Jena, 2014; Lewis et al., 2014). Examination of the presynaptic membrane at the nerve terminal using high resolution AFM (Cho et al., 2004), EM (Cho et al., 2004), and SAXS studies (Kovari et al., 2014), demonstrate the presence of approximately 15 nm cup-shaped porosomes, each possessing a central plug (Figures 4, 5). The outer rim of the porosome opening to the outside is lined by eight equally spaced protein densities (Figure 4D, E). The eight protein densities are observed both in the native neuronal porosome complex (Figure 4D top left) as well as in isolated porosomes reconstituted in lipid bilayers (Figure 4D top right). Similar to AFM micrographs, approximately 8 interconnected protein densities are observed in EM micrographs of purified neuronal porosome preparations (Figure 4E). Electron density and contour mapping, and the resultant 3D topology profiles of the neuronal porosome complex provide further details of the arrangement of proteins, and their interconnection to the central plug region of the complex via distinct spoke-like elements (Figure 4E lower left). The 3D topology of the porosome complex (Figure 4E lower right) obtained from electron density maps, show in greater detail, the circular profile of the porosome complex and a central plug connected via spokes as in a cart wheel. AFM micrographs of inside-out presynaptic membrane, demonstrate inverted cup-shaped porosomes facing the cytosol, some with docked synaptic vesicles at the porosome base (Figure 4F, G). AFM, EM, and photon correlation spectroscopy (Figure 4H, I) demonstrate isolated porosomes to range in size from 12–17 nm. High-resolution AFM micrographs of the neuronal porosome complex present

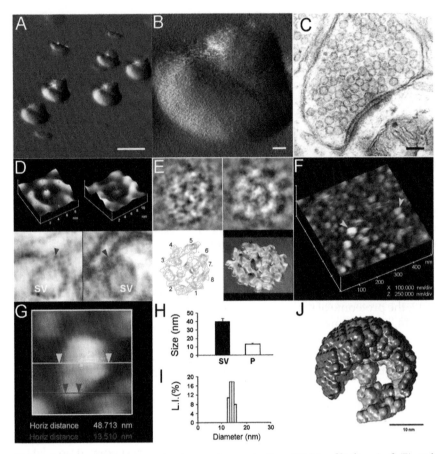

Figure 4: Neuronal porosome structure and organization. (A) Low [Scale = 1 μ] (B) and high-resolution [Scale = 100 nm] atomic force microscope (AFM) micrographs of isolated rat-brain nerve terminals or synaptosomes in buffer. (**C**) Electron microscope (EM) picture of a synaptic terminal [Scale = 100 nm]. (D) AFM micrographs of native neuronal porosome complex at the presynaptic membrane (Fig. D top left), and of an isolated porosome complex reconstituted into lipid membrane (Fig. D top right). Note the native and reconstituted porosomes being morphologically identical. In the electron micrographs below, lower panels are two EM micrographs demonstrating synaptic vesicles (SV) docked at the base of cup-shaped porosome, with a central plug (red arrowhead). (E) EM, electron density, and 3D contour maps provide at nanometer resolution of protein arrangement within the porosome complex. (F) AFM micrograph of the cytosolic compartment of an isolated bouton or synaptosome, demonstrating SV (blue arrow-head) docked at the base of porosomes (red arrow-head). (G) AFM micrograph of a SV docked at a porosome. (H) Measurements (n = 15) using AFM of SV (SV, 40.15 ± 3.14) and porosomes (P, 13.05 ± 0.91) at the presynaptic membrane. (I) Photon correlation spectroscopy performed on isolated neuronal porosomes measure 12–17 nm. (J) X-ray solution scattering (SAXS) of averaged 3-D structure of a SV (purple) docked at the base of a native neuronal porosome (pink) (Kovari et al., 2014). Images of SV-porosome complex using EM, AFM, and SAXS, demonstrate similarity morphology (Cho et al., 2004; Lee et al., 2012; Kovari et al., 2014). ©Bhanu Jena.

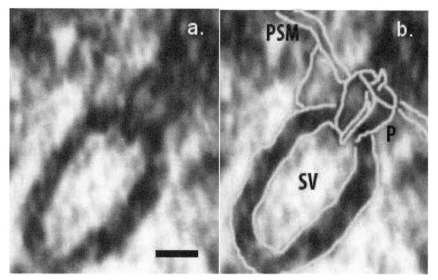

Figure 5: Electron micrograph of a docked synaptic vesicle (SV) at the base of a cup-shaped neuronal porosome complex (P), present at the presynaptic membrane (PSM) of a nerve terminal in a rat brain neuron [Scale = 10 nm]. Micron. (2012) 43: 948–953. *Courtesy of Prof. M. Zhvania.*

at the presynaptic membrane demonstrates the central plug at various conformational states, suggesting the capability of the central plug for vertical motion, and its possible involvement in the rapid opening and closing of the porosome. The neuronal porosome complexed with synaptic vesicle in its native state within synaptosomes, has also been determined using x-ray solution scattering (SASX) (Figure 4J), providing further molecular details of the complex and its interaction with synaptic vesicles (Kovari et al., 2014).

Composition of the Neuronal Porosome Complex

Mass spectrometry (MS) of purified neuronal porosome complex, reveals the presence of approximately 40 proteins within the complex (Lee et al., 2012) (Table 1). Additionally, the dynamic nature of the porosome is reflected from association and dissociation of proteins from the complex during neurotransmission (Lee et al., 2012). Immuno-isolated porosomes from rat brain synaptosome preparations, demonstrate the presence of a number proteins, among them P/Q-type calcium channel, actin, vimentin, the N-ethylmaleimide-sensitive factor (NSF), SNAP-25, syntaxin-1, synaptotagmin-1, alpha subunit of the heterotrimeric GTP-binding $G_{a\alpha'}$ GTPase activating protein (GAP), tubulin, myosin 7b, spectrin beta chain, creatine kinase, dystrophin, langerin, intersectin1, and myosin heavy chain

Table 1: A list of some key proteins composing the neuronal porosome complex. Purified rat brain porosomes from two separate experiments were analyzed by LC-MS/MS on both LTQ and QSTAR XL. Only proteins identified in both samples are reported here all of which had protein confidence ≥ 95% with at least two unique peptides each having 95% confidence or above. Proteins also found in earlier immuno-isolation studies are marked in a separate column, x indicating proteins identified using MALDI-TOF/TOF; * indicating proteins identified using immunoblot analysis (Lee et al., 2012).

Gene symbol	MW	Protein Name	Found in earlier studies
ACTB	42 kDa	Actin, cytoplasmic 1	x, *
AT1A3	112 kDa	Sodium/potassium-transporting ATPase subunit alpha-3	
AT2B1	139 kDa	Plasma membrane calcium-transporting ATPase 1	
AT2B2	137 kDa	Plasma membrane calcium-transporting ATPase 2	
BASP1	22 kDa	Brain acid soluble protein 1	
CAP1	52 kDa	Adenylyl cyclase-associated protein 1	
CN37	47 kDa	2',3'-cyclic-nucleotide 3'-phosphodiesterase	
DPYL2	62 kDa	Dihydropyrimidinase-related protein 2	
DPYL3	62 kDa	Dihydropyrimidinase-related protein 3	
DPYL5	62 kDa	Dihydropyrimidinase-related protein 5	
GLNA	42 kDa	Glutamine synthetase	
GNAO	40 kDa	Guanine nucleotide-binding protein G(o) subunit alpha	x, *
NCAM1	95 kDa	Neural cell adhesion molecule 1	
NSF	83 kDa	Vesicle-fusing ATPase	*
RAB3A	25 kDa	Ras-related protein Rab-3A	
RTN3	102 kDa	Reticulon-3	
RTN4	126 kDa	Reticulon-4	
SNP25	25 kDa	Synaptosomal-associated protein 25	x, *
STX1A	33 kDa	Syntaxin-1A	*
STX1B	33 kDa	Syntaxin-1B	*
STXB1	68 kDa	Syntaxin-binding protein 1	
SYN2	63 kDa	Synapsin-2	
SYPH	33 kDa	Synaptophysin	
SYT1	47 kDa	Synaptotagmin-1	*
TBA1A	50 kDa	Tubulin alpha-1A chain	x
VAMP1	13 kDa	Vesicle-associated membrane protein 1	
VAMP2	13 kDa	Vesicle-associated membrane protein 2	
VATB2	57 kDa	V-type proton ATPase subunit B, brain isoform	

1, and the chloride channel CLC-3 (Lee et al., 2012). A number of these porosome proteins have previously been implicated in neurotransmission and in various neurological disorders (Mikoshiba et al., 1980; Nemhauser and Goldberg, 1985; Reinikainen et al., 1989; Chapman et al., 1996; Yamamoto et al., 1997; Balestrino et al., 1999; Greber et al., 1999; Iino and Maekawa, 1999; Iino et al., 1999; Wu et al., 1999; Cole et al., 2000; Freeman and Field, 2000; Dodson and Charalabapoulou, 2001; Geerlings et al., 2001; Vlkolinsky et al., 2001; Zhang et al., 2002; Flynn et al., 2003; Peirce et al., 2006; Scarr et al., 2006; Jeans et al., 2007; Jensen et al., 2007; Khanna et al., 2007a,b; Kim et al., 2007; Lagow et al., 2007; Scuri et al., 2007; Smith et al., 2007; Sultana et al., 2007; Cao et al., 2009; Garside et al., 2009; Mukaetova-Ladinska et al., 2009; Empson et al., 2010; McKee et al., 2010; Li et al., 2011; Zhao et al., 2011; de Juan-Sanz et al., 2013; Klein et al., 2013; Zhang et al., 2013; Corradini et al., 2014; Sinclair et al., 2015). Furthermore, dynamin was hypothesized to be involved with the porosome complex based on its known association with intersectin and its presence was confirmed by Western blot analysis (Lee et al., 2012). The dynamics of dynamin association-dissociation at the neuronal porosome complex has also been demonstrated since dynamin is increased in porosomes isolated from brain slices following stimulation. In contrast to the increase in dynamin association with the porosome, a dissociation of $G_{a\alpha}$ is observed following stimulation.

As briefly mentioned earlier, a survey of previously published reports reveals the involvement of several porosome proteins in neurotransmission and neurological disorders (Mikoshiba et al., 1980; Nemhauser and Goldberg, 1985; Reinikainen et al., 1989; Chapman et al., 1996; Yamamoto et al., 1997; Balestrino et al., 1999; Greber et al., 1999; Iino and Maekawa, 1999; Iino et al., 1999; Wu et al., 1999; Cole et al., 2000; Freeman and Field, 2000; Dodson and Charalabapoulou, 2001; Geerlings et al., 2001; Vlkolinsky et al., 2001; Zhang et al., 2002; Flynn et al., 2003; Peirce et al., 2006; Scarr et al., 2006; Jeans et al., 2007; Jensen et al., 2007; Khanna et al., 2007a,b; Kim et al., 2007; Lagow et al., 2007; Scuri et al., 2007; Smith et al., 2007; Sultana et al., 2007; Cao et al., 2009; Garside et al., 2009; Mukaetova-Ladinska et al., 2009; Empson et al., 2010; McKee et al., 2010; Li et al., 2011; Zhao et al., 2011; de Juan-Sanz et al., 2013; Klein et al., 2013; Zhang et al., 2013; Corradini et al., 2014; Sinclair et al., 2015), suggesting their interactions within the porosome complex, and their critical role in porosome-mediated neurotransmitter release. Cytoskeletal proteins such as actin and the alpha chain of tubulin are two of the several neuronal porosome proteins. Studies performed in the presence of the actin depolymerizing agent latrunculin A, partially blocks neurotransmitter release at the pre-synaptic terminal of motor neurons (Cole et al., 2000). Similarly, ion-channel proteins such as the alpha sub-unit 3 of the universal Na+/K+ ATPase, identified in the porosome complex, is involved in neuronal secretion. Na+/K+ ATPase activity is blocked by dihydrooubain (DHO) (Balestrino et al., 1999), resulting in an increase in both the amplitude and number of action potentials at the nerve terminal (Scuri et al., 2007). Na+/K+ ATPase inhibition is calcium dependent and increased intracellular; calcium levels inhibit Na+/K+ ATPase, which increases excitability of neurons (Kim et al., 2007). Similarly, the porosome protein plasma membrane calcium ATPases (PMCA) co-localize with synaptohysin (Jensen et al., 2007). Syntaxin-1, also a porosome protein co-localizes with PMCA2 and the glycine transporter 2 (GlyT2) that is found coupled to the Na^+/K^+ pump, suggesting the presence of a protein complex involved in neurotransmission (Geerlings et al., 2001; Garside et al., 2009; de Juan-Sanz et al., 2013). Mutation in the PMCA2 encoding gene is known to result in homozygous deafwaddler mice (dfw/dfw) and they show high levels of calcium accumulation within their synaptic terminals (Dodson and Charalabapoulou, 2001). NAP-22, also known as BASP-1, is a protein found in the neuronal porosome complex, has long been speculated to be involved in synaptic transmission (Yamamoto et al., 1997; Iino and Maekawa, 1999; Iino et al., 1999). NAP-22 is known to bind to the inner leaflet of lipid rafts suggesting interaction with cholesterol. Adenylyl cyclase associated protein-1 (CAP-1) is known to regulate actin polymerization (Freeman and Field, 2000) and both actin and CAP-1 are present in the porosomal complex. In Alzheimer's, the levels of CNPase (2,3-cyclic nucleotide phosphodiesterase) and the heat shock protein 70 (HSP70), are found to increase while the levels of dihydropyrimidinase related protein-2 (DRP-2) decrease (Sultana et al., 2007). Alterations in the levels of SNARE proteins are associated with various neurological disorders. SNAP-25 and synaptophysin are significantly reduced in neurons of patients with Alzheimer's disease (Greber et al., 1999; Mukaetova-Ladinska et al., 2009; Sinclair et al., 2015). Mice that are SNAP-25 (+/–) show disabled learning and memory, and exhibit epileptic like seizures (Corradini et al., 2014). Overexpression of SNAP-25 also results in defects in cognitive function (McKee et al., 2010), and loss of SNAP-25 is also associated with Huntington's disease. Rabphilin3a, another porosome protein is known to be involved in vesicle docking and fusion at the presynaptic membrane (Smith et al., 2007). Increase in synaptophysin levels along with SNAP-25, is also observed in Broddmann's area in the post-mortem brain of patients with bipolar disorder I (Scarr et al., 2006). Reticulons are proteins that contribute to lipid membrane curvature and are found in the neuronal porosome. The presence of reticulons with the porosome and diseases associated with their deregulation lend credence to the role of membrane curvature in modulating synaptic vesicle fusion at the porosome complex. These independent studies reflect on the critical role of various neuronal porosome proteins and their interactions on porosome-mediated neurotransmission.

In the past two decades, our studies demonstrate that membrane-associated t-SNAREs and v-SNAREs interact in a rosette or ring complex, enabling Ca^{+2}-mediated membrane fusion and establishment of the 'fusion pore' (Cho et al., 2002d; Thorn et al., 2004; Jeremic et al., 2004a,b; Cho et al., 2005; Jeremic et al., 2006; Cook et al., 2008; Shin et al., 2010; Issa et al., 2010; Cho et al., 2011). Furthermore, our studies have progressed our understanding of the regulation of secretory vesicle volume, and the requirement of vesicle volume increase for fractional release of intravesicular contents from cells during secretion (Jena et al., 1997; Cho et al., 2002e; Kelly et al., 2004; Jeremic et al., 2005; Lee et al., 2010; Shin et al., 2010; Chen et al., 2011). These results provide the molecular underpinnings of how cells precisely regulate the discharge of a portion of their intravesicular contents during a secretory episode, while retaining full integrity of both the vesicle membrane and the cell plasma membrane. In summary, our studies in the past two decades demonstrate the presence of a new cup-shaped lipoprotein structure at the cell plasma membrane called *'porosome, -the universal secretory portals in cells'*, and elucidate-how the porosome is involved in the regulated fractional release of intravesicular contents from cells with exquisite precision involving membrane fusion and secretory vesicle volume regulation; revealing for the first time the molecular underpinnings of the transient or kiss-and-run mechanism of secretion in cells. We have isolated the porosome from a number of secretory cells including neurons, determined its composition, functionally reconstituted it in lipid membrane, and determined its dynamics and high-resolution structure using a variety of approaches including AFM, EM, and SAXS. Complementing the regulation of the porosome function, our studies have further contributed to our understanding of SNARE and Ca^{+2}-mediated membrane fusion and secretory vesicle volume regulation, both required for the regulated fractional release of intravesicular contents during cell secretion. These results provide for the first time a molecular understanding of the regulated fractional release of intravesicular contents from cells during secretion, and the interactions between porosome proteins and their alterations in disease states.

Acknowledgements

This work has been support in part by grants from the National Institutes of Health DK56212, NS39918 and the National Science Foundation EB00303, CBET1066661 (BPJ).

Conflicts of Interest

The author has no conflicts to declare.

Keywords: Porosome, kiss-and-run, fusion pore, SNARE-rosette, membrane fusion

References

Almers, W., and F.W. Tse. 1990. Transmitter release from synapses: Does a preassembled fusion pore initiate exocytosis? Neuron 4: 813–818.

Aravanis, A.M., J.L. Pyle, and R.W. Tsien. 2003. Single synaptic vesicles fusing transiently and successively without loss of identity. Nature 423: 643–647.

Balestrino, M., J. Young, and P. Aitken. 1999. Block of (Na+,K+)ATPase with ouabain induces spreading depression-like depolarization in hippocampal slices. Brain Res. 838: 37–44.

Cao, F., R. Hata, P. Zhu, M. Niinobe, and M. Sakanaka. 2009. Up-regulation of syntaxin1 in ischemic cortex after permanent focal ischemia in rats. Brain Res. 1272: 52–61.

Ceccarelli, B., W.P. Hurlbut, and A. Mauro. 1973. Turnover of transmitter and synaptic vesicles at the frog neuromuscular junction. J. Cell. Biol. 57: 499–524.

Chapman, E.R., S. An, J.M. Edwardson, and R. Jahn. 1996. A novel function for the second C2 domain of synaptotagmin. Ca2+-triggered dimerization. J. Biol. Chem. 271: 5844–5849.

Chen, Z.-H., J.-S. Lee, L. Shin, W.-J. Cho, and B.P. Jena. 2011. Involvement of β-adrenergic receptor in synaptic vesicle swelling and implication in neurotransmitter release. J. Cell. Mol. Med. 15: 572–576.

Cho, S.-J., A.S. Quinn, M.H. Stromer, S. Dash, J. Cho, D.J. Taatjes, and B.J. Jena. 2002a. Structure and dynamics of the fusion pore in live cells. Cell Biol. Int. 26: 35–42.

Cho, S.-J., K. Jeftinija, A. Glavaski, S. Jeftinija, B.P. Jena, and L.L. Anderson. 2002b. Structure and dynamics of the fusion pores in live GH-secreting cells revealed using atomic force microscopy. Endocrinology 143: 1144–1148.

Cho, S.-J., A. Wakade, G.D. Pappas, and B.P. Jena. 2002c. New structure involved in transient membrane fusion and exocytosis. New York Academy of Science Annals 971: 254–256.

Cho, S.-J., M. Kelly, K.T. Rognlien, J.A. Cho, J.K.H. Horber, and B.P. Jena. 2002d. SNAREs in opposing bilayers interact in a circular array to form conducting pores. Biophys. J. 83: 2522–2527.

Cho, S.-J., A.K.M. Sattar, E.-H. Jeong, M. Satchi, J. Cho, S. Dash, M.S. Mayes, M.H. Stromer, and B.P. Jena. 2002e. Aquaporin 1 regulates GTP-induced rapid gating of water in secretory vesicles. Proc. Natl. Acad. Sci. U.S.A. 99: 4720–4724.

Cho, W.-J., A. Jeremic, K.T. Rognlien, M.G. Zhvania, I. Lazrishvili, B. Tamar, and B.P. Jena. 2004. Structure, isolation, composition and reconstitution of the neuronal fusion pore. Cell Biol. Int. 28: 699–708.

Cho, W.-J., A. Jeremic, and B.P. Jena. 2005. Size of supramolecular SNARE complex: Membrane-directed self-assembly. J. Am. Chem. Soc. 127: 10156–10157.

Cho, W.J., J.-S. Lee, G. Ren, L. Zhang, L. Shin, C.W. Manke, J. Potoff, N. Kotaria, M.G. Zhvania, and B.P. Jena. 2011. Membrane-directed molecular assembly of the neuronal SNARE complex. J. Cell. Mol. Med. 15: 31–37.

Cole, J.C., B.R. Villa, and R.S. Wilkinson. 2000. Disruption of actin impedes transmitter release in snake motor terminals. J. Physiol. 525: 579–586.

Cook, J.D., W.J. Cho, T.L. Stemmler, and B.P. Jena. 2008. Circular dichroism (CD) spectroscopy of the assembly and disassembly of SNAREs: The proteins involved in membrane fusion in cells. Chem. Phys. Lett. 462: 6–9.

Corradini, I., A. Donzelli, F. Antonucci, H. Welzl, M. Loos, R. Martucci, S. De Astis, L. Pattini, F. Inverardi, D. Wolfer, M. Caleo, Y. Bozzi, C. Verderio, C. Frassoni, D. Braida, M. Clerici, H.P. Lipp, M. Sala, and M. Matteoli. 2014. Epileptiform activity and cognitive deficits in SNAP-25 (+/–) mice are normalized by antiepileptic drugs. Cerebral Cortex 24: 364–376.

de Juan-Sanz, J., E. Núñez, L. Villarejo-López, D. Pérez-Hernández, A.E. Rodriguez-Fraticelli, B. López-Corcuera, J. Vázquez, and C. Aragón. 2013. Na+/K+-ATPase is a new interacting partner for the neuronal glycine transporter GlyT2 that downregulates its expression *in vitro* and *in vivo*. J. Neurosci. 33: 14269–14281.

Dodson, H.C., and M. Charalabapoulou. 2001. PMCA2 mutation causes structural changes in the auditory system in deafwaddler mice. J. Neurocytol. 30: 281–292.

Empson, R.M., W. Akemann, and T. Knopfel. 2010. The role of the calcium transporter protein plasma membrane calcium ATPase PMCA2 in cerebellar Purkinje neuron function. Funct. Neuronal. 25: 53–158.

Flynn, S.W., D.J. Lang, A.L. Mackay, V. Goghari, I.M. Vavasour, K.P. Whittall, G.N. Smith, V. Arango, J.J. Mann, A.J. Dwork, P. Falkai, and W.G. Honer. 2003. Abnormalities of myelination in schizophrenia detected *in vivo* with MRI, and post-mortem with analysis of oligodendrocyte proteins. Mol. Psychiatry 8: 811–820.

Freeman, N.L., and J. Field. 2000. Mammalian homolog of the yeast cyclase associated protein, CAP/Srv2p, regulates actin filament assembly. Cell Motil. Cytoskeleton 45: 106–120.

Garside, M.L., P.R. Turner, B. Austen, E.E. Strehler, P.W. Beesley, and R.M. Epson. 2009. Molecular interactions of the plasma membrane calcium ATPase 2 at pre- and post-synaptic sites in rat cerebellum. Neurosci. 162: 383–395.

Geerlings, A., E. Nunez, B. Lopez-Corcuera, and C. Aragon. 2001. Calcium- and syntaxin 1-mediated trafficking of the neuronal glycine transporter GLYT2. J. Biol. Chem. 276: 17584–17590.

Greber, S., G. Lubec, N. Cairns, and M. Fountoulakis. 1999. Decreased levels of synaptosomal associated protein 25 in the brain of patients with Down syndrome and Alzheimer's disease. Electrophoresis 20: 928–934.

Iino, S., and S. Maekawa. 1999. Immunohistochemical demonstration of a neuronal calmodulin-binding protein, NAP-22, in the rat spinal cord. Brain Res. 834: 66–73.

Iino, S., S. Kobayashi, and S. Maekawa. 1999. Immunohistochemical localization of a novel acidic calmodulin-binding protein, NAP-22, in the rat brain. Neurosci. 91: 1435–1444.

Issa, Z., C.W. Manke, B.P. Jena, and J.J. Potoff. 2010. Ca^{2+} bridging of apposed phospholipid bilayer. J. Phys. Chem. 114: 13249–13254.

Jeans, A.F., P.L. Oliver, R. Johnson, M. Capogna, J. Vikman, Z. Molnar, A. Babbs, C.J. Partridge, A. Salehi, M. Bengtsson, L. Eliasson, P. Rorsman, and K.E. Davies. 2007. A dominant mutation in Snap25 causes impaired vesicle trafficking, sensorimotor gating, and ataxia in the blind-drunk mouse. Proc. Natl. Acad. Sci. U.S.A. 104: 2431–2436.

Jena, B.P., S.W. Schneider, J.P. Geibel, P. Webster, H. Oberleithner, and K.C. Sritharan. 1997. Gi regulation of secretory vesicle swelling examined by atomic force microscopy. Proc. Natl. Acad. Sci. U.S.A. 94: 13317–13322.

Jena, B.P., S.-J. Cho, A. Jeremy, M.H. Stromer, and R. Abu-Hamdah. 2003. Structure and composition of the fusion pore. Biophys. J. 84: 1–7.

Jena, B.P. 2014. Neuronal porosome- the secretory portal at the nerve terminal: Its structure-function, composition, and reconstitution. J. Mol. Struct. 1073: 187–195.

Jensen, T.P., A.G. Filoteo, T. Knopfel, and R.M. Empson. 2007. Presynaptic plasma membrane Ca2+ ATPase isoform 2a regulates excitatory synaptic transmission in rat hippocampal CA3. J. Physiol. 579: 85–99.

Jeremic, A., M. Cho, S.-J., Kelly, M.H. Stromer, and B.P. Jena. 2003. Reconstituted fusion pore. Biophys. J. 85: 2035–2043.

Jeremic, A., M. Kelly, J.-H. Cho, S.-J. Cho, J.K.H. Horber, and B.P. Jena. 2004a. Calcium drives fusion of SNARE-apposed bilayers. Cell Biol. Int. 28: 19–31.

Jeremic, A., W.-J. Cho, and B.P. Jena. 2004b. Membrane fusion: what may transpire at the atomic level. J. Biol. Phys. Chem. 4: 139–142.

Jeremic, A., W.-J. Cho, and B.P. Jena. 2005. Involvement of water channels in synaptic vesicle swelling. Exp. Biol. Med. 230: 674–680.

Jeremic, A., A.S. Quinn, W.-J. Cho, D.J. Taatjes, and B.P. Jena. 2006. Energy-dependent disassembly of self-assembled SNARE complex: observation at nanometer resolution using atomic force microscopy. J. Am. Chem. Soc. 128: 26–27.

Kelly, M., W.-J. Cho, A. Jeremic, R. Abu-Hamdah, and B.P. Jena. 2004. Vesicle swelling regulates content expulsion during secretion. Cell Biol. Int. 28: 709–716.

Khanna, R., A. Zougman, and E.F. Stanley. 2007a. A proteomic screen for presynaptic terminal N-type calcium channel (CaV2.2) binding partners. J. Biochem. Mol. Biol. 40: 302–314.

Khanna, R., Q. Li, J. Bewersdorf, and E.F. Stanley. 2007b. The presynaptic CaV2.2 channel-transmitter release site core complex. Eur. J. Neurosci. 26: 547–559.

Kim, J.H., I. Sizov, M. Dobretsov, and H. von Gersdorff. 2007. Presynaptic Ca2+ buffers control the strength of a fast post-tetanic hyperpolarization mediated by the alpha3 Na(+)/K(+)-ATPase. Nature Neurosci. 10: 196–205.

Klein, M.E., T.J. Younts, P.E. Castillo, and B.A. Jordan. 2013. RNA-binding protein Sam68 controls synapse number and local beta-actin mRNA metabolism in dendrites. Proc. Natl. Acad. Sci. U.S.A. 110: 3125–130.

Kovari, L.C., J.S. Brunzelle, K.T. Lewis, W.J. Cho, J.-S. Lee, D.J. Taatjes, and B.P. Jena. 2014. X-ray solution structure of the native neuronal porosome-synaptic vesicle complex: Implication in neurotransmitter release. Micron 56: 37–43.

Lagow, R.D., H. Bao, E.N. Cohen, R.W. Daniels, A. Zuzek, W.H. Williams, G.T. Macleod, R.B. Sutton, and B. Zhang. 2007. Modification of a hydrophobic layer by a point mutation in syntaxin 1A regulates the rate of synaptic vesicle fusion. PLoS Biol. 5: e72.

Lee, J.-S., W.-J. Cho, L. Shin, and B.P. Jena. 2010. Involvement of cholesterol in synaptic vesicle swelling. Exp. Biol. Med. 235: 470–477.

Lee, J.-S., A. Jeremic, L. Shin, W.J. Cho, X. Chen, and B.P. Jena. 2012. Neuronal porosome proteome: Molecular dynamics and architecture. J. Proteomics 75: 3952–3962.

Lewis, K.T., K.R. Maddipati, D.J. Taatjes, and B.P. Jena. 2014. Neuronal porosome lipidome. J. Cell. Mol. Med. 18: 1927–1937.

Li, K.C., F.X. Zhang, C.L. Li, F. Wang, M.Y. Yu, Y.Q. Zhong, K.H. Zhang, Y.J. Lu, Q. Wang, X.L. Ma, J.R. Yao, J.Y. Wang, L.B. Lin, M. Han, Y.Q. Zhang, R. Kuner, H.S. Xiao, L. Bao, X. Gao, and X. Zhang. 2011. Follistatin-like 1 suppresses sensory afferent transmission by activating Na+,K+-ATPase. Neuron 69: 974–987.

McKee, A.G., J.S. Loscher, N.C. O'Sullivan, A. Chadderton, A. Palfi, L. Batti, G.K. Sheridan, S. O'Shea, M. Moran, O. McCabe, A.B. Fernandez, M.N. Pangalos, J.J. O'Connor, C.M. Regan, W.T. O'Connor, P. Humphries, G.J. Farrar, and K.J. Murphy. 2010. AAV-mediated chronic over-expression of SNAP-25 in adult rat dorsal hippocampus impairs memory-associated synaptic plasticity. J. Neurochem. 112: 991–1004.

Mikoshiba, K., E. Aoki, and Y. Tsukada. 1980. 2'-3'-cyclic nucleotide 3'-phosphohydrolase activity in the central nervous system of a myelin deficient mutant (Shiverer). Brain Res. 192: 195–204.

Monck, J.R., and J.M. Fernandez. 1992. The exocytotic fusion pore. J. Cell. Biol. 119: 1395–1404.

Mukaetova-Ladinska, E.B., J.H. Xuereb, F. Garcia-Sierra, J. Hurt, H.J. Gertz, R. Hills, C. Brayne, F.A. Huppert, E.S. Paykel, M.A. McGee, R. Jakes, W.G. Honer, C.R. Harrington, and C.M. Wischik. 2009. Lewy body variant of Alzheimer's disease: Selective neocortical loss of t-SNARE proteins and loss of MAP2 and alpha-synuclein in medial temporal lobe. Sci. World J. 9: 1463–1475.

Neher, E. 1993. Secretion without full fusion. Nature 363: 497–498.

Nemhauser, I., and D.J. Goldberg. 1985. Structural effects in axoplasm of DNase I, an actin depolymerizer that blocks fast axonal transport. Brain Res. 334: 47–58.

Peirce, T.R., N.J. Bray, N.M. Williams, N. Norton, V. Moskvina, A. Preece, V. Haroutunian, J.D. Buxbaum, M.J. Owen, and M.C. O'Donovan. 2006. Convergent evidence for 2',3'-cyclic nucleotide 3'-phosphodiesterase as a possible susceptibility gene for schizophrenia. Arch. Gen. Psychiatry 63: 18–24.

Reinikainen, K.J., A. Pitkanen, and P.J. Riekkinen. 1989. 2',3'-cyclic nucleotide-3'-phosphodiesterase activity as an index of myelin in the post-mortem brains of patients with Alzheimer's disease. Neurosci. Lett. 106: 229–232.

Scarr, E., L. Gray, D. Keriakous, P.J. Robinson, and B. Dean. 2006. Increased levels of SNAP-25 and synaptophysin in the dorsolateral prefrontal cortex in bipolar I disorder. Bipolar Disorders 8: 133–143.

Schneider, S.W., K.C. Sritharan, J.P. Geibel, H. Oberleithner, and B.P. Jena. 1997. Surface dynamics in living acinar cells imaged by atomic force microscopy: Identification of plasma membrane structures involved in exocytosis. Proc. Natl. Acad. Sci. U.S.A. 94: 316–321.

Scuri, R., P. Lombardo, E. Cataldo, C. Ristori, and M. Brunelli. 2007. Inhibition of Na+/K+ ATPase potentiates synaptic transmission in tactile sensory neurons of the leech. Eur. J. Neurosci. 25: 59–167.

Shin, L., W.-J. Cho, J. Cook, T. Stemmler, and B.P. Jena. 2010. Membrane lipids influence protein complex assembly-disassembly. J. Am. Chem. Soc. 132: 5596–5597.

Sinclair, L.I., H.M. Tayler, and S. Love. 2015. Synaptic protein levels altered in vascular dementia. Neuropath. Appl. Neurobiol. 41: 533–543.

Smith, R., P. Klein, Y. Koc-Schmitz, H.J. Waldvogel, R.L. Faull, P. Brundin, M. Plomann, and J.Y. Li. 2007. Loss of SNAP-25 and rabphilin 3a in sensory-motor cortex in Huntington's disease. J. Neurochem. 103: 115–123.

Sultana, R., D. Boyd-Kimball, J. Cai, W.M. Pierce, J.B. Klein, M. Merchant, and D.A. Butterfield. 2007. Proteomics analysis of the Alzheimer's disease hippocampal proteome. J. Alzheimers Dis. 11: 153–164.

Taraska, J.W., D. Perrais, M. Ohara-Imaizumi, S. Nagamatsu, and W. Almers. 2003. Secretory granules are recaptured largely intact after stimulated exocytosis in cultured endocrine cells. Proc. Natl. Acad. Sci. U.S.A. 100: 2070–2075.

Thorn, P., K.E. Fogarty, and I. Parker. 2004. Zymogen granule exocytosis is characterized by long fusion pore openings and preservation of vesicle lipid identity. Proc. Natl. Acad. Sci. U.S.A. 101: 6774–6779.

Vlkolinsky, R., N. Cairns, M. Fountoulakis, and G. Lubec. 2001. Decreased brain levels of 2′,3′-cyclic nucleotide-3′-phosphodiesterase in Down syndrome and Alzheimer's disease. Neurobiol. Aging 22: 547–553.

Wu, M.N., T. Fergestad, T.E. Lloyd, Y. He, K. Broadie, and H.J. Bellen. 1999. Syntaxin 1A interacts with multiple exocytic proteins to regulate neurotransmitter release *in vivo*. Neuron. 23: 593–605.

Yamamoto, Y., Y. Sokawa, and S. Maekawa. 1997. Biochemical evidence for the presence of NAP-22, a novel acidic calmodulin binding protein, in the synaptic vesicles of rat brain. Neurosci. Lett. 224: 127–130.

Zhang, H., P. Ghai, H. Wu, C. Wang, J. Field, and G.-L. Zhou. 2013. Mammalian adenylyl cyclase-associated protein 1 (CAP1) regulates cofilin function, the actin cytoskeleton, and cell adhesion. J. Biol. Chem. 288: 20966–20977.

Zhang, X., M.J. Kim-Miller, M. Fukuda, J.A. Kowalchyk, and T.F. Martin. 2002. Ca2+-dependent synaptotagmin binding to SNAP-25 is essential for Ca2+-triggered exocytosis. Neuron. 34: 599–611.

Zhao, Y.Y., X.Y. Shi, X. Qiu, L. Zhang, W. Lu, S. Yang, C. Li, G.H. Cheng, Z.W. Yang, and Y. Tang. 2011. Enriched environment increases the total number of CNPase positive cells in the corpus callosum of middle-aged rats. Acta Neurobiol. Exp. (Wars.) 71: 322–330.

Section B

Molecular Regulation of Growth/Metabolic Efficiency

CHAPTER-5

Epigenetics and Developmental Programming in Ruminants
Long-Term Impacts on Growth and Development

Lawrence P. Reynolds,[1],* Alison K. Ward[1] and Joel S. Caton[1]

||||||||||||||||||||||||||||||||||||||||||||||||||||||||||||||||||||||||||||||||||||||||||||||||||||||||||||||||||

INTRODUCTION

What is Developmental Programming?

Newborns that are growth-restricted or developmentally compromised in some other way (e.g., altered development of specific organs) have an increased risk of health complications not just as infants but also throughout their lifespan, including a range of metabolic, neurological, behavioral, and reproductive disabilities. Although originally referred to as 'the Barker hypothesis', or 'fetal programming', more recently this concept has been renamed as developmental programming, or developmental origins of health and disease, to reflect the observation that developmental insults during infancy are probably as important as those that occur during fetal life (Barker, 1992, 2004; Paneth and Susser, 1995; Armitage et al., 2004; Wu et al., 2006; Caton and Hess, 2010; Reynolds et al., 2010b; Reynolds and Caton, 2012). The general concept is that a poor environment or 'stress' while *in utero* or during infancy can have long-term effects on the health and well being of that individual.

---

[1] Center for Nutrition and Pregnancy, and Department of Animal Sciences, North Dakota State University, Department 7630, Fargo, ND 58108-6050, USA.
Email: Joel.Caton@ndsu.edu
Email: Alison.Ward@ndsu.edu
* Corresponding author: Larry.Reynolds@ndsu.edu

Epidemiological Underpinnings

Human epidemiological studies worldwide have provided convincing support for the concept of developmental programming. Many have shown a strong association between low birth weight, poor postnatal environment, or other developmental insults such as exposure to stress-related hormones (e.g., corticoids, which are used therapeutically in cases of premature onset of labor to initiate maturation of fetal organ systems in preparation for birth), and the subsequent risk of developing a range of pathologies as adolescents and adults. Such pathologies include cardiovascular disease (Figure 1; Reynolds and Caton, 2012), obesity, type 2 diabetes, poor growth and altered body composition, immune dysfunction, reproductive dysfunction, and behavioral problems including the pervasive developmental disorders (autism, etc.) (Armitage et al., 2004; Barker, 2004; Luther et al., 2005; Wallace et al., 2006; Wu et al., 2006; Caton and Hess, 2010; Reynolds et al., 2010b; Reynolds and Caton, 2012; Reynolds and Vonnahme, 2016).

It is clear that these pathologies have a major impact on the quality of life. Thus, reducing the incidence of low birth weight or poor postnatal environment has the potential to affect the immediate health and survival as well as the lifelong health and productivity of an individual. The potential long-term consequences of developmental programming are

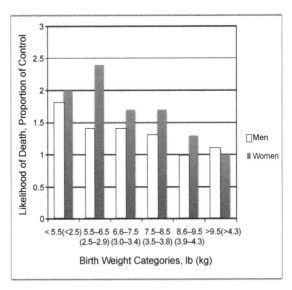

Figure 1: Likelihood of death due to coronary heart disease as adults for various birth weight categories as a proportion of controls (highest birth weight category). Taken from Godfrey and Barker, 2000. Births from Hertfordshire, UK from 1911–1930. N = 1033 deaths for men and 120 deaths for women.

so significant that the UN's World Food Programme recently established the First 1000 Days initiative to provide nutritional support during the critical developmental period from conception to the second birthday (WFP, 2014). Maternal and child nutrition similarly are a focus of the U.S. Government's Feed the Future initiative of the U.S. Agency for International Development (Feed the Future Innovation Lab for Nutrition, 2017).

Developmental Programming in Ruminants

Although the pathologies resulting from developmental programming negatively impact on offspring productivity, perhaps the greatest consequence for livestock production is that the phenotype may not reflect the offspring's genetic potential, thereby leading to poorly informed selection decisions for breeding programs (Reynolds et al., 2010b; Reynolds and Caton, 2012). Unfortunately, there is a dearth of large, long-term data sets relating birth or weaning weights to long-term health and productivity in livestock. Using a data set from the U.S. Sheep Experiment Station for birth and weaning weights of over 82,000 offspring across five decades, we showed highly significant correlations between birth weight, weaning weight, and average daily gain to weaning (Table 1; Reynolds, Vonnahme, Hanna and Taylor, unpublished). Similarly, using records from several thousand Nellore beef cattle, Chud et al. (2014) recently reported a genetic correlation of 0.36 between birth weight and weaning weight and 0.20 between birth weight and accumulated productivity (which they defined as an index of dam efficiency that includes body weight of the calf at weaning [growth trait] and number of offspring of the dam [reproductive trait]).

The concept of developmental programming was based on epidemiological studies in humans, but there was a hint of developmental programming, sometimes referred to as the 'maternal effect', for growth of livestock. For example, the classic crossbreeding experiment of Walton and Hammond with large (Shire) and small (Shetland) horses, showed that uterine environment impacts not only birth weight but also adult size (Figure 2; Hammond, 1927; Walton and Hammond, 1938).

Although less well documented, it seems likely that developmental programing affects livestock production, especially in extensive production systems such as those prevalent in the Intermountain region of the western U.S. and similar environments (e.g., savannahs) throughout the world. In the U.S., for example, livestock are often under a poor nutritional environment during pregnancy due to: (1) breeding of young, often peri-pubertal, dams and the attendant competition for nutrients between the rapidly growing maternal and fetal systems; (2) selection for multiple fetuses (e.g., sheep) in which increased numbers of fetuses are strongly associated with reduced fetal size; (3) selection for increased milk production, in which

Table 1: Correlations of birth weight with weaning weight and average daily gain in four breeds of sheep over 5 decades (1950–1999).*

Breed	Type of birth/weaning	Sex	n	Weaning age, d	Weaning weight, kg	Avg. daily gain, kg	Correlation with birth weight, r**	
							Weaning weight	Average daily gain
Rambouillet	Single/single	M	6521	125.1±8.7	37.2±6.0	0.256±0.044	0.491	0.411
		F	7087	124.5±10.1	33.4±4.9	0.230±0.036	0.487	0.400
	Twin/twin	M	6986	125.9±8.0	32.7±5.3	0.223±0.039	0.376	0.316
		F	7776	125.2±9.5	30.1±4.5	0.206±0.034	0.384	0.333
Columbia	Single/single	M	4175	121.8±8.2	40.8±7.0	0.291±0.051	0.479	0.398
		F	4635	121.6±8.8	37.2±5.8	0.265±0.042	0.522	0.420
	Twin/twin	M	5147	123.1±7.5	35.3±6.2	0.247±0.047	0.391	0.313
		F	5788	123.2±7.8	32.7±5.4	0.228±0.040	0.398	0.322
Targhee	Single/single	M	5416	119.3±9.5	39.3±6.4	0.284±0.048	0.454	0.396
		F	6019	119.1±10.0	35.2±5.5	0.253±0.041	0.458	0.387
	Twin/twin	M	7579	119.2±9.0	33.2±5.6	0.238±0.042	0.307	0.266
		F	8033	118.7±9.5	30.6±4.9	0.219±0.037	0.326	0.289
Polypay†	Single/single	M	532	120.6±11.6	39.8±7.2	0.290±0.056	0.415	0.369
		F	1191	118.2±14.1	35.2±6.3	0.259±0.045	0.366	0.387
	Twin/twin	M	2360	120.6±12.1	34.8±5.5	0.253±0.044	0.317	0.248
		F	3096	119.8±12.8	31.7±5.1	0.231±0.040	0.338	0.274
Overall	Single/single	M	16644	122.2±9.3	38.8±6.6	0.275±0.050	0.469	0.400
		F	18932	121.6±10.4	35.0±5.6	0.248±0.042	0.479	0.402
	Twin/twin	M	22072	122.4±9.2	33.7±5.7	0.237±0.044	0.328	0.265
		F	24693	122.0±10.0	31.1±5.0	0.219±0.038	0.350	0.288

*Data are from the U.S. Sheep Experiment Station (Reynolds, Vonnahme, Hanna and Taylor unpublished).
**For weaning age, weaning weight, and average daily gain, data are presented as means ± SD.
†For Polypay, data include 1970–1999 only.

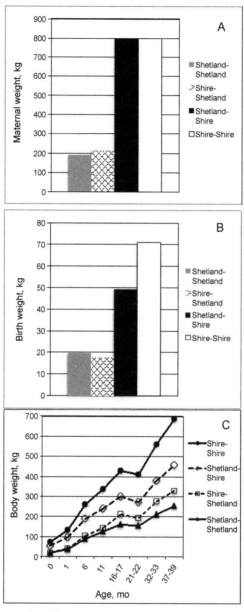

Figure 2: Results of crossbreeding experiment between small (Shetland) and large (Shire) horses. First breed indicates sire and second breed indicates dam. Data taken from Hammond (1927) and Walton and Hammond (1938).

the increased energy demand of lactation competes with the increased energy demand of fetal and placental growth; and (4) breeding of livestock during high environmental temperatures (e.g., summer to early fall) with the subsequent pregnancy during periods of poor forage or poor feed quality (e.g., fall to winter; Wu et al., 2006; Caton and Hess, 2010; Reynolds et al., 2010b; Reynolds and Caton, 2012).

Additionally, in livestock, just as in humans, compromised fetal or neonatal growth has been shown to lead to: (1) increased neonatal morbidity and mortality; (2) altered postnatal growth, including poor feed efficiency, and reduced average daily gain and weaning weight; (3) poor body composition, including increased fat, reduced muscle growth, and reduced meat quality; (4) metabolic disorders, such as poor glucose tolerance and insulin resistance; (5) cardiovascular disease; and (6) dysfunction of specific organs and organ systems, including adipose, brain, cardiovascular, endocrine, gastro-intestinal, immune, kidney, liver, mammary gland, muscle, pancreas, placenta, and reproductive including gonads (Rhind et al., 2001; Sheldon and West, 2004; Wu et al., 2006; Cottrell and Ozanne, 2007; Anway et al., 2008; Gardner et al., 2008, 2009; Caton and Hess, 2010; Du et al., 2010; Reynolds et al., 2010b; Long et al., 2012a,b; Shankar et al., 2011; Bartol and Bagnell, 2012; Connor et al., 2012; Meyer et al., 2012; Reynolds and Caton, 2012; Spencer et al., 2012; Symonds et al., 2012; Jackson et al., 2013; Kilcoyne et al., 2014; Xiong and Zhang, 2013; Cardoso et al., 2014; Schmidt et al., 2014; Zambrano et al., 2014; Meyer and Caton, 2016; Reynolds and Vonnahme, 2016). Thus, nearly all organ systems and bodily processes are affected negatively in various animal models of developmental programming, including livestock (Reynolds et al., 2010b; Reynolds and Caton, 2012).

The initial epidemiological studies that led to the concept of developmental programming focused on individuals with low birth weight, but we now know that birth weight *per se* is only a reflection of an insult or multiple insults to the fetus during development; that is, developmental programming can occur independently of birth weight (Barker, 2004; Reynolds et al., 2010b; Reynolds and Caton, 2012). For example, in humans and animal models, offspring of mothers who experience nutrient restriction early in pregnancy but receive adequate nutrition later in pregnancy, resulting in normal birth weights, still exhibit many of the same phenotypes, including poor growth, increased adiposity, poor glucose tolerance, and dyslipidemia, as offspring from mothers that are undernourished for the whole of pregnancy (Barker, 2004; Ford et al., 2007; Vonnahme et al., 2007; Dong et al., 2013).

These and similar observations emphasize the importance of interventions designed to correct developmental defects during fetal or early postnatal life, because if the organ systems are indeed programmed, then later interventions may be much less effective (Greenwood et al., 2000, 2004; Barker, 2004, 2007; Reynolds et al., 2010b; Reynolds and Caton, 2012). These observations also emphasize the importance of understanding how to manage offspring from compromised pregnancies. For example, in humans and animal models, rapid body weight gain during infancy further impairs body composition, leading to obesity in the offspring (Barker, 2004, 2007; Reynolds et al., 2010b; Reynolds and Caton, 2012). Additionally, it has been shown in humans and in animal models that developmental insults in one generation can have consequences for later generations even in the absence of further insults, and it seems likely these transgenerational effects may depend on epigenetic alterations in gene expression (see the section, Epigenetics and Developmental Programming, below; Anderson et al., 2006; Ismail-Beigi et al., 2006; Wu et al., 2006; Reynolds et al., 2010b; Meyer et al., 2012; Reynolds and Caton, 2012; Reynolds et al., 2013). Lastly, these observations emphasize the critical need for increased research efforts to understand the basis (i.e., the mechanisms) of developmental programming in terms of a variety of maternal stressors, effects at various developmental stages (i.e., pre- as well as postnatally), intergenerational consequences, and potential therapeutic interventions (Barker, 2007; Caton and Hess, 2010; Reynolds et al., 2010b; Reynolds and Caton, 2012).

Based on epidemiological studies in humans and corroborated by numerous controlled studies in animal models, such as those described above, the most likely explanation for the long-term effects of various insults during fetal or postnatal life is two-fold: (1) irreversible alterations in tissue and organ structure (i.e., *a structural defect*), and (2) permanent changes in tissue function (i.e., a permanent change in gene expression leading to *a functional defect*).

The first mechanism, a structural defect, can include, for example, changes in brown adipose deposition (Symonds et al., 2012), altered nephron number leading to renal dysfunction (Richter et al., 2016), and altered pancreatic β-cell mass (Martin-Gronert and Ozanne, 2012; Gatford and Simmons, 2013). However, developmental programming of a structural defect is perhaps best exemplified by muscle. This is because the number of individual muscle cells (myocytes), or muscle fibers, is established before birth; after that, muscle fiber size can increase by the addition of nuclei [from muscle satellite cells] and subsequent hypertrophy, but no new muscle fibers can be added and muscle growth is therefore limited by the number of myocytes at birth (Du et al., 2010, 2011). Thus, factors affecting fetal muscle development can lead to permanent, irreversible changes in muscle structure and its growth potential (Du et al., 2011).

Muscle fiber number in offspring is affected by maternal nutrient restriction during pregnancy (Du et al., 2010), and it has been argued that developing muscle is especially vulnerable to nutrient availability because of its low priority in terms of nutrient partitioning during fetal development, due primarily to its lower metabolic demands compared with tissues such as brain, gut, and placenta, a concept first articulated by Sir Joseph Barcroft (Barcroft, 1946). Maternal nutrient restriction, or other factors that limit nutrient availability such as multiple fetuses, also affect development of other muscle cells in addition to myocytes, including intramuscular adipocytes (which regulate intramuscular fat, or marbling) and fibroblasts (connective tissue-producing cells), leading to alterations in not only muscle size but also muscle marbling and connective tissue content in the offspring (Du et al., 2011). As with most other organ systems, the period of pregnancy during which nutrient restriction is experienced determines which of the cell types (i.e., myocytes, adipocytes, or fibroblasts) is most affected (Du et al., 2011).

The second mechanism, a functional defect due to altered gene expression, is best explained by a relatively novel concept termed 'epigenetics' (see the section, Epigenetics and Developmental Programming, below; Sinclair et al., 2010; Meyer et al., 2012; Reynolds and Caton, 2012; Wang et al., 2012; Reynolds et al., 2013; Skinner, 2014). Of course, in reality structural and functional defects are intimately inter-related and their effects are difficult to separate in most cases.

Effects of Maternal Nutrition

Maternal nutritional status is a major factor in developmental programming events and ultimately offspring outcomes (Wallace, 1948; Wallace et al., 1999; Wu et al., 2006; Caton et al., 2007; Caton and Hess, 2010; Reynolds and Caton, 2012; Funston et al., 2012; Robinson et al., 2013; Vonnahme et al., 2015; Meyer and Caton, 2016; Reynolds and Vonnahme, 2016). Prenatal growth trajectory is responsive to maternal nutrient intake from the early stages of embryonic life, when nutrient requirements for conceptus growth are reported to be negligible (NRC, 1996, 2007; Robinson et al., 1999). Growth restricted neonates are at risk of immediate postnatal complications and may also exhibit poor growth and development, with significant negative consequences later in life (Wu et al., 2006; Caton and Hess, 2010; Funston et al., 2012; Reynolds and Caton, 2012).

As mentioned, within ruminant livestock production systems, there is real potential for periods of undernutrition (extensive grazing, drought, winter dominancy, multiple fetuses, or high milk output). Decreased growth rate and suboptimal carcasses cost feedlot producers millions of dollars annually. Fetal growth restriction and maternal undernutrition are implicated in negative impacts upon growth efficiency and body composition (Greenwood et al., 1998, 2000; Wu et al., 2006; Caton et al., 2007;

Larson et al., 2009; Robinson et al., 2013). Even though evidence indicates that maternal nutrient restriction can significantly alter composition of offspring growth in the absence of birth weight differences (Reynolds and Caton, 2012), birth weights in cattle are still highly correlated to postnatal growth performance (Robinson et al., 2013). Permanent changes in postnatal metabolism induced by maternal nutritional perturbations present a significant challenge to livestock producers because nutritional management decisions are often based on average body weight of a given group of animals. Therefore, information regarding factors contributing to animal inefficiencies, such as developmental programming, has the potential to improve efficiency and profitability of ruminant livestock programs, which in turn will help meet the grand challenge of nearly doubling livestock production by 2050 to feed a global population of 9.6 billion by the year 2050 (Elliot, 2013; Gerland et al., 2014; Reynolds et al., 2015; United Nations, 2015).

In ruminant livestock, compromised maternal nutrition can present in many different ways. Generally, because of extensive production systems associated with the parent population, seasonal changes in forage quality, and traditional approaches to supplementation, maternal nutrient restriction is usually thought of as the most likely nutritional issue facing producing females. However, during periods of confinement or lush forage growth, overconsumption and nutritional excess can be an issue, and data from some laboratories clearly show negative consequences for the offspring in response to excess maternal nutrition. In addition to general nutrient restriction or excess, specific nutrient imbalances in the maternal diet can also have negative consequences for the developing offspring. In the sections below, we will expand on the impacts of maternal nutrient restriction, nutrient excess, and selected specific nutrients on developmental programing of offspring.

For the purpose of this discussion, nutrient restriction includes any series of events that reduces fetal and/or perinatal nutrient supply during critical windows of development (Caton and Hess, 2010; Reynolds and Caton, 2012). Nutrient restriction can result from altered maternal nutrient supply, placental insufficiency, deranged maternal metabolism, physiological extremes, and environmental conditions, among many other scenarios. From a practical standpoint, compromised maternal nutrient supply, adverse environmental conditions, and other "events" leading to stress responses are the most likely observed causes of nutrient restriction in ruminant livestock. These also include multiple births in some breeds, which may contribute to physiological extremes and result in reduced nutrient supply to developing fetuses and offspring. In this section we will focus primarily on maternal dietary-induced nutrient restriction.

Due to the pattern of placental growth in relation to fetal growth during gestation (Redmer et al., 2004), it is important to realize that the effects of nutrient restriction during pregnancy may depend on the timing, level, and (or) length of nutrient restriction (Reynolds and Caton, 2012; Reynolds et al., 2013; Vonnahme et al., 2015; Zhang et al., 2015). Luther et al. (2005) suggested that maternal nutrient restriction in sheep through mid-pregnancy could reduce placental size and function, while having minimal impacts on fetal body weight near term. In addition, nutrient restriction during late pregnancy often reduces fetal weight. Recent data (Reed et al., 2007; Swanson et al., 2008; Vonnahme et al., 2015) demonstrate that maternal nutrient restriction during the last two thirds of pregnancy in sheep can reduce late term fetal and offspring birth weights. Robinson et al. (2013) indicated that nutrient restriction in beef cattle often reduces birth weights and results in compromised growth later in life. Sletmoen-Olson et al. (2000) demonstrated that both low and high levels of metabolizable protein supplementation to mature beef cows reduce birth weights relative to controls fed at the projected requirement. In contrast, protein supplementation of cows during the last trimester of pregnancy has been reported to have little effect on birth weights (Martin et al., 2007; Larson et al., 2009). Other data (Spitzer et al., 1995; Stalker et al., 2007) indicate that increasing body condition before parturition can increase calf birth weight. Taken together, the available data are interpreted to indicate that birth weight in sheep is more susceptible to maternal nutrient restriction than is birth weight in beef cattle. This is likely a result of differential placental growth patterns between cattle and sheep (Reynolds et al., 2005; Vonnahme and Lemley, 2012). In addition, in both sheep and cattle, multiple fetuses are associated with not only reduced birth weight but also reduced uterine and umbilical blood flows (Christenson and Prior, 1978; Ferrell and Reynolds, 1992); in addition, total capillary volume available for transplacental exchange is reduced in multiple fetuses due to a smaller placental size (Vonnahme et al., 2008).

Maternal nutrient excess during gestation can also have detrimental impacts on the developing offspring. It has been shown that extreme over-nourishment in singleton-bearing adolescent ewes throughout gestation results in rapid maternal growth and fat deposition at the expense of growth of the gravid uterus (Wallace et al., 1996, 1999, 2001; Redmer et al., 2012). In this sheep model, rapid maternal growth results in placental growth restriction, and premature delivery of low-birth weight, metabolically compromised lambs compared with moderately nourished ewes of equivalent age and genetic background (Wallace et al., 2012). Data from a more moderate maternal over-nourishment model (Swanson et al., 2008; Vonnahme et al., 2010) demonstrates that birth weights are decreased when adolescent ewes are fed 140% of recommended nutrient intake requirements from day 40 of pregnancy until parturition, which indicates that moderate over-nutrition during the last two-thirds of pregnancy can cause moderate fetal growth restriction. In addition, low birth weight lambs from the above studies were metabolically compromised later in life (Vonnahme et al., 2010; Yunusova et al., 2013). Placental and fetal growth restriction are generally seen in adolescent, but not mature, over-nourished ewes; however, gestation length and colostrum yield are negatively impacted in adult ewes that are over-nourished (Wallace et al., 2005), indicating that the health and growth of offspring may also be altered by maternal over-nutrition in mature ewes.

Whereas the above discussion focuses primarily on maternal nutrient restriction or excess in terms of total nutritional supply, which is most often achieved by altering dry matter intake, considerable research exists concerning specific nutrients and developmental programming. Across species, nutrients within the major classes of carbohydrates, protein, lipids, vitamins, and minerals have been investigated in the context of developmental programming paradigms, and in each case examples exist where maternal nutrient supply can impact offspring outcomes. In many cases, offspring changes induced by maternal nutrient supply are long-term. In other examples, nutrient realimentation and/or biological plasticity allows for compensation and or protection from measureable adverse outcomes of developmental programming events.

When developmental events are impacted by inappropriate maternal nutrition, offspring are often both metabolically and functionally different from those of adequately fed mothers. Research with beef cattle in Nebraska (Funston et al., 2012) indicates that maternal nutrition and protein supplementation during gestation can have long-term effects on the offspring, including changes in weaning weight, carcass characteristics, and reproductive traits. Funston et al. (2012) also showed that long-term functional differences may not always be foreshadowed by early life measures like birth weight. Radunz et al. (2012) reported that prepartum maternal dietary energy source in beef cows can alter offspring adipose tissue development, glucose metabolism, insulin sensitivity, and long-term intramuscular fat deposition. Others (Lan et al., 2013) demonstrated maternal dietary starch levels during gestation in sheep can affect fetal DNA methylation and gene expression. In follow-up research with cattle, Wang et al. (2015) recently reported that expression of imprinted genes and DNMT (DNA methyltransferase) in offspring are influenced by maternal nutrition (specifically starch in this case), indicating epigenetic mechanisms may underpin offspring responses to changes in maternal nutrition. In a review paper assessing the impacts of developmental programming in cattle, Robinson et al. (2013) concluded that fetal programming and related maternal effects were pronounced and may explain the considerable variation in growth- and production-related traits in cattle such as body weight, feed intake, carcass and muscle weight, and lean, fat, and bone weights.

Although inappropriate maternal macronutrient (energy, protein, and fat) supply can clearly have impacts on the offspring, maternal micronutrient (vitamins, trace minerals, and specific amino acids) supply can also impact the offspring responses, including having long-term consequences. Research from our laboratories using supranutritional levels of maternal dietary selenium (Caton et al., 2007; Ward et al., 2008; Vonnahme et al, 2010; Camacho et al., 2012; Meyer et al., 2013, Yunusova et al., 2013; Caton et al., 2014a,b) has demonstrated changes in offspring birth weight, growth, nutrient digestion, glucose metabolism and insulin sensitivity, visceral fat, intestinal vascularity, and endocrine profiles in lambs in some but not all studies.

Research investigating ruminally-protected arginine supplementation in pregnant ewes fed either adequate or nutrient restricted diets has demonstrated increased growth trajectory of lambs (Peine et al., 2013, 2014). Research from New Zealand (McCoard et al., 2013) demonstrated that provision of L-arginine intravenously to adequately fed, twin bearing ewes from d 100 of gestation until birth increased birth weight of female but not male lambs and increased brown adipose stores of all lambs at birth. Penagaricano et al. (2013) reported that maternal methionine supplementation in Holstein cows caused significant changes in the transcriptome of the resulting flushed embryos; interestingly, genes with altered expression included those involved in embryonic development and immune responses.

Sinclair et al. (2007) provided some of the first strong evidence in ruminants that dietary supply of 1-carbon metabolism precursors during the periconceptional period could result in adult offspring that were heavier, fatter, insulin resistant, and had altered immune responses and elevated blood pressure compared with offspring from control fed ewes. Dietary treatments of reduced cobalt and sulfur resulted in decreased ruminal synthesis of vitamin B12 and methionine, which concomitantly resulted in reduced plasma B12, folate, and methionine and elevated plasma homocysteine concentrations. Additionally, the work of Sinclair et al. (2007) indicated the dietary treatments resulted in altered methylation status of approximately 4% of 1,400 CpG islands examined.

More recently, Kwong et al. (2010) examined the effects of maternal B-vitamin status at conception on genes coding for regulatory enzymes in the methionine-folate cycles in bovine oocytes, somatic cells and embryos. Their data suggest that maternal folate status drives embryonic folate metabolism because of the heavy reliance of the embryo on recycled folate. Additionally, Juchem et al. (2012) reported that dairy cows supplemented with ruminally-protected B vitamins have improved conception rates at 42 days after first breeding, which is consistent with the conclusions of Laanpere et al. (2010), who stated that "The results of previous studies clearly emphasize that imbalances in folate metabolism and related gene variants may impair female fecundity as well as compromise implantation and the chance of a live birth." These data coupled with the finding of Sinclair et al. (2007) indicating dietary 1-carbon precursors provided during the periconceptional period could have long-term effects on offspring clearly indicate a strong need for additional work in this area.

Other 'Stressors' and Developmental Programming

Although many studies have confirmed the importance of maternal nutrition pre- and postnatally (see above), a host of other factors are probably also involved in developmental programming. Such factors include lifestyle choices such as smoking or alcohol consumption; maternal factors including not only nutrition/malnutrition but also maternal age, ethnicity (or breed in animals), poor maternal health or poor access to health care, sedentary lifestyle, maternal stress including relational stress, and glucocorticoids and pre-term birth; and environmental exposures not only to various endocrine disrupting compounds but also to smoke, drugs, temperature and humidity, high altitude, etc. (Table 2). In addition, these various factors can be exclusive to humans (e.g., lifestyle choices) or can impact both humans and animals (e.g., maternal age, poor maternal health, sedentary lifestyle, maternal stress, environmental exposures; Table 2).

Placental Programming

Not only is placental vascular development, and consequently placental function (e.g., vascular function, blood flow and nutrient delivery), altered in many cases of developmental programming, but placental size also is often affected (Mayhew et al., 2004; Redmer et al., 2004; Reynolds et al., 2006, 2010a; Jansson and Powell, 2007; Belkacemi et al., 2010; Leach, 2011; Lemley et al., 2012; Shukla et al., 2014; Reynolds and Vonnahme, 2016). These observations have led to the concept of 'placental programming' as the basis of alterations of fetal growth and of development of the various fetal organ systems (Borowicz and Reynolds, 2010; Vonnahme and Lemley, 2012; Vonnahme et al., 2013a). Defects in placental growth and developmental, including vascular defects, precede altered fetal growth and development (Redmer et al., 2005, 2009; Reynolds et al., 2010a; Vonnahme and Lemley, 2012; Vonnahme et al., 2013a; Reynolds and Vonnahme, 2016). In addition, we have recently shown that placental vascular development is defective very early (within the first three weeks after mating) in highly compromised pregnancies (Reynolds et al., 2013, 2014).

One of the most interesting concepts that has emerged is the close relationship among very simple measures of placental size/development and the risk of developing various chronic diseases later in life in humans. Such simple measures include not only placental weight, but also placental width and length, and placental thickness (Thornburg et al., 2010; Barker et al., 2012; Barker and Thornburg, 2013). Whether such simple measures of placental size and shape are related to developmental programming in livestock has not been established. Nevertheless, it seems clear that placental programming is a major factor contributing to developmental programming.

Table 2: Risk factors for low birth weight/developmental programming.

Risk Factor	Reference(s)*
Lifestyle Choices (Humans only)	
Smoking – primarily associated with pre-term birth	Hellemans et al., 2010; Dupont et al., 2012; Aizer and Currie, 2014; Been et al., 2014; Juul et al., 2014
Alcohol consumption – low birth weight/developmental disorders	
Maternal Factors (Humans and/or Animals)	
Poverty/Lack of education/malnutrition	Armitage et al., 2004; Sheldon and West, 2004; Fowden et al., 2006; Nathanielsz, 2006; Fowden and Forhead, 2009; Reynolds et al., 2010b; Alfaradhi and Ozanne, 2011; Hochberg et al., 2011; Li et al., 2011; Barker et al., 2012; Dupont et al., 2012; Brunton, 2013; Harris et al., 2013; Reynolds et al., 2013; Vonnahme et al., 2013a, 2013b; Aizer and Currie, 2014; Juul et al., 2014; Schneider et al., 2014; Seneviratne et al., 2014; Zambrano et al., 2014
Maternal age	
Ethnicity/breed (genetic background)	
Pre-term birth and glucocorticoids	
Poor maternal health and/or health-care status	
Sedentary lifestyle	
Maternal stress (e.g., relational)	
Marital status	
Environmental Exposures (Humans and Animals)	
Herbicides, pesticides, and fungicides	Fowden et al., 2006; Richter et al., 2007; Anway et al., 2008; Crain et al., 2008; Fowden and Forhead, 2009; Bellingham et al., 2010; Reynolds et al., 2010b; Bonacasa et al., 2011; Fowler et al., 2012; McLachlan et al., 2012; Spencer et al., 2012; Evans et al., 2014; Galbally et al., 2014; Juul et al., 2014; Skinner, 2014
Other environmental exposures – e.g., human waste/fertilizer, smoke, phytosteroids, drugs, temperature and humidity, high altitude, etc.	

* See References Cited for full reference.

Epigenetics and Developmental Programming

What is Epigenetics?

Epigenetics, which literally translates to "above genetics", is defined as heritable changes in gene expression without alteration of the genetic code. Epigenetic alterations are heritable through mitosis within an individual as well as transgenerationally through reproduction. Patterns of epigenetic alterations play a key role in the programming of pluripotent stem cells as they differentiate into terminal cell types. It is also responsible for X-inactivation and imprinting (whereby one allele is selectively silenced depending on whether it was maternally or paternally inherited).

Epigenetic changes and the resultant changes in gene expression serve as a more rapid response in adaptation to environmental selection pressures than evolution via genetic mutation (i.e., natural selection). Perturbations during fetal development, such as reduced nutrient availability, can induce epigenetic changes that promote an energetically "thrifty" phenotype. These alterations can persist through multiple generations, as exemplified by the pioneering studies by Barker and colleagues of children born during the Dutch Famine (also called the Dutch Hunger Winter) of 1944–1945. Reduced intrauterine growth due to caloric restriction during gestation was significantly associated with type 2 diabetes and glucose tolerance in adulthood (Ravelli et al., 1998).

Within the nucleus, DNA exists within a three-dimensional chromatin structure. The basic repeating unit of chromatin, the nucleosome, is an octamer of histones (two each of H2A, H2B, H3, and H4) that form a core around which oligonucleotide strands are wound. Each histone subunit consists of a globular portion that forms the core and an N-terminal tail, which can undergo various post-translational modifications (Figure 3). Chromatin is present in

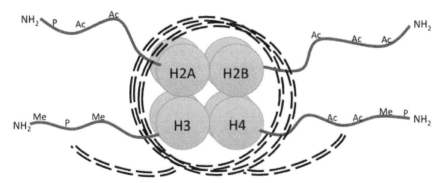

Figure 3: The basic unit of chromatin structure consists of DNA wrapped around an octamer core of histones (H2A, H2B, H3, and H4). Posttranslational modifications present on the N-terminus histone tails include acetylation (Ac), methylation (Me), and phosphorylation (P).

two forms: heterochromatin, which is compact and transcriptionally silent, and euchromatin, which is "loose" or "open" and is transcriptionally active. Epigenetic modifications alter the state of chromatin (from euchromatin to heterochromatin or vice versa) to effect changes in gene expression.

Histone Modification Posttranslational modifications of histone tails, including methylation, acetylation, phosphorylation, and ubiquitylation, alter chromatin structure and subsequently gene expression. The type and location of modification and their effect on gene expression collectively form the histone code. Standard nomenclature for histone modification is listed in order as histone, amino acid, residue number, and modification. For example, H3K18ac refers to acetylation of lysine 18 of histone 3.

Histone acetylation occurs at lysine residues of histone tails. Histone acetyl transferases (HATs) catalyze the transfer of an acetyl group from acetyl CoA to the ε-amino group of lysine (Sterner and Berger, 2000). This modification can be reversed by the action of histone deacetylases (HDACs). In general, histone acetylation is associated with transcriptionally active euchromatin and histone deacetylation with transcriptionally inactive heterochromatin. Transcription co-activators, such as ACTR, often have HAT activity (Chen et al., 1997).

Histone methylation occurs at lysine and arginine residues, primarily within the tails of H3 and H4 (Shilatifard, 2006). The effect of histone methylation on chromatin structure differs dependent upon which residue is modified. For example, methylation of H3K4 and H3K79 are associated with transcriptionally active regions; however, methylation of H3K9, H3K27, and H4K20 are associated with transcriptional repression (Dillon et al., 2005; Delage and Dashwook, 2008). Histone methyltransferases (HMTases) catalyze the transfer of a methyl group from the donor S-adenosylmethionine (SAM) to nitrogen atoms within the side chain or at the N-terminal (Clarke, 1993). Multiple families comprise HMTases, including type I and type II arginine methyltransferases as well as SET domain- and non-SET domain-containing lysine methyltransferases. Lysine may be mono, di, or trimethylated and arginine may be mono or dimethylated. Additionally, dimethylated arginine can further be classified by symmetrical or asymmetrical conformation (Shilatifard, 2006). Histone methylation is generally a stable modification, but it can be reversed by enzymatic demethylation as well as deimination of arginine to citrulline (Bannister and Kouzarides, 2011).

Histone phosphorylation occurs on serine, tyrosine, and threonine residues, primarily on histone tails but also at the H3Y41 residue, located within the H3 core (Dawson et al., 2009). Histone phosphorylation is dynamic, with a high rate of turnover. Kinases catalyze the transfer of a phosphoryl group from ATP to the hydroxyl group of the amino acid side chain. The reverse reaction is catalyzed by phosphatases (Oki et al., 2007;

Bannister and Kouzarides, 2011). Histone phosphorylation is generally associated with transcription activation, especially genes involved in the regulation of proliferation (Bungard et al., 2010; Rossetto et al., 2012). Conversely, phosphorylation of H2BY37 is associated with transcription suppression (Mahajan et al., 2012). Additionally, phosphorylation of serine 139 of the histone variant H2AX (commonly referred to as γH2AX) plays a key role in DNA damage repair, recruiting DNA damage repair and signaling factors (Rossetto et al., 2012).

Ubiquitin is a polypeptide 76 amino acids long that is covalently bound to lysine residues by the action of E3 ubiquitin ligases (Hershko and Ciechanover, 1998; Bannister and Kouzarides, 2011). This modification is dynamic and can be reversed by the action of de-ubiquitin. Histones can be mono or polyubiquitinated. Ubiquitination is associated with gene activation at some locations, such as H2BK123ub1 (Kim et al., 2009) and with transcriptional silencing at other residues, such as H2AK119ub1 (Wang et al., 2004). It also plays a key role in the regulation of other histone modifications. For example, monoubiquitination of H2 by the enzyme Rad6 is required for both SET and non-SET domain-mediated lysine methylation (Dover et al., 2002; Sun and Allis, 2002; Wood et al., 2003).

Several other histone modifications have also been observed, including sumoylation, ADP ribosylation, and biotinylation. Sumoylation is similar to ubiquitination, and involves the attachment of polypeptides to the lysine residues of histones. It is associated with transcriptional repression and is antagonistic to acetylation and ubiquitination, as it competes for lysine binding sites (Shiio and Eisenman, 2003). Glutamate and arginine residues may be mono or poly-ADP ribosylated. This is a dynamic and reversible modification, but little is known about its effect(s) on chromatin structure and gene expression (Hassa et al., 2006). Lastly, lysine residues of histones may be modified by the addition of biotin by biotinidase or holocarboxylase synthetase. Histone biotinylation has been associated with transcriptional silencing, the G1 phase of the cell cycle, and in response to DNA damage (Kothapalli et al., 2005).

DNA Methylation

In addition to histone modification, epigenetic control of gene expression is also enacted through DNA methylation. The cytosine-guanidine dinucleotide motif (commonly referred to as a CpG) is present in clusters throughout the genome in CpG 'islands', most frequently located in the promoter region and first exon of genes (Larsen et al., 1992). These motifs can be methylated at the 5-carbon position of the cytosine, forming 5-methylcytosine. This palindromic motif is methylated on both the forward and reverse strands of double-stranded DNA. The reaction is catalyzed by

DNA methyltransferases (DNMTs), utilizing S-adenosylmethionine (SAM) as a methyl donor (Dhe-Paganon et al., 2011). During mitosis, DNMT1 fully methylates all hemi-methylated residues on the daughter strand, thereby maintaining the fidelity of methylation patterns through cell division. DNA methylation is reversed by the action demethylases (Ramchandani et al., 1999).

DNA methylation, particularly in the promoter region of a gene, is associated with transcriptional repression. DNA methylation prevents transcription through two mechanisms: blocking the binding of transcription factors and creating binding sites for repressors. Additionally, DNA methylation and histone modifications interact to promote and maintain transcriptional silencing. For example, methylcytosine-binding protein (MeCP) binds methylated CpG, creating binding cites for HDAC and thereby facilitating histone deacetylation (Jones et al., 1998). Similarly, methylation of H3K9 creates a binding site for DNMT, which subsequently methylate CpG doublets (Tamaru and Selker, 2001).

Modes of Epigenetic Alterations in Offspring

Imprinting

Imprinting is the phenomenon by which one copy of a gene is turned off dependent upon the parent of origin. Genes may be maternally or paternally imprinted, where the gene inherited from the mother or father, respectively, is epigenetically silenced. Approximately one hundred genes are imprinted in mammals, comprising 1–2% of genes (Barlow, 2011). DNA methylation patterns associated with imprinting are resistant to reprogramming in early embryogenesis, but are reset and reprogrammed in the germ line of each generation (Youngson and Whitelaw, 2008).

Imprinting is controlled by the methylation status of control regions, known as differentially methylated domains or imprinting control regions (ICR), located near the imprinted genes. Depending upon the gene, methylation of the ICR may silence the gene by blocking the binding of transcription factors or enhancers, or activate the gene by blocking the binding of insulators, which repress transcription (Pfeifer, 2000). Additionally, imprinted genes are commonly present in clusters, and therefore ICR may control more than one gene (Barlow, 2011).

Arguably, the most notable example of imprinting is that of insulin-like growth factor 2 (IGF2). It is maternally imprinted; i.e., the gene inherited from the father is expressed while the one from the mother is silenced (DiChiara et al., 1991). The IGF2 gene is methylated in oocytes and conversely unmethylated in sperm cells, resulting in its differential expression with parent of origin (Gerbert et al., 2006). The conflict theory of genomic imprinting (also known as the kinship theory) argues that increased growth and resource utilization by the fetus favors paternal evolutionary fitness over maternal fitness, and therefore there is selection pressure to maternally imprint genes that promote growth and conversely to paternally imprint genes which inhibit growth (Wilkins and Haig, 2003).

An example of a phenotypic trait inherited through imprinting is callipyge in sheep. Callipyge, which translates to "beautiful buttocks", is a form of muscle hypertrophy that significantly increases muscle development in the hindquarters in sheep (Jackson and Green, 1993). The causative mutation is located between two imprinted genes and subsequently for the trait to be expressed the sheep must be heterozygous (one normal allele and one mutant allele) with the mutant allele from the paternal side (Freking et al., 2002).

Maternal Factors and Uterine Environment

As we have discussed, conditions during embryonic and fetal development, such as nutrient availability, shape the potential for growth and development through developmental programming. Epigenetic modification is a mechanism through which developmental programming effects are executed. Factors such as maternal obesity, over-nutrition, and nutrient deprivation alter the trajectory of embryonic and fetal development, creating lasting effects that persist through to adulthood.

Maternal obesity is associated with metabolic dysregulation in the offspring, including insulin intolerance, obesity, and type II diabetes (Shankar et al., 2008, 2010). Studies in mice (Yang et al., 2013) and rats (Borengasser et al., 2013) have implicated epigenetic remodeling as the mechanism through which adipocyte differentiation is altered, ultimately leading to metabolic dysfunction later in development. These studies found that maternal obesity altered the expression of genes regulating the development of adipocytes from precursor cells. Reduced promoter DNA methylation as well as H3K27 trimethylation resulted in greater expression of the pro-adipogenic transcription factor Zfp423, leading to increased adipocyte differentiation within fetal tissues.

In cattle, maternal over-nutrition during gestation is associated with increased fetal expression of markers of intramuscular adipogenesis. Fetuses collected at mid to late gestation from cows fed 150% of the maintenance requirements displayed increased expression of Zfp423, C/EBPα, and PPARγ, all of which are transcription factors involved in adipocyte differentiation and proliferation (Duarte et al., 2014). Feeding a high-starch diet (versus a low-starch, isocaloric and isonitrogenous control) from mid to late gestation resulted in calves with increased skeletal muscle expression of the DNA methyltransferase DNMT3a as well the imprinted genes H19, IGF2F and MEG8, implying that maternal energy source alters epigenetic modulation (Wang et al., 2015).

Maternal nutrient restriction is also associated with epigenetic changes that result in developmental programming. Rats that were feed-restricted by 50% during gestation produced pups with lower birth weight that displayed rapid compensatory growth and developed obesity and metabolic syndrome as adults (Tosh et al., 2010). Continuation of maternal feed restriction through lactation delayed compensatory gain and prevented the development of obesity and metabolic syndrome at maturity. The obese, gestationally feed-restricted rats reared by ad libitum-fed dams had significantly greater hepatic H3K4 trimethylation of exon 1 and 2 of the IGF1 gene, and subsequently greater IGF1 expression than the non-obese, feed-restricted reared and control rats.

Protein restriction during pregnancy has been associated with epigenetically mediated dysregulation of glucocorticoid metabolism. Protein restriction during gestation and lactation resulted in hypomethylation and increased expression of the hepatic glucocorticoid receptor in rats (Lillycrop et al., 2007). Expression of methyl CpG-binding protein and its binding to the promoter of the glucocorticoid receptor gene was reduced, and multiple histone modifications (H3K9ac, H4K9ac, H3K9me) were significantly increased, implicating epigenetic alterations as the mechanism responsible for glucocorticoid dysregulation.

Paternal Factors

While most research has focused on the maternal impact on epigenetic programming, there is now increasing interest in paternal effects. Offspring of male mice fed a low-protein, high-sucrose diet were shown to have increased hepatic expression of genes involved in lipid and cholesterol synthesis (Carone et al., 2010). This was associated with increased liver saturated fatty acid and decreased cholesterol content. The transcription factor PPARα exhibited increased DNA methylation and decreased gene expression, and accounted for nearly 14% of the global differences in gene expression between control and low protein-sired pups. In contrast, female offspring of male rats fed a high-fat diet developed impaired pancreatic β-cell function, resulting in reduced insulin secretion and glucose tolerance (Ng et al., 2010). Expression of 642 pancreatic islet genes was altered, and hypomethylation of the gene Il12ra2, which had the greatest increase in expression, was observed. Further research is needed to unravel the effect(s) of paternal environmental factors, including nutrition, on developmental programming and epigenetic modifications.

Epigenetic Rescue by Diet and Other Regulators

One-Carbon Metabolism

One-carbon metabolism refers to the transfer of methyl groups from donor molecules through metabolic reactions. Methyl transfer is a key process in epigenetic modulation of gene expression, as both histone and DNA methylation are essential epigenetic modifications. Methyl donors, historically referred to as lipotropic factors, include methionine, folate, betaine, choline, and vitamin B12. These factors serve as precursors for the generation of S-adenosylmethionine (SAM), the substrate for histone and DNA methylation (Figure 4; Clarke, 1993; Dhe-Paganon et al., 2011). Therefore, modulating the availability of methyl donors (and consequently SAM) could conceivably alter the kinetics of DNA and histone methylation, thereby changing the epigenetic landscape.

The classic example of the effect of one-carbon metabolism on epigenetic alterations is that of agouti variable yellow (A^{vy}) mice. The agouti protein controls the expression of black (eumelanin) versus yellow (phaeomelanin) pigment production. Wild-type agouti is characterized by

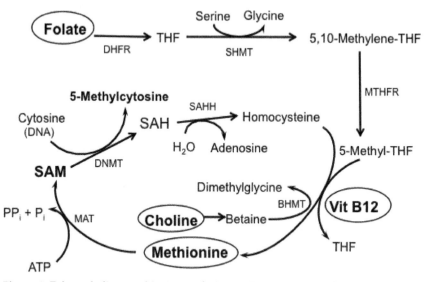

Figure 4: Folate, choline, methionine, and vitamin B12 serve as methyl donors in one-carbon metabolism. Abbreviations: DHFR, dihydrofolate reductase; THF, tetrahydrofolate; SHMT, serine hydroxymethyl transferase; MTHFR, methylenetetrahydrofolate reductase; BHMT, betaine hydroxymethyl transferase; MAT, methionineadenosyl transferase; SAM, S-adenosylmethionine; DNMT, DNA methyltransferase; SAH, S-adenosylhomocysteine; SAHH, S-adenosylhomocysteine hydrolase.

development of a black hair with a yellow band at the base, which results in a brown coat. The A^{vy} allele contains an intracisternal A particle (IAP) retrotransposon inserted near the agouti gene promoter (Dolinoy, 2008). The transposon acts as a controlling element, overtaking control from the agouti gene promoter, causing ectopic overexpression of the agouti protein, which results in a yellow phenotype as well as obesity, type II diabetes, and increased susceptibility to cancer. The IAP contains a CpG island that when methylated ablates its action as a controlling element and reestablishes normal expression of the agouti gene and a normal phenotype, dubbed pseudo-agouti. Supplementing the maternal diet during gestation with methyl donors (choline, betaine, folic acid, vitamin B12, and methionine) significantly increased the proportion of pups that were pseudo-agouti (normal) with methylated IAP versus yellow (obese) pups with unmethylated IAP (Cooney et al., 2002). These results indicate that altering the availability of dietary methyl donors during gestation can significantly alter the epigenome and subsequently phenotype of the offspring.

Methyl donor supplementation during gestation also can mitigate the effects of maternal obesity on developmental programming. Mice born to dams fed a high fat diet during gestation and lactation were significantly heavier at birth and weaning and displayed altered feeding behavior, having increased preference for sugar and fat (Vucetic et al., 2010). This was associated with increased expression and promoter hypomethylation of genes involved in the reward center of the brain as well as global hypomethylation of brain DNA. In a subsequent study, the addition of methyl donors to the high fat diet eliminated pre-weaning weight differences, reduced the preference for high fat foods, and ameliorated gene specific and global hypomethylation (Carlin et al., 2013). This research highlights the potential of methyl donor supplementation to reduce or eliminate the deleterious effects of developmental programming during gestation.

Maternal methyl donor supplementation is also associated with reduced susceptibility to cancer. Resistance to mammary tumor development was found in response to a carcinogenic challenge in female rats born to dams supplemented with methyl donors during gestation (Cho et al., 2012). Incidence of tumor development was lower and delayed in maternally supplemented offspring. Expression of HDAC1 and methyl CpG-binding protein 2 within the tumors was significantly reduced; however, no change in global DNA methylation was observed.

There also is evidence to suggest that supplementation of methyl donors later in life can have a significant effect on phenotype. Supplementing the diets of feedlot cattle with "lipotropic factors" such as choline, folic acid, and vitamin B12 increased yield grade and marbling score and decreased back fat thickness, resulting in a greater carcass value (Smith et al., 1974). Lipotropic factors may also reduce cancer risk through epigenetic modulation. Supplementation of female rats with five times the basal level of methyl donors during puberty reduced mammary tumor development (size and number) following injection of the carcinogenic agent N-nitroso-N-methylurea (Cho et al., 2014). Expression of HDAC1 and DNMT1 was reduced, suggesting that the protective effects of methyl donor supplementation were achieved via an epigenetic mechanism.

Butyrate

Butyrate is a short chain fatty acid produced by fiber fermentation in the rumen and colon. It is a potent HDAC inhibitor, and consequently significantly increases histone acetylation (Davie, 2003) as well as decreases histone methylation through epigenetic cross talk (Marinova et al., 2011). Treatment of bovine kidney epithelial cells *in vitro* with sodium butyrate increased H3 acetylation and induced apoptosis and cell cycle arrest (Li and Elsasser, 2005). In a follow-up study it was observed that sodium butyrate altered the expression of 450 genes in bovine kidney epithelial cells in culture, the majority of which were down-regulated and involved in cell cycle control (Li and Li, 2006).

In a depression model in rats, sodium butyrate treatment exhibited significant antidepressant effects and altered the expression of several genes within the hippocampus through increased H4 acetylation (Yamawaki et al., 2012). It also delayed the development of Huntington's disease and increased lifespan in mice and was associated with increased H3 and H4 acetylation (Ferrante et al., 2003). Circulating levels of butyrate can be altered, as patients treating impaired glucose tolerance with acarbose, an α-glucosidase inhibitor, had increased serum butyrate from 3.3 to 4.2 µM after 4 weeks of treatment (Wolver and Chiasson, 2000). Therefore altering butyrate levels can potentially be used as an agent to alter the epigenetic landscape and is a potential avenue for treatment for multiple neurological and other disorders.

Implications and Future Directions

Although we are beginning to understand the basis of developmental programming, there also is much that remains to be discovered. For example, the preponderance of data indicating transmission of developmentally induced alterations in germ line gene expression in a variety of animals suggests this is a universal phenomenon that occurs across species (Anway et al., 2008; Grazul-Bilska et al., 2012; Spencer et al., 2012; Brunton, 2013; Juul et al., 2014; Skinner, 2014). Although many controlled and relatively short-term studies in livestock have confirmed that developmental programming affects the fetus *in utero* as well as the offspring during infancy, few studies have examined whether these effects persist into adulthood and affect lifetime, or even transgenerational, productivity, primarily because of the expense of such studies in large animals with relatively long inter-generational intervals. In one study, overfeeding of ewes during pregnancy resulted in altered glucose and insulin tolerance as well as increased adiposity in both the offspring (F1) and their offspring (F2) (Shasa et al., 2015). As we have mentioned, there also is a need for much better epidemiological data linking birth or weaning weights to short- or long-term productivity in livestock.

At the same time, although a host of mechanisms have been described, an integrated view of the mechanisms of developmental programming is completely lacking. For example in sheep, over- and underfeeding during the peri-conceptual period negatively affect oocyte quality, and consequently the oocyte's ability to be fertilized or develop into a normal embryo (Borowczyk et al., 2006; Grazul-Bilska et al., 2012). Similarly, underfeeding of ewes from 8 weeks before until 30 days after mating alters the timing of birth as well as fetal maturation at the end of pregnancy, leading to a high rate of postnatal mortality (Kumarasamy et al., 2005). However, the mechanisms by which altered dietary intake during the peri-conceptual period alters not only oocyte function and embryonic development but also fetal development and postnatal survival is unknown. Particularly acute, due to the profound effects of placental development and function on embryo-fetal development, is the need for a much better understanding of the mechanisms regulating early placental development, and the many factors that can influence it both positively and negatively.

Data relevant to intelligent design of therapeutic or management strategies to overcome the negative consequences of developmental programming also are lacking, although there are some hints in the literature. For example, dietary protein supplementation of heifers or cows during pregnancy has been shown to increase calf birth and weaning weights as well as pregnancy rates of their heifer calves (Sletmoen-Olson et al., 2000; Caton et al., 2007; Funston et al., 2010). Altering dietary intake during the last third of gestation in a model of maternal overfeeding was able to partially restore placental development but was unable to 'rescue' fetal growth (Redmer et al., 2012). In other studies, supplementing diet-restricted ewes with Se during pregnancy has been shown to impact vascularity and mass of maternal small intestine as well as milk production and, in some cases, lamb birth weight (Vonnahme et al., 2015). Finally, subjecting developing heifers to a 'stair-step' nutritional regimen (alternating dietary restriction and ad-libitum feeding) shortened the time to puberty (Cardoso et al., 2014). These few studies emphasize the importance of examining much more carefully the effects of dietary interventions in developmental programming, and reiterate the importance of moving beyond the phenomenology to the mechanisms of developmental programming.

Some therapeutic interventions have been shown to overcome the negative consequences of developmental programming. We have already mentioned the need for more studies of the benefits of dietary lipotropes, or methyl donors. As another example, parenteral injection of L-Arg, which serves as a precursor for nitric oxide, an important vasodilator, as well as for polyamines, which are important for cellular growth and replication, periconceptually reduced embryonic loss and improved litter size in sheep (Luther et al., 2009; Saevre et al., 2011). Injection of L-Arg late in pregnancy also increased fetal protein accretion and lamb birth weight (De Boo et al., 2005). Similarly, injection of L-Arg during the last third of pregnancy enhanced fetal survival and growth in sheep carrying multiple fetuses (Lassala et al., 2011). In ewes undernourished beginning on day 28 post-mating, injection of L-Arg during the last 60% of pregnancy was able to restore lamb birth weights to those of adequately nourished ewes (Lassala et al., 2010). Using the same model of maternal dietary restriction as Lassala et al. (2010) and Satterfield et al. (2010) showed that treatment with sildenafil, which enhances nitric oxide levels, beginning on day 28 of gestation resulted in increased fetal weight in both restricted and adequately-fed ewes.

These studies highlight the importance of continued investigation of dietary, management and therapeutic strategies to improve postnatal, adult, and transgenerational outcomes in animals subjected to the negative consequences of developmental programming. Based on our current understanding, we would suggest focusing on the consequences of these various strategies in terms not only of their effects on whole animal, organ system and cellular structure and function, but also their effects on gene expression via alterations in the epigenetic landscape. Perhaps in the future, we might even use this knowledge to integrate epigenetic status, or 'epimutations', into livestock breeding programs.

Acknowledgements

The authors wish to acknowledge their appreciation for the invaluable contributions of their many mentors, colleagues, and graduate and undergraduate students to this work; many of these individuals are cited in the References Cited. We also wish to acknowledge the funding provided for this work by the many funding agencies, including North Dakota State University and the North Dakota Agricultural Experiment Station.

Keywords: Developmental programming, epigenetics, livestock, stressors, pregnancy, placenta

References

Aizer, A., and J. Currie. 2014. The intergenerational transmission of inequality: Maternal disadvantage and health at birth. Science 344: 856–861.

Alfaradhi, M.Z., and S.E. Ozanne. 2011. Developmental programming in response to maternal overnutrition. Frontiers in Genetics 2: 1–13.

Anderson, C.M., F. Lopez, A. Zimmer, and J.N. Benoit. 2006. Placental insufficiency leads to developmental hypertension and mesenteric artery dysfunction in two generations of Sprague-Dawley rat offspring. Biol. Reprod. 74: 538–544.

Anway, M.D., S.S. Rekow, and M.K. Skinner. 2008. Transgenerational epigenetic programming of the embryonic testis transcriptome. Genomics 91: 30–40.

Armitage, J.A., I.Y. Khan, P.D. Taylor, P.W. Nathanielsz, and L. Poston. 2004. Developmental programming of the metabolic syndrome by maternal nutritional imbalance: how strong is the evidence from experimental models in mammals? J. Physiol. 561: 355–377.

Bannister, A.J., and T. Kouzarides. 2011. Regulation of chromatin by histone modifications. Cell Res. 21: 381–395.

Barcroft, J. 1946. Researches on Pre-Natal Life. Blackwell, Oxford.

Barker, D.J.P. 1992. Fetal and Infant Origins of Adult Disease. BMJ Publishing Group, London.

Barker, D.J.P. 2004. Developmental origins of well-being. Philos. Trans. Royal Soc. London 359: 1359–1366.

Barker, D.J.P. 2007. Introduction: The window of opportunity. Symposium: Novel concepts in the developmental origins of adult health and disease. J. Nutr. 137: 1058–1059.

Barker, D.J.P., G. Larsen, C. Osmond, K.L. Thornburg, E. Kajantie, and J.G. Eriksson. 2012. The placental origins of sudden cardiac death. Internat. J. Epidemiol. 41: 1394–1399.

Barker, D.J.P., and K.L. Thornburg. 2013. Placental programming of chronic diseases, cancer and lifespan: A review. Placenta 34: 841–845.

Barlow, D.P. 2011. Genomic imprinting: A mammalian epigenetic discovery model. Annu. Rev. Genet. 45: 379–403.

Bartol, F.F., and C.A. Bagnell. 2012. Lactocrine programming of female reproductive tract development: Environmental connections to the reproductive continuum. Molec. Cell. Endocrinol. 354: 16–21.

Been, J.V., U.B. Nurmatov, B. Cox, T.S. Nawrot, C.P. van Schayck, and A. Sheikh. 2014. Effect of smoke-free legislation on perinatal and child health: A systematic review and meta-analysis. The Lancet 383: 1549–1560.

Belkacemi, L., D.M. Nelson, M. Desai, and M.G. Ross. 2010. Maternal undernutrition influences placental-fetal development. Biol. Reprod. 83: 325–331.

Bellingham, M., P.A. Fowler, M.R. Amezaga, C.M. Whitelaw, S.M. Rhind, C. Cotinot, B. Mandon-Pepin, R.M. Sharpe, and N.P. Evans. 2010. Foetal hypothalamic and pituitary expression of gonadotrophin-releasing hormone and galanin systems is disturbed by exposure to sewage sludge chemicals via maternal ingestion. J. Neuroendocrinol. 22: 527–533.

Bonacasa, B., R.C.M. Siow, and G.E. Mann. 2011. Impact of dietary soy isoflavones in pregnancy on fetal programming of endothelial function in offspring. Microcirculation 18: 270–285.

Borengasser, S.J., Y. Zhong, P. Kang, F. Lindsey, M.J.J. Ronis, T.M. Badger, H. Gomez-Acevedo, and K. Shankar. 2013. Maternal obesity enhances white adipose tissue differentiation and alters genome-scale DNA methylation in male rat offspring. Endocrinology 154: 4113–4125.

Borowczyk, E., J.S. Caton, D.A. Redmer, J.J. Bilski, R.M. Weigl, K.A. Vonnahme, P.P. Borowicz, J.D. Kirsch, K.C. Kraft, L.P. Reynolds, and A.T. Grazul-Bilska. 2006. Effects of plane of nutrition on *in vitro* fertilization and early embryonic development in sheep. J. Anim. Sci. 84: 1593–1599.

Borowicz, P., and L.P. Reynolds. 2010. 'Placental programming': More may still be less. J. Physiol. 588: 393.

Brunton, P.J. 2013. Effects of maternal exposure to social stress during pregnancy: Consequences for mother and offspring. Reproduction 146: R175–R189.

Bungard, D., B.J. Fuerth, P.Y. Zheng, B. Faubert, N.L. Mass, B. Viollet, D. Carling, C.B. Thomoson, R.G. Jones, and S.L. Berger. 2010. Signaling kinase AMPK activates stress-promoted transcription via histone H2B phosphorylation. Science 329: 1201–1205.

Camacho, L.E., A.M. Meyer, T.L. Neville, C.J. Hammer, D.A. Redmer, L.P. Reynolds, J.S. Caton, and K.A. Vonnahme. 2012. Neonatal hormone changes and growth in lambs born to dams receiving differing nutritional intakes and selenium supplementation during gestation. Reproduction 144: 23–35.

Cardoso, R.C., B.R.C. Alves, L.D. Prezotto, J.F. Thorson, L.O. Tedeschi, D.H. Keisler, C.S. Park, M. Amstalden, and G.L. Williams. 2014. Use of a stair-step compensatory gain nutritional regimen to program the onset of puberty in beef heifers. J. Anim. Sci. 92: 292–299.

Carlin, J., R. George, and T.M. Reyes. 2013. Methyl donor supplementation blocks the adverse effects of maternal high fat diet on offspring physiology. PLoS ONE 8: e63549.

Carone, B.R., F. Fauquier, N. Habib, J.M. Shea, C.E. Hart, R. Li, C. Block, C. Li, H. Gu, P.D. Zamore, A. Meissner, Z. Weng, H.A. Hofmann, N. Friedman, and O.J. Rando. 2010. Paternally induced transgenerational environmental reprogramming of metabolic gene expression on mammals. Cell 143: 1084–1096.

Caton, J., K. Vonnahme, J. Reed, T. Neville, C. Effertz, C. Hammer, J. Luther, D. Redmer, and L. Reynolds. 2007. Effects of maternal nutrition on birth weight and postnatal nutrient metabolism. Proc. International Symposium on Energy and Protein Metabolism. Vichy, France. EAAP Publication No. 124: 101–102.

Caton, J.S., and B.W. Hess. 2010. Maternal plane of nutrition: Impacts on fetal outcomes and postnatal offspring responses. Invited Review. pp. 104–122. *In*: B.W. Hess, T. Delcurto, J.G.P. Bowman, and R.C. Waterman (eds.). Proc. 4th Grazing Livestock Nutrition Conference. Proc. West Sect. Am. Soc. Anim. Sci., Champaign, IL.

Caton, J.S., A.M. Meyer, R.D. Yunusova, P.P. Borowicz, L.P. Reynolds, D.A. Redmer, C.J. Hammer, and K. Vonnahme. 2014a. Effects of maternal selenium supply and nutritional plane on offspring intestinal biology. pp. 14. *In*: Proc. 15th International Symposium on Trace Elements in Man and Animals (TEMA-15) University of Florida, Orlando.

Caton, J.S., T.L. Neville, L.P. Reynolds, C.J. Hammer, K.A. Vonnahme, A.M. Meyer, and J.B. Taylor. 2014b. Biofortification of maternal diets with selenium: Postnatal growth outcomes. pp. 159–161. *In*: G.S. Banuelos, Z.-Q. Lin, and X. Yin (eds.). Selenium in the Environment and Human Health. CRC Press, Taylor & Francis Group, London.

Chen, H., R.J. Lin, R.L. Schiltz, D. Chakravarti, A. Nash, L. Nagy, M.L. Privalsky, Y. Nakatani, and R.M. Evans. 1997. Nuclear receptor coactivator ACTR is a novel histone acetyltransferase and forms a multimeric activation complex with P/CAF and CBP/p300. Cell. 90: 569–580.

Cho, K., L. Mabasa, S. Bae, M.W. Walters, and C.S. Park. 2012. Maternal high-methyl diet suppresses mammary carcinogenesis in female rat offspring. Carciongenesis 33: 1106–1112.

Cho, K., W.S. Choi, C.L. Crane, and C.S. Park. 2014. Pubertal supplementation of lipotropes in female rats reduces mammary cancer risk by suppressing histone deacetylase 1. Eur. J. Nutr. 53: 1139–1143.

Christenson, R.K., and R.L. Prior. 1978. Uterine blood flow and nutrient uptake during late gestation in ewes with different number of fetuses. J. Anim. Sci. 46: 189–200.

Chud, T.C.S., S.L. Caetano, M.E. Buzankas, D.A. Grossi, D.G.F. Guidolin, G.B. Nascimento, J.O. Rosa, R.B. Lobo, and D.P. Munari. 2014. Genetic analysis for gestation length, birth weight, weaning weight, and accumulated productivity in Nellore beef cattle. Livest. Sci. 170: 16–21.

Clarke, S. 1993. Protein methylation. Curr. Opin. Cell Biol. 5: 977–983.

Connor, K.L., M.H. Vickers, J. Beltrand, M.J. Meaney, and D.M. Sloboda. 2012. Nature, nurture, or nutrition? Impact of maternal nutrition on maternal care, offspring development and reproductive function. J. Physiol. 590.9: 2167–2180.

Cooney, C.A., A.A. Dave, and G.L. Wolff. 2002. Maternal methyl supplements in mice affect epigenetic variation and DNA methylation of offspring. J. Nutr. 132: 2393S–2400S.

Cottrell, E.C., and S.M. Ozanne. 2007. Developmental programming of energy balance and the metabolic syndrome. Proc. Nutr. Soc. 66: 198–206.

Crain, D.A., S.J. Janssen, T.M. Edwards, J. Heindel, S.-M. Ho, P. Hunt, T. Iguchi, A. Juul, J.A. McLachlan, J. Schwartz, N. Skakkebaek, A.M. Soto, S. Swan, C. Walker, T.K. Woodruff, T.J. Woodruff, L.C. Guidice, and L.J. Guillette. 2008. Female reproductive disorders: The role of endocrine-disrupting compounds and developmental timing. Fertil. Steril. 90: 911–940.

Davie, J.R. 2003. Inhibition of histone deacetylase activity by butyrate. J. Nutr. 133: 2485S–2493S.

Dawson, M.A., A.J. Bannister, B. Gottgens, S.D. Foster, T. Bartke, A.R. Green, and T. Kouzarides. 2009. JAK2 phosphorylated histone H3Y31 and excluded HP1α from chromatin. Nature 461: 819–822.

De Boo, H.A., P.L. van Zijl, D.E. Smith, W. Kulik, H.N. Lafeber, and J.E. Harding. 2005. Arginine and mixed amino acids increase protein accretion in the growth-restricted and normal ovine fetus by different mechanisms. Pediatr. Res. 58: 270–277.

Delage, B., and R.H. Dashwook. 2008. Dietary manipulation of histone structure and function. Annu. Rev. Nutr. 28: 347–366.

Dhe-Paganon, S., F. Syeda, and L. Park. 2011. DNA methyl transferase 1: Regulatory mechanisms and implications in health and disease. Int. J. Biochem. Mol. Biol. 2: 58–66.

DiChiara, T.M., E.J. Roberstson, and A. Efstratiadis. 1991. Paternal imprinting of the mouse insulin-like growth factor II gene. Cell 64: 849–859.

Dillon, S.C., X. Zhang, R.C. Trievel, and X. Cheng. 2005. The SET-domain protein superfamily: Protein lysine methyltransferases. Genome Biol. 6: 227.

Dolinoy, D.C. 2008. The agouti mouse model: An epigenetic biosensor for nutritional and environmental alterations on the fetal epigenome. Nutr. Rev. 66: S7–S11.

Dong, M., Q. Zheng, S.P. Ford, P.W. Nathanielsz, and J. Ren. 2013. Maternal obesity, lipotoxicity and cardiovascular diseases in offspring. J. Molec. Cell. Cardiology 55: 111–116.

Dover, J., J. Schneider, M.A. Tawia-Boateng, A. Wood, K. Dean, M. Johnston, and A. Shilatifard. 2002. Methylation of histone H3 by COMPASS requires ubiquitination of histone H2 by Rad6. J. Biol. Chem. 277: 28368–28371.

Du, M., J. Tong, J. Zhao, K.R. Underwood, M. Zhu, S.P. Ford, and P.W. Nathanielsz. 2010. Fetal programming of skeletal muscle development in ruminant animals. J. Anim. Sci. 88: E51–60.

Du, M., J.X. Zhao, X. Yan, Y. Huang, L.V. Nicodemus, L. Yue, R.J. McCormick, and M.J. Zhu. 2011. Fetal muscle development, mesenchymal multipotent cell differentiation and associated signaling pathways. J. Anim. Sci. 89: 583–590.

Duarte, M.S., M.P. Gionbelli, P.V.R. Paulino, N.V.L. Serão, C.S. Nascimento, M.E. Botelho, T.S. Martins, S.C.V. Filho, M.V. Dodson, and M. Du. 2014. Maternal overnutrition enhances mRNA expression of adipogenic markers and collagen deposition in skeletal muscle of beef cattle fetuses. J. Anim. Sci. 92: 3846–3854.

Dupont, C., A.G. Cordier, C. Junien, B. Mandon-Pepin, R. Levy, and P. Chavatte-Palmer. 2012. Maternal environment and reproductive function of the offspring. Theriogenology 78: 1405–1414.

Elliot, I. 2013. Meat output must double by 2050. Feedstuffs. Accessed 1/15/2013: http://feedstuffsfoodlink.com/story-meat-output-must-double-by-2050-71-66920.

Evans, N.P., M. Bellingham, R.M. Sharpe, C. Cotinot, S.M. Rhind, C. Kyle, H. Erhard, S. Hombach-Klonisch, P.M. Lind, and P.A. Fowler. 2014. Reproduction Symposium: Does grazing on biosolids-treated pasture pose a pathophysiological risk associated with increased exposure to endocrine disrupting compounds? J. Anim. Sci. 92: 3185–3198.

Feed the Future Innovation Lab for Nutriton. USAID and Tufts University, Boston, MA. http://www.nutritioninnovationlab.org (last accessed 06 January 2017).

Ferrante, R.J., J.K. Kubilus, J. Lee. H. Ryu, A. Beesen, B. Zucker, K. Smith, N.W. Kowall, R.R. Ratan, R. Luthi-Carter, and S.M. Hersch. 2003. Histone deacetylase inhibition by sodium butyrate chemotherapy ameliorates the neurodegenerative phenotype in Huntington's disease mice. J. Neurosci. 23: 9418–9427.

Ferrell, C.L., and L.P. Reynolds. 1992. Uterine and umbilical blood flows and net nutrient uptake by fetuses and uteroplacental tissues of cows gravid with either single or twin fetuses. J. Anim. Sci. 70: 426–433.

Ford, S.P., B.W. Hess, M.M. Schwope, M.J. Nijland, J.S. Gilbert, K.A. Vonnahme, W.J. Means, H. Han, and P.W. Nathanielsz. 2007. Maternal undernutrition during early to mid-gestation in the ewe results in altered growth, adiposity, and glucose tolerance in male offspring. J. Anim. Sci. 85: 1285–1294.

Fowden, A.L., D.A. Giussani, and A.J. Forhead. 2006. Intrauterine programming of physiological systems: Causes and consequences. Physiology 21: 29–37.

Fowden, A.L., and A.J. Forhead. 2009. Hormones as epigenetic signals in developmental programming. Exp. Physiol. 94(6): 607–625.

Fowler, P.A., M. Bellingham, K.D. Sinclair, N.P. Evans, P. Pocar, B. Fischer, K. Schaedlich, J.-S. Schmidt, M.R. Amezaga, S. Bhattachary, S.M. Rhind, and P.J. O'Shaughnessy. 2012. Impact of endocrine-disrupting compounds (EDCs) on female reproductive health. Molec. Cell. Endocrinol. 355: 231–239.

Freking, B.A., S.K. Murphy, A.A. Wylie, S.J. Rhodes, J.W. Keele, K.A. Leymaster, R.L. Jirtle, and T.P.L. Smith. 2002. Identification of the single base change causing callipyge muscle hypertrophy phenotype, the only known example of polar overdominance in mammals. Genome Res. 12: 1496–1506.

Funston, R.N., D.M. Larson, and K.A. Vonnahme. 2010. Effects of maternal nutrition on conceptus growth and offspring performance: Implications for beef cattle production. J. Anim. Sci. 88(E. Suppl.): E205–E215.

Funston, R.N., A.F. Summers, and A.J. Roberts. 2012. Alpharma Beef Cattle Nutrition Symposium: Implications of nutritional management for beef cow-calf systems. J. Anim. Sci. 90: 2301–2307. doi:10.2527/jas2011-4568.

Galbally, M., M. Snellen, and J. Power. 2014. Antipsychotic drugs in pregnancy: A review of their maternal and fetal effects. Therapeutic Adv. Drug Safety 52: 100–109.

Gardner, D.S., R.G. Lea, and K.D. Sinclair. 2008. Developmental programming of reproduction and fertility: What is the evidence? Animal 2: 1128–1134.

Gardner, D.S., S.E. Ozanne, and K.D. Sinclair. 2009. Effect of the early-life nutritional environment on fecundity and fertility of mammals. Phil. Trans. R. Soc. B 364: 3419–3427.

Gatford, K.L., and R.A. Simmons. 2013. Prenatal programming of insulin secretion in intrauterine growth restriction. Clin. Obstet. Gynecol. 56: 520–528.

Gerbert, C., C. Wrenzycki, D. Herrmann, D. Gröger, R. Reinhardt, P. Hajkova, A. Lucas-Hahn, J. Carnwath, H. Lehrach, and H. Niemann. 2006. The bovine IGF2 gene is differentially methylated in oocyte and sperm DNA. Genomics 88: 222–229.

Gerland, P., A.E. Raftery, H. Sevcikova, N. Li, D. Gu, T. Spoorenberg, L. Alkema, B.K. Fosdick, J. Chunn, N. Lalic, G. Bay, T. Buettner, G.K. Heilig, and J. Wilmoth. 2014. World population stabilization unlikely this century. Science 346: 234–237.

Grazul-Bilska, A.T., E. Borowczyk, J.J. Bilski, L.P. Reynolds, D.A. Redmer, J.S. Caton, and K.A. Vonnahme. 2012. Overfeeding and underfeeding have detrimental effects on oocyte quality measured by *in vitro* fertilization and early embryonic development in sheep. Domestic Anim. Endocrinol. 43: 289–298.

Greenwood, P.L., A.S. Hunt, J.W. Hermanson, and A.W. Bell. 1998. Effects of birth weight and postnatal nutrition on neonatal sheep: I. Body growth and composition, and some aspects of energetic efficiency. J. Anim. Sci. 76: 2354–2367.

Greenwood, P.L., A.S. Hunt, J.W. Hermanson, and A.W. Bell. 2000. Effects of birth weight and postnatal nutrition on neonatal sheep: II. Skeletal muscle growth and development. J. Anim. Sci. 78: 50–61.

Greenwood, P.L., A.S. Hunt, and A.W. Bell. 2004. Effects of birth weight and postnatal nutrition on neonatal sheep: IV. Organ growth. J. Anim. Sci. 82: 422–428.

Hammond, J. 1927. The Physiology of Reproduction in the Cow. Cambridge University Press, Cambridge, UK.

Harris, E.K., E.P. Berg, E.L. Berg, and K.A. Vonnahme. 2013. Effect of maternal activity during gestation on maternal behavior, fetal growth, umbilical blood flow, and farrowing characteristics in pigs. J. Anim. Sci. 91: 734–744.

Hassa, P.O., S.S. Haenni, M. Elser, and M.O. Hottiger. 2006. Nuclear ADP-ribosylation reactions in mammalian cells: Where are we today and where are we going? Microbiol. Mol. Bio. Rev. 70: 789–829.

Hellemans, K.G.C., J.H. Sliwowska, P. Verma, and J. Weinberg. 2010. Prenatal alcohol exposure: Fetal programming and later life vulnerability to stress, depression and anxiety disorders. Neurosci. Biobehav. Rev. 34: 791–807.

Hershko, A., and A. Ciechanover. 1998. The ubiquitin system. Annu. Rev. Biochem. 67: 425–479.

Hochberg, Z., R. Feil, M. Constancia, M. Fraga, C. Junien, J.-C. Carel, P. Boileau, Y. Le Bouc, C.L. Deal, K. Lillycrop, R. Scharfmann, A. Sheppard, M. Skinner, M. Szyf, R.A. Waterland, D.J. Waxman, E. Whitelaw, K. Ong, and K. Albertsson-Wikland. 2011. Child health, developmental plasticity, and epigenetic programming. Endocr. Rev. 32: 159–224.

Ismail-Beigi, F., P.M. Catalano, and R.W. Hanson. 2006. Metabolic programming: Fetal origins of obesity and metabolic syndrome in the adult. Am. J. Physiol. 291: E439–E440.

Jackson, L.M., A. Mytinger, E.K. Roberts, T.M. Lee, D.L. Foster, V. Padmanabhan, and H.T. Jansen. 2013. Developmental programming: Postnatal steroids complete prenatal steroid actions to differentially organize the GnRH surge mechanism and reproductive behavior in female sheep. Endocrinology 154: 1612–1623.

Jackson, S.P., and R.D. Green. 1993. Muscle trait inheritance, growth performance and feed efficiency of sheep exhibiting a muscle hypertrophy phenotype. J. Anim. Sci. 71: 241.

Jansson, T., and T.L. Powell. 2007. Role of the placenta in fetal programming: Underlying mechanisms and potential interventional approaches. Clin. Sci. 113: 1–13.

Jones, P.L., G.J. Veenstra, P.A. Warde, D. Vermaak, S.U. Kass, N. Landsberger, J. Strouboulis, and A.P. Wolffe. 1998. Methylated DNA and MeCP2 recruit histone deacetylase to repress transcription. Nature Genet. 19: 187–191.

Juchem, S.O., P.H. Robinson, and E. Evans. 2012. A fat based rumen protection technology post-ruminally delivers a B vitamin complex to impact performance of multiparous Holstein cows. Anim. Feed Sci. Technol. 174: 68–78.

Juul, A., K. Almstrup, A.-M. Andersson, T.K. Jensen, N. Jorgensen, K.M. Main, E. Rajpert-De Meyts, and N.E. Toppari J, Skakkebaek. 2014. Possible fetal determinants of male infertility. Nature Rev. Endocrinol. 10: 553–562.

Kilcoyne, K.R., L.B. Smith, N. Atanassova, S. Macpherson, C. McKinnell, S. van den Driesche, M.S. Jobling, T.J.G. Chambers, K. De Gendt, G. Verhoeven, L. O'Hara, S. Platts, L. Renato de Franca, N.L.M. Lara, R.A. Anderson, and R.M. Sharpe. 2014. Fetal programming of adult Leydig cell function by androgenic effects on stem/progenitor cells. Proc. Natl. Acad. Sci. U.S.A. 111: E1924–E1932.

Kim, J., M. Guermah, R.K. McGinty, J.S. Lee, Z. Tang, T.A. Milne, A. Shilatifard, T.W. Muir, and R.G. Roeder. 2009. RAD6-mediated transcription-coupled H2B ubiquitylation directly stimulates K3K4 methylation in human cells. Cell 137: 459–471.

Kothapalli, N., G. Camporeale, A. Kueh, Y.C. Chew, A.M. Oommen, J.B. Griffin, and J. Zempleni. 2005. Biological functions of biotinylated histones. J. Nutr. Biochem. 16: 446–448.

Kumarasamy, V., M.D. Mitchell, F.H. Bloomfield, M.H. Oliver, M.E. Campbell, J.R. Challis, and J.E. Harding. 2005. Effects of periconceptional undernutrition on the initiation of parturition in sheep. Am. J. Physiol. Regul. Integr. Comp. Physiol. 288: R67–R72.

Kwong, W.Y., S.J. Adamiak, A. Gwynn, R. Singh, and K.D. Sinclair. 2010. Endogenous folates and single-carbon metabolism in the ovarian follicle, oocyte and pre-implantation embryo. Reproduction 139: 705–715.

Laanpere, M., S. Altmäe, A. Stavreus-Evers, T.K. Nilsson, A. Yngve, and A. Salumets. 2010. Folate-mediated one-carbon metabolism and its effect on female fertility and pregnancy viability. Nutr. Rev. 68: 99–113.

Lan, X., E.C. Cretney, J. Kropp, K. Khateeb, M.A. Berg, F. Peñagaricano, R. Magness, A.E. Radunz, and H. Khatib. 2013. Maternal diet during pregnancy induces gene expression and DNA methylation changes in fetal tissues in sheep. Front. Genet. 4: 49.

Larsen, F., G. Gundersen, R. Lopez, and H. Prydz. 1992. CpG islands as gene markers in the human genome. Genomics 13: 1095–1107.

Larson, D.M., J.L. Martin, D.C. Adams, and R.N. Funston. 2009. Winter grazing system and supplementation during late gestation influence performance of beef cows and steer progeny. J. Anim. Sci. 87: 1147–1155.

Lassala, A., F.W. Bazer, T.A. Cudd, S. Datta, D.H. Keisler, M.C. Satterfield, T.E. Spencer, and G. Wu. 2010. Parenteral administration of L-arginine prevents fetal growth restriction in undernourished ewes. J. Nutr. 140: 1242–1248.

Lassala, A., F.W. Bazer, T.A. Cudd, S. Datta, D.H. Keisler, M.C. Satterfield, T.E. Spencer, and G. Wu. 2011. Parenteral administration of L-arginine enhances fetal survival and growth in sheep carrying multiple fetuses. J. Nutr. 141: 849–855.

Leach, L. 2011. Placental vascular dysfunction in diabetic pregnancies: Intimations of fetal cardiovascular disease? Microcirculation 18: 263–269.

Lemley, C.O., A.M. Meyer, L.E. Camacho, T.L. Neville, D.J. Newman, J.S. Caton, and K.A. Vonnahme. 2012. Melatonin supplementation alters uteroplacental hemodynamics and fetal development in an ovine model of intrauterine growth restriction. Am. J. Physiol. Regul. Integr. Comp. Physiol. 302: R454–R467.

Li, C.J., and T.H. Elsasser. 2005. Butyrate-induced apoptosis and cell cycle arrest in bovine kidney epithelial cells: Involvement of caspase and proteasome pathways. J. Anim. Sci. 83: 89–97.

Li, M., D.M. Sloboda, and M.H. Vickers. 2011. Maternal obesity and developmental programming of metabolic disorders in offspring: Evidence from animal models. Exp. Diabetes Res. 2011: 1–9 (doi:10.1155/2011/592408).

Li, R.W., and C.J. Li. 2006. Butyrate induces profound changes in gene expression related to multiple signal pathways in bovine kidney epithelial cells. BMC Genomics 7: 234.

Lillycrop, K.A., J.L. Slater-Jefferies, M.A. Hanson, K.M. Godfrey, A.A. Jackson, and G.C. Burdge. 2007. Induction of altered epigenetic regulation of the hepatic glucocorticoid receptor in the offspring of rats fed a protein-restricted diet during pregnancy suggests that reduced DNA methyltransferase-1 expression is involved in impaired DNA methylation and changes in histone modification. Br. J. Nutr. 97: 1064–1073.

Long, N.M., C.B. Tousley, K.R. Underwood, S.I. Paisley, W.J. Means, B.W. Hess, M. Du, and S.P. Ford. 2012a. Effects of early- to mid-gestational undernutrition with or without protein supplementation on offspring growth, carcass characteristics, and adipocyte size in beef cattle. J. Anim. Sci. 90: 197–206.

Long, N.M., P.W. Nathanielsz, and S.P. Ford. 2012b. The impact of maternal overnutrition and obesity on hypothalamic-pituitary-adrenal axis in response to offspring stress. Domestic Anim. Endocrinol. 42: 195–202.

Luther, J.S., D.A. Redmer, L.P. Reynolds, and J.M. Wallace. 2005. Nutritional paradigms of ovine fetal growth restriction: implications for human pregnancy. Human Fertility 8: 179–187.

Luther, J.S., E.J. Windorski, J.S. Caton, G. Wu, J.D. Kirsch, K.A. Vonnahme, L.P. Reynolds, and C.S. Schauer. 2009. Effects of arginine supplementation on reproductive performance in Rambouillet ewes. http://www.ag.ndsu.edu/HettingerREC/sheep/2009-Sheep-Report.pdf.

Mahajan, K., B. Fang, J.M. Koomen, and N.P. Mahajan. 2012. H2B Tyr37 phosphorylation suppresses expression of replication-dependent core histone genes. Nat. Struct. Mol. Biol. 19: 930–937.

Marinova, Z., Y. Leng, P. Leeds, and D.M. Chuang. 2011. Histone deacetylase inhibition alters histone methylation associated with heat shock protein 70 promoter modifications in astrocytes and neurons. Neuropharmacology 60: 1109–1115.

Martin, J.L., K.A. Vonnahme, D.C. Adams, G.P. Lardy, and R.N. Funston. 2007. Effects of dam nutrition on growth and reproductive performance of heifers calves. J. Anim. Sci. 85: 841–847.

Martin-Gronert, M.S., and S.E. Ozanne. 2012. Metabolic programming of insulin action and secretion. Diabetes, Obesity and Metabolism 14(3): 29–39.

Mayhew, T.M., D.S. Charnock-Jones, and P. Kaufmann. 2004. Aspects of human fetoplacental vasculogenesis and angiogenesis. III. Changes in complicated pregnancies. Placenta 25: 127–139.

McCoard, S., F. Sales, N. Wards, Q. Sciascia, M. Oliver, J. Koolaard, and D. van der Linden. 2013. Parenteral administration of twin-bearing ewes with L-arginine enhances the birth weight and brown fat stores in sheep. SpringerPlus 2: 684.

McLachlan, J.A., S.L. Tilghman, M.E. Burow, and M.R. Bratton. 2012. Environmental signaling and reproduction: A comparative biological and chemical perspective. Molec. Cell. Endocrinol. 354: 60–62.

Meyer, A.M., J.J. Reed, T.L. Neville, J.B. Taylor, L.P. Reynolds, D.A. Redmer, K.A. Vonnahme, and J.S. Caton. 2012. Effects of nutritional plane and selenium supply during gestation on visceral organ mass and indices of intestinal growth and vascularity in primiparous ewes at parturition and during early lactation. J. Anim. Sci. 90: 2733–2749.

Meyer, A.M., T.L. Neville, J.J. Reed, J.B. Taylor, L.P. Reynolds, D.A. Redmer, C.J. Hammer, K.A. Vonnahme, and J.S. Caton. 2013. Maternal nutritional plane and selenium supply during gestation impacts visceral organ mass and intestinal growth and vascularity of neonatal lamb offspring. J. Anim. Sci. 91: 2628–2639.

Meyer, A.M., and J.S. Caton. 2016. The role of the small intestine in developmental programming: Impact of maternal nutrition on the dam and offspring. Advances in Nutrition (in press).

Ng, S.F., R.C.Y. Lin, D.R. Laybutt, R. Barres, J.A. Owens, and M.J. Morris. 2010. Chronic high-fat diet in fathers programs b-cell dysfunction in female rat offspring. Nature 467: 963–967.

NRC. 1996. Nutrient Requirements of Beef Cattle, 7th Rev. Ed. National Academy Press, Washington, DC.

NRC. 2007. Nutrients Requirements of Small Ruminants, Sheep, Goats, Cervids and New World Camelids. The National Academies Press, Washington, DC.

Oki, M., H. Aihara, and T. Ito. 2007. Role of histone phosphorylation in chromatin dynamics and its implications in diseases. Subcell Biochem. 41: 319–336.

Paneth, N., and M. Susser. 1995. Early origin of coronary heart disease (the "Barker hypothesis"). Brit. Med. J. 1310: 411–412.

Peine, J.L., G.Q. Jia, M.L. Van Emon, T.L. Neville, J.D. Kirsch, C.J. Hammer, S.T. O'Rourke, L.P. Reynolds, and J.S. Caton. 2013. Effects of maternal nutrition and rumen-protected arginine supplementation on ewe and postnatal lamb performance. Proc. West. Sec. Amer. Soc. Anim. Sci. 64: 80–83.

Peine, J., G. Gia, M. Van Emon, T. Neville, J. Kirsch, C. Hammer, S. O'Rourke, L. Reynolds, and J. Caton. 2014. PPO.03 Effects of maternal nutrition and ruminal-protected arginine supplementation on offspring growth. Arch. Dis. Child Fetal Neonatal Ed. 99(1A): 151. doi: 10.1136/archdischild-2014-306576.444.

Penagaricano, F., A.H. Souza, P.D. Carvalho, A.M. Driver, R. Gambra, J. Kropp, K.S. Hackbart, D. Luchini, R.D. Shaver, M.C. Wiltbank, and H. Khatib. 2013. Effect of maternal methionine supplementation on the transcriptome of bovine preimplantation embryos. PLoS ONE 8: e72302. doi:10.1371/journal.pone.0072302.

Pfeifer, K. 2000. Mechanisms of genomic imprinting. Am. J. Hum. Genet. 67: 777–787.

Radunz, A.E., F.L. Fluharty, A.E. Relling, T.L. Felix, L.M. Shoup, H.N. Zerby, and S.C. Loerch. 2012. Prepartum dietary energy source fed to beef cows: II. Effects on progeny postnatal growth, glucose tolerance, and carcass composition. J. Anim. Sci. 2012.90: 4962–4974. doi: 10.2527/jas2012-5098.

Ramchandani, S., S.K. Bhattacharya, N. Cervoni, and M. Szyf. 1999. DNA methylation is a reversible biological signal. Proc. Natl. Acad. Sci. U.S.A. 96: 6107–6112.

Ravelli, A.C.J., J.H.P. van der Meulen, R.P.J. Michels, C. Osmond, D.J.P. Barker, C.N. Hales, and O.P. Bleker. 1998. Glucose tolerance in adults after prenatal exposure to famine. Lancet 351: 173–177.

Redmer, D.A., J.M. Wallace, and L.P. Reynolds. 2004. Effect of nutrient intake during pregnancy on fetal and placental growth and vascular development. Domest. Anim. Endocrinol. 27: 199–217.

Redmer, D.A., R.P. Aitken, J.S. Milne, L.P. Reynolds, and J.M. Wallace. 2005. Influence of maternal nutrition on messenger RNA expression of placental angiogenic factors and their receptors at mid-gestation in adolescent sheep. Biol. Reprod. 72: 1004–1009.

Redmer, D.A., J. Luther, J. Milne, R. Aitken, M. Johnson, P. Borowicz, M. Borowicz, L.P. Reynolds, and J. Wallace. 2009. Fetoplacental growth and vascular development in overnourished adolescent sheep at day 50, 90 and 130 of gestation. Reproduction 137: 749–757.

Redmer, D.A., J.S. Luther, J.S. Milne, R.P. Aitken, M.L. Johnson, P.P. Borowicz, L.P. Reynolds, J.S. Caton, and J.M. Wallace. 2012. Decreasing maternal nutrient intake during the final third of pregnancy in previously overnourished adolescent sheep: Effects on maternal nutrient partitioning, feto-placental outcomes. Placenta 33: 114–121.

Reed, J.J., M.A. Ward, K.A. Vonnahme, T.L. Neville, S.L. Julius, P.P. Borowicz, J.B. Taylor, D.A. Redmer, A.T. Grazul-Bilska, L.P. Reynolds, and J.S. Caton. 2007. Effects of selenium supply and dietary restriction on maternal and fetal body weight, visceral organ mass, cellularity estimates, and jejunal vascularity in pregnant ewe lambs. J. Anim. Sci. 85: 2721–2733.

Reynolds, L.P., P.P. Borowicz, K.A. Vonnahme, M.L. Johnson, A.T. Grazul-Bilska, J.M. Wallace, J.S. Caton, and D.A. Redmer. 2005. Animal models of placental angiogenesis. Placenta 26: 689–708.

Reynolds, L.P., J.S. Caton, D.A. Redmer, A.T. Grazul-Bilska, K.A. Vonnahme, P.P. Borowicz, J.S. Luther, J.M. Wallace, G. Wu, and T.E. Spencer. 2006. Evidence for altered placental blood flow and vascularity in compromised pregnancies. Invited (Topical) review. J. Physiol. 572.1: 51–58.

Reynolds, L.P., P.P. Borowicz, J.S. Caton, K.A. Vonnahme, J.S. Luther, D.S. Buchanan, S.A. Hafex, A.T. Grazul-Bilska, and D.A. Redmer. 2010a. Uteroplacental vascular development and placental function: An update. Int. J. Dev. Biol. 54(2-3): 355–366.

Reynolds, L.P., P.P. Borowica, J.S. Caton, K.A. Vonnahme, J.S. Luther, C.J. Hammer, K.R. Maddock Carlin, A.T. Grazul-Bilska, and D.A. Redmer. 2010b. Developmental programming: The concept, large animal models, and the key role of uteroplacental vascular development. J. Anim. Sci. 88(13): E61–72.

Reynolds, L.P., and J.S. Caton. 2012. Role of the pre- and post-natal environment in developmental programming of health and productivity. Invited review. Molecular and Cellular Endocrinology, Special Issue 'Environment, Epigenetics and Reproduction' M.K. Skinner (ed.). Mol. Cell Endocrinol. 354: 54–59.

Reynolds, L.P., K.A. Vonnahme, C.O. Lemley, D.A. Redmer, A.T. Grazul-Bilska, P.P. Borowicz, and J.S. Caton. 2013. Maternal stress and placental vascular function and remodeling. Curr. Vasc. Pharmacol. 11: 564–593.

Reynolds, L.P., P.P. Borowicz, C. Palmieri, and A.T. Grazul-Bilska. 2014. Placental vascular defects in compromised pregnancies: effects of assisted reproductive technologies and other maternal stressors. pp. 193–204. In: L. Zhang, and C.A. Ducsay (eds.). Advances in Fetal and Neonatal Physiology. Advances in Experimental Medicine and Biology 814. Springer Science+Business Media, NY.

Reynolds, L.P., M. Wulster-Radcliffe, D.K. Aaron, and T.A. Davis. 2015. Issue and opinions: Importance of animals in agricultural sustainability and food security. J. Nutr. 145: 1377–79; doi:10.3945/jn.115.212217.

Reynolds, L.P., and K.A. Vonnahme. 2016. Triennial Reproduction Symposium: Developmental programming of fertility. J. Anim. Sci. (submitted).

Rhind, S.M., M.T. Rae, and A.N. Brooks. 2001. Effects of nutrition and environmental factors on the fetal programming of the reproductive axis. Reproduction 122: 205–214.

Richter, C.A., L.S. Birnbaum, F. Farabollini, R.R. Newbold, B.S. Rubin, C.E. Talsness, J.G. Vandenbergh, D.R. Walser-Kuntz, and F.S. vom Saal. 2007. *In vivo* effects of bisphenol A in laboratory rodent studies. Reprod. Toxicol. 24: 199–224.

Richter, V.F.I., J.F. Briffa, K.M. Moritz, M.E. Wlodek, and D.H. Hryciw. 2016. The role of maternal nutrition, metabolic function and the placental in developmental programming of renal dysfunction. Clin. Exp. Pharmacol. Physiol. 43: 135–141.

Robinson, D.L., L.M. Cafe, and P.L. Greenwood. 2013. Meat science and muscle biology symposium: Developmental programming in cattle: Consequences for growth, efficiency, carcass, muscle, and beef quality characteristics. J. Anim. Sci. 91: 1428–1442. doi:10.2527/jas2012-5799.

Robinson, J.J., K.D. Sinclair, and T.G. McEvoy. 1999. Nutritional effects on foetal growth. Anim. Sci. 68: 315–331.

Rossetto, D., N. Avvakumov, and J. Côté. 2012. Histone phosphorylation: a chromatin modification involved in diverse nuclear events. Epigenetics 7: 1098–1108.

Saevre, C., A.M. Meyer, M.L. Van Emon, D.A. Redmer, J.S. Caton, J.D. Kirsch, J.S. Luther, and C.S. Schauer. 2011. Impacts of arginine on ovarian function and reproductive performance at the time of maternal recognition of pregnancy in ewes. NDSU Sheep Res. Rep. 52: 13–16.

Satterfield, M.C., F.W. Bazer, T.E. Spencer, and G. Wu. 2010. Sildenafil citrate treatment enhances amino acid availability in the conceptus and fetal growth in an ovine model of intrauterine growth restriction. J. Nutr. 140: 251–258.

Schmidt, K.L., E.A. MacDougall-Shakleton, K.K. Soma, and S.A. MacDougall-Shakleton. 2014. Developmental programming of the HPA and HPG axes by early-life stress in male and female song sparrows. Gen. Comp. Endocrinol. 196: 72–80.

Schneider, J.E., J.M. Brozek, and E. Keen-Rinehart. 2014. Our stolen figures: The interface of sexual differentiation, endocrine disruptors, maternal programming, and energy balance. Horm. Behav. 66: 110–119.

Seneviratne, S.N., G.K. Parry, L.M.E. McCowan, E. Ekeroma, Y. Jiang, S. Gusso, G. Peres, R.O. Rodrigues, S. Craigie, W.S. Cutfield, and P.L. Hofman. 2014. Antenatal exercise in overweight and obese women and its effects on offspring and maternal health: Design and rationale of the IMPROVE (Improving Maternal and Progeny Obesity Via Exercise) randomized controlled trial. BMC Pregnancy Childbirth 14: 148 http://www.biomedcentral.com/1471-2393/14/148.

Shankar, K., A. Harrell, X. Liu, J.M. Gilchrist, M.J. Ronis, and T.M. Badger. 2008. Maternal obesity at conception programs obesity in the offspring. Am. J. Physiol. Regul. Integr. Comp. Physiol. 294: R528–R538.

Shankar, K., P. Kang, A. Harrell, Y. Zhong, J.C. Marecki, M.J. Ronis, and T.M. Badger. 2010. Maternal overweight programs insulin and adiponectin signaling in the offspring. Endocrinology 151: 2577–2589.

Shankar, K., Y. Zhong, P. Kang, F. Lau, M.L. Blackburn, J.-R. Chen, S.J. Borengasser, M.J.J. Ronis, and T.M. Badger. 2011. Maternal obesity promotes a proinflammatory signature in rat uterus and blastocyst. Endocrinology 152: 4158–4170.

Shasa, D.R., J.F. Odhiambo, N.L. Long, N. Tuersunjiang, P.W. Nathanielsz, and S.P. Ford. 2015. Multi-generational impact of maternal overnutrition/obesity in sheep on the neonatal leptin surge in granddaughters. Int. J. Obesity 39: 695–701.

Sheldon, B.C., and S.A. West. 2004. Maternal dominance, maternal condition, and offspring sex ratio in ungulates. Am. Naturalist 163: 40–54.

Shiio, Y., and R.N. Eisenman. 2003. Histone sumoylation is associated with transcriptional repression. PNAS 100: 13225–13230.

Shilatifard, A. 2006. Chromatin modifications by methylation and ubiquitinaion: Implications in the regulation of gene expression. Annu. Rev. Biochem. 75: 243–269.

Shukla, P., C.O. Lemley, N. Dubey, A.M. Meyer, S.T. O'Rourke, and K.A. Vonnahme. 2014. Effect of maternal nutrient restriction and melatonin supplementation from mid to late gestation on vascular reactivity of maternal and fetal placental arteries. Placenta 35: 461–466.

Sinclair, K.D., C. Allegrucci, R. Singh, D.S. Gardner, S. Sebastian, J. Bispham, A. Thurston, J.F. Huntley, W.D. Rees, C.A. Maloney, R.G. Lea, J. Craigon, T.G. McEvoy, and L.E. Young. 2007. DNA methylation, insulin resistance, and blood pressure in offspring determined by maternal periconceptional B vitamin and methionine status. PNAS 104: 19351–19356.

Sinclair, K.D., A. Karamitri, and D.S. Gardner. 2010. Dietary regulation of developmental programming in ruminants: epigenetic modifications in the germline. Soc. Reprod. Fertil. 67: 59–72.

Skinner, M.K. 2014. Endocrine disruptor induction of epigenetic transgenerational inheritance of disease. Molec. Cell. Endocrinol. 398: 4–12.

Sletmoen-Olson, K.E., J.S. Caton, L.P. Reynolds, and K.C. Olson. 2000. Undegraded intake protein supplementation. I. Effects on forage utilization and performance of periparturient beef cows fed low-quality hay during gestation and lactation. J. Anim. Sci. 78: 449–455.

Smith, G.S., J.W. Chambers, A.L. Neumann, E.E. Ray, and A.B. Nelson. 1974. Lipotropic factors for beef cattle fed high-concentrate diets. J. Anim. Sci. 38: 627–633.

Spencer, T.E., K.A. Dunlap, and J. Filant. 2012. Comparative developmental biology of the uterus: Insights into mechanisms and developmental disruption. Molec. Cell. Endocrinol. 354: 34–53.

Spitzer, J.C., D.G. Morrison, R.P. Wettemann, and L.C. Faulkner. 1995. Reproductive responses and calf birth and weaning weight as affected by body condition at parturition and postpartum weight gain in primiprous beef cows. J. Anim. Sci. 73: 1251–1257.

Stalker, L.A., L.A. Ciminski, D.C. Adams, T.J. Klopfenstein, and R.T. Clark. 2007. Effects of weaning date and prepartum protein supplementation on cow performance and calf growth. Rangeland Ecol. Management 60: 578–587.

Sterner, D.E., and S.L. Berger. 2000. Acetylation of histones and transcription-related factors. Microbiol Mol. Biol. Rev. 64: 435–459.

Sun, Z.W., and C.D. Allis. 2002. Ubiquitination of histone H2B regulates H3 methylation and gene silencing in yeast. Nature 418: 104–108.

Swanson, T.J., C.J. Hammer, J.S. Luther, D.B. Carlson, J.B. Taylor, D.A. Redmer, T.L. Neville, J.J. Reed, L.P. Reynolds, J.S. Caton, and K.A. Vonnahme. 2008. Effects of gestational plane of nutrition and selenium supplementation on mammary development and colostrum quality in pregnant ewe lambs. J. Anim. Sci. 86: 2415–2423.

Symonds, M.E., M. Pope, D. Sharkey, and H. Budge. 2012. Adipose tissue and fetal programming. Diabetologia 55: 1597–1606.

Tamaru, H., and E.U. Selker. 2001. A histone H3 methyltransferase controls DNA methylation in *Neurospora crassa*. Nature 414: 277–83.

Thornburg, K.L., P.F. O'Tierney, and S. Louey. 2010. The placenta is a programming agent for cardiovascular disease. Placenta 31: S54–S59. doi:10.1016/j.placenta.2010.01.002.

Tosh, D.N., Q. Fu, C.W. Callaway, R.A. McKnight, T.C. McMillin, M.G. Ross, R.H. Lane, and M. Desai. 2010. Epigenetics of programmed obesity: Alteration of IUGR rat hepatic IGF1 mRNA expression and histone structure in rapid vs. delayed postnatal catch-up growth. Am. J. Physiol. Gastrointest. Liver Physiol. 299: G1023–G1029.

United Nations. 2015. Revision of World Population Prospects (https://esa.un.org/unpd/wpp/), UN Department of Economic and Social Affairs, Population Division. Accessed 12/28/2015.

Vonnahme, K.A., M.J. Zhu, P.P. Borowicz, T.W. Geary, B.W. Hess, L.P. Reynolds, J.S. Caton, W.J. Means, and S.P. Ford. 2007. Effect of early gestational undernutrition on angiogenic factor expression and vascularity in the bovine placentome. J. Anim. Sci. 85: 2464–2472.

Vonnahme, K.A., J. Evoniuk, M.L. Johnson, P.P. Borowicz, D.A. Redmer, L.P. Reynolds, and A.T. Grazul-Bilska. 2008. Placental vascularity and growth factor expression in singleton, twin, and triplet pregnancies in the sheep. Endocrine 33: 53–61.

Vonnahme, K.A., J.S. Luther, L.P. Reynolds, C.J. Hammer, D.B. Carlson, D.A. Redmer, and J.S. Caton. 2010. Impacts of maternal selenium and nutritional level on growth, adiposity, and glucose tolerance in female offspring in sheep. Domest. Anim. Endocrinol. 39: 240–248.

Vonnahme, K.A., and C.O. Lemley. 2012. Programming the offspring through altered uteroplacental hemodynamics: How maternal environment impacts uterine and umbilical blood flow in cattle, sheep and pigs. Reprod. Fertil. Develop. 24: 97–104.

Vonnahme, K.A., C.O. Lemley, P. Shukla, and S.T. O'Rourke. 2013a. Placental programming: How the maternal environment can impact placental function. J. Anim. Sci. 91: 2467–2480.

Vonnahme, K.A., T.L. Neville, G.A. Perry, D.A. Redmer, L.P. Reynolds, and J.S. Caton. 2013b. Maternal dietary intake alters organ mass and endocrine and metabolic profiles in pregnant ewe lambs. Anim. Reprod. Sci. 141: 131–141.

Vonnahme, K.A., C.O. Lemley, J.S. Caton, and A.M. Meyer. 2015. Impacts of maternal nutrition on vascularity of nutrient transferring tissues during gestation and lactation. Nutrients 7: 3497–3523. doi:10.3390/nu7053497.

Vucetic, Z., J. Kimmel, K. Totoki, E. Hollenbeck, and T.M. Reyes. 2010. Maternal high-fat diet alters methylation and gene expression of dopamine and opioid-related genes. Endocrinology 151: 4756–4764.

Wallace, J.M., R.P. Aitken, and M.A. Cheyne. 1996. Nutrient partitioning and fetal growth in rapidly growing adolescent ewes. J. Reprod. Fertil. 107: 183–190.

Wallace, J.M., D.A. Bourke, and R.P. Aitken. 1999. Nutrition and fetal growth: Paradoxical effects in the overnourished adolescent sheep. J. Reprod. Fertil. 54(1): 385–399.

Wallace, J.M., D.A. Bourke, P. Da Silva, and R.P. Aitken. 2001. Nutrient partitioning during adolescent pregnancy. Reproduction 122: 347–357.

Wallace, J.M., J.S. Milne, and R.P. Aitken. 2005. The effect of overnourishing singleton-bearing adult ewes on nutrient partitioning to the gravid uterus. Br. J. Nutr. 94: 533–539.

Wallace, J.M., J.S. Luther, J.S. Milne, R.P. Aitken, D.A. Redmer, L.P. Reynolds, and W.W. Hay, Jr. 2006. Nutritional modulation of adolescent pregnancy outcome—a review. Placenta 27(a): S61–S68.

Wallace, J.M., J.S. Milne, C.L. Adam, and R.P. Aitken. 2012. Adverse metabolic phenotype in low-birth-weight lambs and its modification by postnatal nutrition. Br. J. Nutr. 107: 510–522.

Wallace, L.R. 1948. The growth of lambs before and after birth in relation to the level of nutrition. J. Agric. Sci. 38: 243–300.

Walton, A., and J. Hammond. 1938. The maternal effects on growth and conformation in Shire horse-Shetland pony crosses. Proc. R. Soc. Lond. B. 125: 311–335.

Wang, H., L. Wang, H. Erdjument-Bromage, M. Vidal, P. Tempst, R.S. Jones, and Y. Zhang. 2004. Role of histone H2A ubiquitination in polycomb silencing. Nature 431: 873–878.

Wang, J., Z. Wu, D. Li, N. Li, S.V. Dindot, M.C. Satterfield, F.W. Bazer, and G. Wu. 2012. Nutrition, epigenetics, and metabolic syndrome. Antioxicants Redox Signaling 17: 282–301.

Wang, X., X. Lan, A.E. Radunz, and H. Khatib. 2015. Maternal nutrition during pregnancy is associated with differential expression of imprinted genes and DNA methyltransferases in muscle of beef cattle offspring. J. Anim. Sci. 93: 35–40.

Ward, M.A., T.L. Neville, J.J. Reed, J.B. Taylor, D.M. Hallford, S.A. Soto-Navarro, K.A. Vonnahme, D.A. Redmer, L.P. Reynolds, and J.S. Caton. 2008. Effects of selenium supply and dietary restriction on maternal and fetal metabolic hormones in pregnant ewe lambs. J. Anim. Sci. 86: 1254–1262.

WFP. 2014. World Food Program, United Nations. http://wfpusa.org/blog/march-4-nutrition-nourishing-future-first-1000-days.

Wilkins, J.F., and D. Haig. 2003. What good is genomic imprinting: The function of parent-specific gene expression. Nat. Rev. Genet. 4: 359–368.

Wolver, T.M.S., and J.L. Chiasson. 2000. Acarbose raises serum butyrate in human subjects with impaired glucose tolerance. Br. J. Nutr. 84: 57–61.

Wood, A., N.J. Krogan, J. Dover, J. Schneider, J. Heidt, M.A. Boateng, K. Dean, A. Golshani, Y. Zhang, J.F. Greenblatt, M. Johnston, and A. Shilatifard. 2003. Bre1, an E3 ubiquitin ligase required for recruitment and substrate selection of Rad6 at a promoter. Mol. Cell. 11: 267–274.

Wu, G., F.W. Bazer, J.M. Wallace, and T.E. Spencer. 2006. Intrauterine growth retardation: Implications for the animal sciences. J. Anim. Sci. 84: 2316–2337.

Xiong, F., and L. Zhang. 2013. Role of hypothalamic-pituitary-adrenal axis in developmental programming of health and disease. Front. Neuroendol. 34: 27–46.

Yamawaki, Y., M. Fuchikama, S. Morinobu, M. Segawa, T. Matsumoto, and S. Yamawaki. 2012. Antidepressant-like effect of sodium butyrate (HDAC inhibitor) and its molecular mechanism of action in the rat hippocampus. World J. Biol. Psychiatry 13: 458–467.

Yang, Q.Y., J.F. Liang, C.J. Rogers, J.X. Zhao, M.J. Zhu, and M. Du. 2013. Maternal obesity induces epigenetic modifications to facilitate Zfp423 expression and enhance adipogenic differentiation in fetal mice. Diabetes 62: 3727–3735.

Youngson, N.A., and E. Whitelaw. 2008. Transgenerational epigenetic effects. Annu. Rev. Genomics Hum. Genet. 9: 233–257.

Yunusova, R., T.L. Neville, K.A. Vonnahme, C.J. Hammer, J.J. Reed, J.B. Taylor, D.A. Redmer, L.P. Reynolds, and J.S. Caton. 2013. Impacts of maternal selenium supply and nutritional plane on visceral tissues and intestinal biology in offspring. J. Anim. Sci. 91: 2229–2242.

Zambrano, E., C. Guzman, G.L. Rodriguez-Gonzalez, M. Durand-Carbajal, and P.W. Nathanielsz. 2014. Fetal programming of sexual development and reproductive function. Molec. Cell. Endocrinol. 382: 538–549.

Zhang, S.T., R.H. Reguault, P.L. Barker, K.J. Botting, I.C. McMillen, C.M. McMillan, C.T. Roberts, and J.L. Morrison. 2015. Placental adaptations in growth restriction. Nutrients 7: 360–389. doi:10.3390/nu7010360.

CHAPTER-6

Molecular Physiology of Feed Efficiency in Beef Cattle

Claire Fitzsimons, * *Mark McGee, Kate Keogh,*
Sinéad M. Waters and *David A. Kenny**

||||||||||||||||||||||||||||||||||||||||||||||||||||||||||||||||||||||||||||||||||||||||||||||||||||||||||||||||||||||||||||||||||||||||||||||||||||||

INTRODUCTION

Globally there is an enormous challenge to feed a growing human population within 'planetary boundaries', especially those relating to land use and climate change (Gill, 2013). Within this resource-constrained setting, rising world demand for food, including meat, must be met with minimal environmental impact.

A consequence of dietary transition towards greater meat consumption worldwide is an increase in demand for animal feed (Gill, 2013). Increasing global demand for cereal grains from competitive forces such as human population growth, declining availability of arable land and, more recently, biofuel production (McNeill, 2013) has highlighted the substantial use of cereals in animal feeds. Compared with pigs and poultry, on a life-cycle basis, beef cattle have the capacity to consume large amounts of low-cost, low-quality forages relative to higher-cost concentrates (Nielsen et al., 2013). Furthermore, these forages are often grown on land unsuitable for arable crop production. At an animal level, meat production by ruminants is less efficient than pigs and poultry however, when compared on the basis of human-edible inputs (cereals and proteins) the ruminant has a clear efficiency advantage (Reynolds et al., 2011).

As feed provision is the single largest direct cost incurred by beef producers, it is a major economic factor influencing profitability of beef

Teagasc Agriculture and Grassland Research and Innovation Centre, Grange, Dunsany, Co. Meath, Ireland.
* Corresponding author: david.kenny@teagasc.ie

farming enterprises (Finneran et al., 2010). Conversion of feeds into food of animal origin is associated with energy and nutrient losses, and environmental pollution, including the excretion of nitrogen, phosphorus, trace elements, carbon dioxide and methane (Niemann et al., 2011). In the context of climate change and more restrictive environmental legislation, the environmental footprint of beef production must be minimised. Consequently, there is considerable interest in improved feed efficiency as a means of augmenting the economic and environmental sustainability of beef production systems.

At the animal-level, many alternative definitions of feed efficiency exist, each differing in their application and benefits (Arthur et al., 2001; Berry and Crowley, 2013). Traditionally, feed conversion ratio (FCR) (i.e., feed:gain) or its mathematical inverse, feed conversion efficiency (i.e., gain:feed), was widely used. More recently, residual feed intake (RFI) defined as the difference between observed and predicted intake — calculated as the residuals from a multiple regression model of intake on the various energy expenditures (e.g., weight, growth, etc.) — has become the preferred measurement (Savietto et al., 2014). Because RFI is independent of level of animal production, it is a very useful trait to examine the biological mechanisms associated with variation in feed efficiency.

There is significant inherent inter-animal variation in feed efficiency where cattle are offered feed to appetite. For example, we have shown phenotypic differences in RFI in the region of 20% or greater within populations of growing cattle (Kelly et al., 2010a,b; Crowley et al., 2010; Lawrence et al., 2012; Fitzsimons et al., 2013). Concomitant with this, there is also significant genetic variance for the RFI trait (Crowley et al., 2010).

In addition to natural inherent variation between animals fed to appetite, feed efficiencies may be 'imposed' on cattle through nutritional management practices such as those associated with compensatory (or catch-up) growth. Compensatory growth may be defined as a physiological process whereby an organism accelerates its growth after a period of restricted development, usually due to reduced feed (nutrient) intake, in order to reach the weight of animals maintained on a continuous, unimpeded growth trajectory (Hornick et al., 2000). The process has likely evolved as a coping mechanism for animals and in particular, herbivores such as ruminants to withstand periods of alternating food availability, without compromising overall lifetime growth and development (Dmitriew, 2011). It is widely exploited in beef production systems, particularly those based on seasonal grass growth, as a method to reduce feed costs during 'winter' periods, when feed is expensive (Ashfield et al., 2014). Compensatory growth is associated with increased feed efficiency in cattle and the greatest increment of accelerated growth and indeed greater efficiency typically occurs during the first two months of re-alimentation (Hornick et al., 2000; Keady, 2011; Keogh et al., 2015a).

Research to date has clearly shown that feed efficiency is a complex multifaceted trait, under the control of many biological processes (Niemann et al., 2011). Thus a fundamental understanding of the biological and molecular mechanisms regulating the trait is essential in order to harness the potential of genomic selection to augment genetic improvement in the trait (Paradis et al., 2015).

This review aims to discuss the main biological processes that govern feed efficiency and nutrient utilisation of cattle and the observed variation cited in the published literature to-date. We propose to focus on both inherent inter-animal variation in situations where feed is available on an *ad libitum* basis as well as where feed nutrient supply is managed, i.e., to exploit compensatory growth.

Appetite and Feed Intake

The voluntary feed intake of cattle is regulated by a combination of physical and metabolic mechanisms, with the relative importance of these two processes dependent on prevailing dietary characteristics. Physical distension of the rumen and gastrointestinal tract; hepatic metabolism of products of digestion; and endocrine products of the gastrointestinal tract (GIT), pancreas, adipose tissue, and, possibly, muscle, all contribute signals that are processed centrally in the brain to control hunger, satiety, and energy expenditure (Roche et al., 2008; Hills et al., 2015). While ruminal distention may dominate control of feed intake when ruminants consume low-energy fibrous diets or when energy requirements are high, fuel-sensing by tissues is likely to control appetite when fuel supply exceeds requirements (Allen, 2014). Furthermore, regulation of food intake should be considered over different time scales. For example, the biochemical regulation of short-term prandial events such as eating impetus or meal size will differ from longer term control of feed intake to sustain normal physiological homeostasis (Roche et al., 2008; Hills et al., 2015).

Within the central nervous system (CNS) the nuclei of the hypothalamus are important mediators in the regulation of appetite and energy homeostasis, with the role of the arcuate nucleus (ARC), in particular, well established. The ARC contains proopiomelanocortin/cocaine- and amphetamine-regulated transcript (*POMC/CART*)-expressing neurons, whose activation suppresses feeding. In contrast, the activation of a second population of arcuate neurons, neuropeptide Y/agouti-related peptide- (NPY/AgRP) expressing neurons stimulates feeding (Allen, 2014).

Additional peripheral tissues are also involved in appetite regulation including adipose tissue, where the appetite suppressant leptin is synthesised which acts to inhibit appetite with the hypothalamus. Additionally, insulin signalling to the hypothalamus results in the suppression of appetite and inhibition of gluconeogenesis in the liver (Pliquett et al., 2006). Furthermore, leptin and insulin appear to be counter-regulatory, affecting peripheral tissues through partitioning of substrate utilization (Strat et al., 2005). Indeed differential utilization of lipid and carbohydrate has been suggested to contribute to variation in feed efficiency in beef cattle (Welch et al., 2012).

Lines et al. (2014) suggested that variation in appetite was a major factor contributing to differences in feed intake between heifers that were divergently selected for RFI. Similarly, greater feed intakes were associated with improved feed efficiency in cattle undergoing compensatory growth (Hornick et al., 1998a; Keady, 2011; Keogh et al., 2015a). However, the literature describing the endocrine control of appetite and voluntary feed intake (VFI) regulation, within the context of effects on variation in feed efficiency potential of cattle, is scant. While Richardson and Herd (2004) reported a positive correlation between leptin and RFI ranking, other studies have failed to establish such an association (Brown, 2005; Kelly et al., 2010a, 2011a). Despite this, however, Perkins et al. (2014) found that mRNA expression of leptin was up-regulated in adipose tissue of low RFI steers, suggesting that greater leptin expression may be contributing to improved feed efficiency. However, it must be noted that variation in gene transcription does not always translate to differences in protein abundance. Data from our own group, using feed restricted cattle, provides evidence for a down-regulation of the gene coding for leptin in skeletal muscle of cattle undergoing a period of dietary restriction concomitant with poorer feed efficiency (Doran, 2013). This is consistent with the findings for the leptin receptor, *LEPR*, in hepatic tissue of beef cattle managed in a similar way (Keogh et al., 2016a). Indeed the data of Keogh et al. (2015a) show that intake per unit of bodyweight was higher in animals undergoing re-alimentation after a period of sub-optimal growth, suggesting that an increased appetite may be driving compensatory growth and may also be contributing to the improved nutrient utilisation typically evident during re-alimentation. Attendant with this, the authors also observed lower systemic concentrations of leptin in these animals (Keogh et al., 2015b). While systemic concentrations of leptin generally correlate with body adiposity, some authors have identified alterations in serum leptin concentrations as a result of changes in energy intake and growth rate independent of adiposity (Foster and Nagatani, 1999; Delavaud et al., 2000). Both Vega et al. (2009) and Keogh et al. (2015b) reported lower concentrations of leptin during diet restriction, which extended well beyond the initiation of realimentation. As leptin functions to suppress appetite by binding to NPY neurons in the hypothalamus and inhibiting the appetite stimulatory effect of NPY, lower systemic concentrations of leptin may increase hunger signals in the brain leading to the greater intake of feed observed in some studies (Vega et al., 2009;

Keogh et al., 2015b) where lower systemic concentrations of leptin were associated with improved feed efficiency in cattle undergoing compensatory growth (Ahima, 2005).

Feeding behaviour is determined by the integration of central and peripheral signals in brain feeding centres (Allen, 2014). These include stimulatory (orexigenic) signals that increase hunger as well as inhibitory (anorexigenic) signals that increase satiety (Steinert et al., 2013). In contrast to the anorexigenic and more long term homeostatic effects of leptin the peptide hormone ghrelin is representative of a shorter term appetite stimulant. For example, the absence of digesta in the GIT may lead to secretion of ghrelin from the cells of the abomasum in ruminants, which can signal to specific receptors in the hypothalamus (Hayashida et al., 2001). Ghrelin is a putative signal of energy insufficiency and its concentration increases with fasting and decreases upon feeding (Wertz-Lutz et al., 2008). Foote et al. (2014) examined the relationship between total ghrelin concentrations and growth- and carcass-related traits across both steers and heifers offered a finishing diet and reported a negative association with both dry matter intake (DMI) and feed efficiency, respectively. While there are few studies that have measured systemic concentrations of ghrelin in cattle within a feed efficiency context, Sherman et al. (2008) identified a single nucleotide polymorphism (SNP) in the gene that codes for ghrelin which had a tendency for a negative association with RFI and FCR and a negative association with partial efficiency of growth (PEG) further establishing a role for appetite or satiety in the control of feed efficiency. Similar to ghrelin, cholecystokinin is a peptide hormone also produced in the GIT and synthesised primarily in the mucosal lining of the duodenum. However, whilst ghrelin stimulates appetite, cholecystokinin functions in appetite suppression. Abo-Ismail et al. (2013) reported a negative association between a SNP in the cholecystokinin B receptor (*CCKBR*) gene and DMI consistent with the anorexigenic properties of the hormone. However, a negative association was also evident between this particular polymorphism and RFI in that study (Abo-Ismail et al., 2013) however plasma concentrations of cholecystokinin were not measured. As outlined above, VFI or appetite is regulated centrally by the hypothalamus via the integration of physical feedback mechanisms and hormonal feedback from peripheral tissues (Roche et al., 2008). The NPY and POMC expressing neurons, primarily found in the hypothalamus function as metabolic sensors and are critical mediators of energy homeostasis (Lenard and Berthoud, 2008). Perkins et al. (2014) reported lower mRNA expression of *NPY* and relaxin-3 (*RNL3*), in hypothalamic ARC tissue of efficient beef cattle, both of which have orexigenic properties, and greater expression of *POMC* which decreases feed intake. In agreement, Alam et al. (2012) reported a negative correlation between RFI and *POMC* mRNA expression in the duodenum of Holstein-Friesian and Jersey cross cows. Conversely though, Sherman et al. (2008)

found no association between a SNP in the *POMC* gene, however, three polymorphisms in the *NPY* gene were associated with feed conversion ratio (FCR), with two SNPs positively associated and one negatively associated with FCR.

These studies suggest that polymorphisms and differential mRNA expression of genes involved in the regulation of appetite and satiety in the hypothalamus may underlie variation in feed efficiency in beef cattle however, additional targeted studies are required to confirm this theory. Feed intake is a function of meal size and frequency. Initiation of meals is likely to be determined when inhibitory signals from the previous meal subside and stimulatory signals increase, and conversely, feeding is likely to continue until inhibitory signals intensify and stimulatory signals diminish (Allen, 2014). It is unclear if hunger is primarily the absence of satiety or if it is dependent on mechanisms that specifically stimulate feeding (Allen, 2014). A schematic representation of the metabolic neuro-endocrine control of appetite in cattle is presented in Figure 1. Expression of other

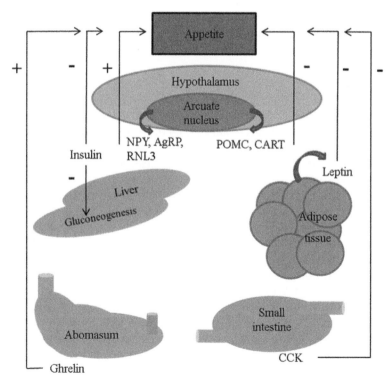

Figure 1: Proposed neuro-endocrine sources of variation in appetite regulation of beef cattle. Schematic synthesised on the basis of the findings of Abo-Ismail et al. (2013), Doran, 2013, Keogh et al. (2016a), Perkins et al. (2014) and Sherman et al. (2008).

genes regulating appetence and satiety that may underlie variation in feed efficiency include gastrin, gastrin-releasing peptide, gastric inhibitory polypeptide in the stomach, secretin in the duodenum and somatostatin in the jejunum, all of which warrant further investigation.

In addition to its fundamental importance to energy homeostasis, Susenbeth et al. (2004) suggested that activity associated with eating was a significant energy sink in beef cattle. For example, for low-quality roughage, it was estimated that the energy cost of eating accounted for approximately 26.5% of metabolisable energy intake (Susenbeth et al., 1998).

Montanholi et al. (2010) reported that feeding behaviour accounted for 18% of variation associated with RFI. The observed lower DMI of feed efficient cattle (Lancaster et al., 2009; Kelly et al., 2010a; Montanholi et al., 2010; Durunna et al., 2012; McGee et al., 2014) suggests greater impetus to feed in inefficient cattle coupled with greater energy expenditure associated with higher feed intake. Indeed, physical feedback mechanisms may directly result in the lower feed intake of feed efficient cattle, as suggested by the findings of Fitzsimons et al. (2014a), who reported a smaller reticulo-rumen complex (8% lighter) in low compared with high RFI bulls, consistent with the difference (12%) in DMI observed. In conclusion, it seems that further investigation into the metabolic and neuroendocrine factors regulating appetite may enhance our understanding of variation in feed efficiency.

Digestion

The reticulo-rumen is central to the profile of nutrients available for absorption in cattle (Dijkstra et al., 2007). For example, typically in excess of 90% of neutral detergent fibre digestion (Huhtanen et al., 2006) and between 25% and almost 100% of starch digestion (Nozière et al., 2010) occurs in the rumen. Volatile fatty acids (VFA) and microbial protein are the primary end-products resulting from ruminal digestion. The VFA provide approximately 80% of the metabolizable energy needs of the animal (Krehbiel, 2014), whereas microbial protein typically comprises 50 to 80% of the crude protein supply that reaches the small intestine (Owens et al., 2014).

Studies have suggested a genetic basis for variation in digestion of feed, over and above feeding level and diet type in ruminants (Smuts et al., 1995; Herd and Arthur, 2009). Increasing feed intake usually decreases diet digestibility, as a consequence of a shorter ruminal residency time (Huhtanen et al., 2006). Thus, lower apparent digestibility *per se* would be expected with high RFI animals. However, studies that have examined diet DM digestibility (DMD) in high and low RFI beef cattle have reported contrasting results. For example, Richardson et al. (1996), Nkrumah et al. (2006), Krueger et al. (2009) and McDonnell et al. (2016) all reported either higher, or a strong tendency towards higher digestibility coefficients for various dietary nutrients in low RFI cattle, whereas Richardson and Herd (2004), Brown (2005), Cruz et al. (2010), Lawrence et al. (2011, 2012), Gomes et al. (2013) and Fitzsimons et al. (2013, 2014a) failed to observe any difference in ability to digest feed between cattle of divergent RFI status. Additionally, increases in digestibility were also reported during greater feed efficiency in cattle undergoing re-alimentation (Grimaud et al., 1998). In some instances, the absence of differences may be related to the nature of the diet offered, as the effect of feed intake on digestion is of lower magnitude with forage than with concentrate diets (Chilliard et al., 1995). For example, McGee et al. (2005) reported that DMD (total faecal collection) of grass silage determined using steers fed either 0.95 or 0.75 of *ad libitum* intake, was unaffected by level of feeding. Additionally, the indirect marker methodologies employed in the majority of studies may not be sufficiently sensitive, to identify modest differences in digestive potential, should they exist. It is unclear whether any association between digestion and RFI is an inherent efficiency or mainly related to a higher passage rate of digesta through the rumen. Nevertheless, there is evidence to suggest that there is a genetic basis to variation in digestion between cattle of varying energetic efficiency potential (Channon et al., 2004). For example, Abo-Ismail et al. (2013) found that SNPs in trypsin 2 (*PRSS2*) and cholecystokinin B receptor (*CCKBR*) genes, which are involved in digestion and metabolic processes, were associated with RFI and FCR in beef cattle.

The GIT has been shown to be one of the most reactive tissues to accelerated growth (Ryan et al., 1993; Yambayamba et al., 1996a; Keogh et al., 2015a) during realimentation, where increased feed efficiency potential is also typically evident. Improved energetic efficiency during compensatory growth may be the consequence of an increase in the size of components of the GIT, for example through increasing the surface area of rumen papaillae and intestinal villi. Indeed, greater expression of genes involved in cellular adhesion, protein folding and cytoskeleton in rumen papillae has been observed in cattle during compensatory growth, with the inverse of these processes apparent following a prior period of dietary restriction and poor feed efficiency (Keogh et al., 2016b). These results suggest that the structural state of the GIT may play an important role in governing feed efficiency, with an increase in the structural integrity of the rumen papillae during re-alimentation potentially contributing to improvements during periods of accelerated growth. Indeed, following a period of dietary restriction in goats, Sun et al. (2013) observed reduced rumen papillae height, width and surface area compared to their non-restricted contemporaries, which was subsequently reversed upon re-alimentation. Greater structural integrity of the rumen papillae may facilitate greater absorption of nutrients and contribute to improved feed efficiency evident during compensatory growth. O'Shea et al. (2016) observed lower expression of *DSG1*, in the rumen papillae of cattle following a period of dietary restriction and subsequently greater *DSG1* expression in animals undergoing compensatory growth. *DSG1* encodes an adhesion molecule involved in maintaining the structural integrity of epithelial cells including rumen epithelium. This provides further evidence for alterations to rumen papillae structure as a consequence of dietary restriction and subsequent compensatory growth. Overall, alterations in GIT absorptive capacity, following a period of dietary restriction, may allow for an increase in nutrient absorption during re-alimentation, potentially mediated through replenishment of GIT epithelial cells/papillae/villi.

In general, from the studies conducted to-date, few obvious differences in the primary rumen fermentation variables measured are evident between high and low RFI cattle and, in cases where variance was observed, results have been ambivalent. For example, there are indications that low RFI cattle have higher concentrations of propionate and a lower acetate:propionate ratio in ruminal fluid compared to those of high RFI (Lawrence et al., 2011; Fitzsimons et al., 2013). However, Krueger et al. (2009) found the opposite, while other authors (Fitzsimons et al., 2014a,b; Hernandez-Sanabria et al., 2012) failed to find any difference in VFA proportions between cattle of varying energetic efficiency. In gestating beef cows, Fitzsimons et al. (2014b) reported no differences in rumen VFA or lactic acid concentrations between high and low RFI groups, however, rumen pH was higher and ammonia concentrations were lower in low RFI cows compared to their high RFI counterparts. Similarly, with lactating dairy cows, previously selected for divergence in RFI, no differences in rumen pH and fermentation parameters were detected (Rius et al., 2012; Thornhill et al., 2014), with the exception of rumen ammonia concentrations, which were lower for high RFI cows (Rius et al., 2012). Additionally, lower feed efficiency has been associated with reduced VFA absorption across ruminal epithelium (Albornoz et al., 2013; Zhang et al., 2013). Data from our own laboratory (O'Shea et al., 2016) provide evidence for elevated concentrations of n-butyric acid in the rumen digesta of animals undergoing a period of restricted growth and this may in turn be involved in maintaining ruminal papillae and consequently VFA absorption (O'Shea et al., 2016). In their review of published gene expression data within segments of the gastrointestinal tract, Connor et al. (2010a) discuss the role of monocarboxylate transporters in absorption of short chain fatty acids (SCFA) across the rumen wall. Considering the potential differences in VFA production between cattle of high and low RFI, it would not be unreasonable to suggest that differences in absorption potential of SCFA may exist between animals varying in energetic efficiency potential. Cumulatively, these findings suggest that some inherent differences in the efficiency of the ruminal fermentation process may exist between animals of high and low RFI but that such differences are potentially dependent on the chemical composition of the prevailing diet.

Rumen Microbiome and Methanogenesis

The duration of residency of digesta in the rumen is known to affect fermentation patterns (Janssen, 2010). Equally, feeding a lower volume of the same diet alters both the bacterial and archael communities within the bovine rumen (McCabe et al., 2015). While diet has been shown to have a major influence on the rumen microbiome (de Menezes et al., 2011; Carberry et al., 2012), there is also evidence of a strong host-microbiota specificity in the ruminant (Kittelmann et al., 2014; Henderson et al., 2015). Therefore, considering that efficient animals consume less feed per unit of bodyweight compared to their inefficient contemporaries, it is not incomprehensible that host feed efficiency phenotype directly affects the composition of the rumen microbiome thus improving overall nutrient utilisation from ingested feed. Indeed previous research has provided evidence for associations between the rumen microbiome and RFI phenotypes in growing beef cattle (Guan et al., 2008; Hernandez-Sanabria et al., 2012; Carberry et al., 2012). Molecular bacterial profile analysis of rumen digesta from steers offered a high-energy finishing diet suggest a link between VFA composition, rumen bacterial profiles and RFI phenotype (Hernandez-Sanabria et al., 2012; Guan et al., 2008). However, since diet is a major factor in determining community structure and fermentation patterns within the rumen, the effect of RFI phenotype on ruminal community structure across contrasting diets has also been investigated in beef cattle. In our own studies, the association between RFI ranking and bacterial profiles was stronger when a forage only (grass silage) as opposed to a cereal based diet was offered (Carberry et al., 2012). It is likely however, that a core bacterial community exists within the rumen regardless of host energetic efficiency potential but that differences between efficient and inefficient animals may be reflected in proportions of minor microbial populations, the importance of which will vary in accordance with prevailing diet.

Prevotella has been described as one of the most dominant bacterial genus within the rumen microbiome irrespective of diet offered (Pitta et al., 2010). Our own work (Carberry et al., 2012) has shown that *Prevotella* abundance was higher in inefficient *Bos taurus* heifers, an observation that was confirmed by McCann et al. (2014) using Brahman bulls. While *Prevotella* abundance may directly reflect feed intake, it is also possible that this micro-organism contributes to inefficiencies in nutrient digestion and utilisation.

Methane production in the rumen is an energetically wasteful process, accounting for a loss of up to 15% of dietary gross energy (Van Nevel and Demeyer, 1996). Variation in rumen microbial populations are associated with differences in methane production between individual animals (Moss et al., 2000) and work from our group suggests that feed efficient cattle emit less methane (Fitzsimons et al., 2013). Research has shown that while dietary manipulation can influence the abundance of total and specific methanogen species (Carberry et al., 2014a), abundance of either total or specific methanogenic species did not differ between animals divergent in RFI (Carberry et al., 2014a,b). However, at the genotype level, various genotypes of *Methanobrevibacter smithii* were more abundant in cattle of high compared to low RFI status, regardless of diet composition (Carberry et al., 2014b) demonstrating that a core group of methanogens exist across feed efficiency phenotypes, but significant differences exist in the distribution of genotypes within species and may contribute to the observed changes in methane emissions between efficient and inefficient animals. A more recent study from our group on the rumen solid digesta fraction of Holstein Friesian bulls suggests an as yet uncultured *Succinivibrionaceae* species is associated with the lower methane emissions and higher concentrations of propionate in the rumen typically evident with higher consumption of highly digestible feed (McCabe et al., 2015).

The preceding narrative suggests that differences in rumen digesta turnover time, microbial fermentation efficiency as well as the morphology and structural integrity of ruminal epithelial tissue may all contribute to observed variation in feed efficiency between cattle. However, further work is required to establish the relative importance and independence of these processes and the underlying biochemical control of mechanisms involved.

Intestinal Absorption and Cell Morphology

Despite the rumen being the major site of feed digestion and nutrient absorption, up to 70% of starch escaping rumen fermentation is converted to glucose, the majority of which is subjected to absorption across the intestinal wall (Nozière et al., 2010). From their review, Nozière et al. (2010) suggest that variation in intestinal absorption between individual animals exists but can often be influenced by amylase, surface exposure of starch to enzymes in the small intestine and feed characteristics. Additionally, absorption of glucose within the small intestine *per se* is variable, with greatest absorptive capacity in the jejunum and least in the ileum (Nozière et al., 2010). Given such observed variation, it is possible that enhanced intestinal absorption of nutrients may be a factor underlying inter-animal variation in feed efficiency. This hypothesis is supported by the findings of Meyer et al. (2014) who reported statistically significant correlations between jejunal mucosal density and both RFI ($r = -0.33$) and FCR ($r = 0.42$) in steers and heifers. These data suggest that feed efficient cattle may have a greater ability to absorb nutrients in the small intestine than their inefficient contemporaries.

Consistent with this hypothesis, Montanholi et al. (2013) found that cell number in the duodenum and ileum of low RFI was higher than that of high RFI steers also suggesting a more metabolically active small intestine in feed efficient cattle. The authors also reported relationships between crypt area, perimeter and cell size and RFI however, these relationships were inconsistent and require further investigation (Montanholi et al., 2013). Similarly, Levesque et al. (2015) reported increases in jejunal and ileal villi height in pigs during compensatory growth, which may contribute to greater surface area for absorption. Moreover, up-regulation of transcripts for the genes *S100A2* and *LTF* both of which are involved in cellular growth and differentiation was apparent during re-alimentation and increased efficiency in jejunal epithelial cells (Keogh et al., 2016c). These results further suggest that alterations in the absorptive capacity of intestinal epithelial cells may influence feed efficiency potential. Consistent with this, following a period of dietary restriction of goats, jejunal villi height and width were reduced which was subsequently reversed upon re-alimentation (Sun et al., 2013).

Meyer et al. (2014) investigated the effect of feed efficiency on the expression of genes pertaining to angiogenesis in the jejunum of heifers and steers but found no relationships with the exception of a positive association between mRNA expression of kinase inert domain receptor (*KDR*) and a tendency (*P* = 0.06) for a positive association between endothelial nitric oxide synthase 3 (*NOS3*) and feed intake. Evidence for a potential role of differential intestinal nutrient absorption in variation in feed efficiency in beef cattle was also demonstrated at a genomic level by Serao et al. (2013) who reported SNPs that mapped to genes involved in fat digestion and absorption and small intestine transport of phospholipids and cholesterol from the epithelium to the intracellular space.

Further exploration of the molecular control of intestinal nutrient absorption is justified to fully characterise the contribution of this key process to variation in feed efficiency.

Nutrient Partitioning and Composition of Growth

In addition to their importance as determinants of the value of beef cattle (Conroy et al., 2010) muscle and fat tissue make a significant contribution to overall body energy homeostasis. Thus any discussion on the regulation of feed efficiency must take cognisance of the underlying biological control of these economically important tissues. Currently, there is much equivocation in the published literature on the contribution of differences in body composition to variation in feed efficiency status. For example, to-date positive (Kelly et al., 2010a), negative (Crowley et al., 2011) or neutral (Basarab et al., 2003; Fitzsimons et al., 2013, 2014a) relationships between RFI and ultrasonically measured longissimus muscle size have been reported for growing beef cattle. Similar confusion exists between phenotypic based reports for carcass muscularity score and lean tissue content. Additionally, at a genetic level, Crowley et al. (2011) found carcass muscularity to be positively associated with improved feed efficiency status, while, in contrast, Mao et al. (2013) failed to establish a genetic correlation between RFI and carcass lean meat yield.

Consistent with the published data for lean tissue growth, there is disagreement between studies on the relationship between feed efficiency status and measures of body fat content. Some studies that used the base model (metabolic body weight and average daily gain) to predict dry matter intake have reported either a positive (Richardson et al., 2001; Basarab et al., 2003; Basarab et al., 2007) or no relationship (Castro Bulle et al., 2007; Bouquet et al., 2010; Gomes et al., 2012; Fitzsimons et al., 2014a) between feed efficiency and carcass fatness traits. However, where others have included adjustments for body composition in the base model a trend for a positive relationship (Basarab et al., 2003), no relationship (Basarab et al., 2007), or a negative relationship (McGee et al., 2013) between RFI and measures of body fat were reported.

While the energetic efficiency of body fat deposition is inferior to that of protein (Herd et al., 2004) the energy requirements to maintain existing fat depots is less than that of protein (Welch et al., 2012). Indeed, there is considerable inter-animal variation in efficiency of lean gain, most likely due to differing rates of protein turnover and degradation between organs and tissues (McBride and Kelly, 1990; Herd et al., 2004). Consistent with this premise, it is not unreasonable to suggest that differences may exist between animals of varying feed efficiency potential, in biochemical processes governing nutrient partitioning and fat and protein accretion and maintenance (Herd and Arthur, 2009).

Variation in nutrient partitioning and subsequent implications for differences in body composition between cattle divergent for feed efficiency may be influenced through differing intermediary metabolism (Randel and Welsh, 2013). For example, insulin's role in nutrient uptake and utilisation renders insulin sensitivity and its associated signalling pathway a prime candidate that potentially contributes to variation in feed efficiency. Additionally, given the key role of the somatotropic axis in cellular growth, proliferation, differentiation and the maintenance of homeostasis in mammals, examination of the functioning of this axis in cattle divergent for feed efficiency may provide more information on the control of this trait.

Insulin Sensitivity

Insulin is a highly metabolically active hormone responsible for many anabolic processes as well as nutrient homeostasis (Taniguchi et al., 2005; Cheng et al., 2010). Insulin sensitivity has been suggested as a contributor towards variation in energetic efficiency (Welch et al., 2012) as it mediates nutrient uptake and utilization, principally glucose homeostasis, within tissues with a large energy demand such as skeletal muscle and liver (Cheng et al., 2010). Insulin increases muscle energy requirements by increasing transport of amino acids and protein synthesis with a concomitant decrease in muscle protein breakdown (Hocquette et al., 1998) suggesting that animals with increased systemic concentrations of insulin have increased maintenance energy requirements. Conversely, enhanced skeletal muscle insulin sensitivity is associated with a decrease in basal insulin concentrations coupled with an increase in carcass growth and decreased carcass fat (Hocquette et al., 1998) as confirmed by Istasse et al. (1990) and Hocquette et al. (1999) who compared double muscled and non-double muscled cattle. Collectively, this suggests that animals with improved feed efficiency may have decreased basal plasma insulin concentrations coupled with enhanced skeletal muscle insulin sensitivity.

The literature on the contribution of glucoregulatory mechanisms to variation in phenotypic RFI is conflicting in that circulating concentrations of insulin have been reported to be both greater (Kelly et al., 2011a) and lower (Richardson et al., 2004) in low compared to high RFI cattle. Additionally, the limited sampling regimen used in most studies would not have been adequate to determine true hormone or metabolite status (Richardson et al., 2002). To this end, Fitzsimons et al. (2014c) specifically examined the insulinogenic response and potential to clear systemic glucose in response to a glucose tolerance test (GTT), which employs a comprehensive temporal sampling protocol. These authors found no differences in either glucose or insulin response between heifers divergent for RFI. This is in agreement with the findings of GTTs conducted by Redden et al. (2011) and Ramos (2011) who reported similar temporal patterns and area under the curve (AUC) response data for glucose and insulin concentrations for ewe lambs and steers, respectively, irrespective of RFI phenotype. Despite this however, Shafer (2011) demonstrated a lower insulinogenic index, in response to a GTT, in low RFI cattle, suggesting greater insulin sensitivity in feed efficient cattle. Similarly, time to clear 50% of the maximum glucose concentration from circulation tended to be lower in low RFI compared to high RFI ewe lambs (Redden et al., 2011). This suggests that improved feed efficiency may lead to more efficient glucose utilization, possibly through enhanced insulin responsiveness. In agreement, Keogh et al. (2015c) using a GTT, reported cattle exhibiting improved feed efficiency during compensatory growth required lower concentrations of insulin to clear systemic glucose than their non-feed restricted counterparts, see Figure 2. This suggests that cattle undergoing compensatory growth with improved feed efficiency may have enhanced insulin sensitivity, contributing to the increased growth, development and nutrient homeostasis potential of these animals.

In the study of Keogh et al. (2015c), it was also suggested that the improved insulin response or sensitivity of cattle undergoing compensatory

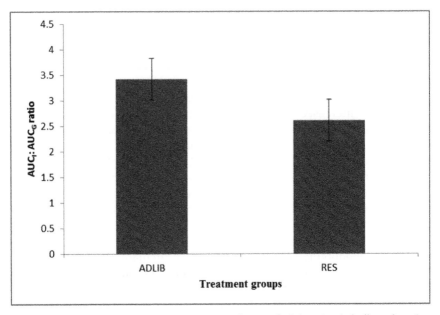

Figure 2: Lower insulin response to intravenous glucose administration in bulls undergoing compensatory growth and exhibiting an improved feed efficiency phenotype (Keogh et al., 2015c).

AUC = area under the response curve.

Adlib = bulls fed to *ad libitum* feed intake.

Res = bulls initially offered a restricted feed intake and subsequently offered feed *ad libitum*.

growth may be contributing to reduced lipogenesis and adipose deposition. Enhanced insulin sensitivity may contribute to improved feed efficiency by allowing more energy from feed to be deposited as protein as opposed to adipose tissue (Keogh et al., 2015c). In agreement, Richardson et al. (2004) found that the steer progeny of high RFI parents had higher end of test plasma concentrations of insulin and greater body fat composition compared to their low RFI contemporaries. The similar back fat depth of the high and low RFI heifers in the study of Fitzsimons et al. (2014c) supports the absence of an effect of phenotypic RFI on insulin sensitivity. Kolath et al. (2006a) and Lawrence et al. (2012) also reported that growing beef cattle divergent for phenotypic RFI had similar plasma insulin concentrations coupled with equal subcutaneous fat thickness, suggesting no differences in insulin sensitivity between cattle divergent for feed efficiency.

As glucose is an important energy source for skeletal muscle tissue and as insulin is involved in the control of nutrient uptake and storage and regulation of glucose homeostasis at the cellular level, differential expression of genes in the insulin signaling pathway (ISP) may identify potential mechanisms contributing to variation in RFI. This was demonstrated by the genome wide association study and subsequent pathway analysis of

Rolf et al. (2011), who found a gene adjacent to regions predicted to harbour quantitative trait loci associated with RFI resided within the ISP. Using a more targeted approach, Fitzsimons et al. (2014c) examined the transcription of genes within the ISP in *M. longissimus dorsi* tissue, which is a major consumer of glucose (Hocquette et al., 1998), of bulls divergent for RFI. In that study, no relationships between mRNA expression of genes in the ISP and RFI with the exception of *SREBP1c*, a transcription factor that mediates insulin-stimulated fatty acid synthesis in bovine *M. longissimus dorsi* (Han et al., 2013), was detected (Fitzsimons et al., 2014c). The positive correlation between mRNA expression of *SREBP1c* and RFI, coupled with the greater final subcutaneous fat depth in high RFI bulls in the study of Fitzsimons et al. (2014c), supports the role of this gene in lipid homeostasis (Horton et al., 2002). Therefore, it can be suggested that genes under transcriptional control of *SREBP1c* such as fatty acid synthase (*FASN*) and acetyl-CoA carboxylase (*ACCA*; Horton et al., 2002) may be underlying variation in feed efficiency. The findings of Welch et al. (2013) confirm this hypothesis whereby mRNA expression of *FASN* was higher in skeletal muscle tissue of feed efficient cattle, implicating fatty acid synthesis as a factor that may be involved in the control of feed efficiency. Differences in lipid metabolism between feed efficiency phenotypes in beef cattle were also confirmed by Abo-Ismail et al. (2014) who reported a SNP the *GTF2F2* gene to have an effect on RFI. Additionally, previous research has shown that mRNA expression of genes involved in fatty acid transport were upregulated in cattle exhibiting an improved feed efficiency phenotype further implicating the role of fatty acid synthesis and transport in feed efficiency in beef cattle (see Table 1).

Differential expression of some genes in the ISP in cattle with improved feed efficiency during compensatory growth was reported by Keogh et al. (2015c). Genes involved in lipogenesis (*ACLY*), the MAPK signalling cascade (*GRB10*) and mediators of insulin and growth factor signalling (*FOXO1, O3* and *O4*) were upregulated in feed efficient cattle undergoing compensatory growth. Interestingly, upregulation of *FOXO* genes during compensatory growth and increased nutrient intake agreed with the positive correlation between mRNA expression of *FOXO1* and DMI in the study of Fitzsimons et al. (2014c). Upregulation of *FOXO* genes in response to increased feed intake suggests that this family of genes may assist muscle cells to counteract possible increased reactive oxygen species as a consequence of enhanced cellular metabolism (Keogh et al., 2015c). However, in contrast to these studies, Chen et al. (2011) reported no differential expression of genes involved in the ISP in liver tissue of cattle divergent for RFI.

Together, the results of the aforementioned studies suggest that insulin-mediated metabolism of glucose and associated pathways such as glycolysis and cell proliferation and growth of skeletal muscle tissue are most likely

Table 1: Transporter genes involved in fatty acid synthesis and transport that were upregulated in several tissues in feed efficient cattle relative to their inefficient contemporaries.

Gene ID	Name	Fold change	Tissue
SLC27A2	Solute carrier family 27, member 2	2.72	Liver[1]
FADS1	Fatty acid desaturase 1	2.70	Liver[2]
SLC27A6	Solute carrier family 27, member 6	4.69	Liver[1]
ELOVL5	ELOVL family member 5, elongation of long chain fatty acids (FEN1/Elo2, SUR4/Elo3-like)	2.09	Liver[1]
FADS6	Fatty acid desaturase domain family, member 6	2.56	Liver[1]
FABP1	Fatty acid binding protein 1	1.61	Liver[3]
FADS2	Fatty acid desaturase 2	1.87	Liver[3]
SLC27A6	Solute carrier family 27, member 6	2.77	Skeletal muscle[4]
FASN	Fatty acid synthase	1.91	Skeletal muscle[4]
FABP3	Fatty acid binding protein 3, muscle and heart	1.89	Skeletal muscle[4]
FADS3	Fatty acid desaturase 3	2.35	Skeletal muscle[4]
FASN	Fatty acid synthase	0.51	Skeletal muscle[5]
ELOVL4	ELOVL fatty acid elongase 4	2.59	Rumen papillae[6]
FADS3	Fatty acid desaturase 3	2.25	Rumen papillae[6]

1: Connor et al. (2010b), 2: Keogh et al. (2016a), 3: Tizioto et al. (2015), 4: Keogh et al. (2016d), 5: Welch et al. (2013), 6: Keogh et al., 2016b.

not biological processes underlying variation in feed efficiency. However, lipogenesis *per se*, may be a potential biological mechanism involved in the control of feed efficiency that requires further exploration across a range of tissues such as skeletal muscle and liver. Indeed, differential gene expression analysis of adipose tissue itself may elucidate some of the biological processes underpinning variation in feed efficiency in beef cattle.

Somatotropic Axis

The somatotropic axis is an evolutionarily conserved signalling pathway across mammalian species; it is involved in a number of fundamental biological processes, including cellular growth, proliferation, differentiation and the maintenance of homeostasis in animals (Clemmons, 2009; Duan et al., 2010). The pathway consists of growth hormone (GH), IGF-I and IGF-II, associated binding proteins (IGFBP1-6, GHBP) and receptors (IGFR, GHR). The well documented control of the somatotropic axis on growth and muscle metabolism (Florini et al., 1996; Oksbjerg et al., 2004; Duan et al., 2010) indicates that the system could have substantial effects on the overall energetic efficiency of feed efficient animals. Indeed, positive genetic correlations have been reported between RFI and systemic IGF-I concentrations (Johnston et al., 2002; Moore et al., 2005). Despite this, the published literature is inconsistent at the phenotypic level, with Nascimento et al. (2015) reporting higher IGF-I concentrations in low RFI animals, Brown (2005) and Lancaster et al. (2008) reporting lower concentrations and other studies failing to establish any relationship between RFI status and systemic concentrations of IGF-1 (Kelly et al., 2010a, 2011a; Lawrence et al., 2012; Welch et al., 2013). This inconsistency may reflect, to some extent, differences in the prevailing dietary management employed between studies (Brown, 2005; Welch et al., 2013). In situations where feed supply is managed, for example during a period of imposed dietary restriction, signalling within the somatotropic axis becomes refractory such that although systemic concentrations of GH typically increase, this is not mirrored in greater hepatic synthesis of IGF-I (Breier, 1999; Keogh et al., 2015d). This has resulted in studies (Yambayamba et al., 1996b; Hornick et al., 1998b; Cabaraux et al., 2003; Keogh et al., 2015b), reporting lower circulating IGF-I concentrations in cattle offered a restricted diet compared to their unrestricted contemporaries. The typical rise in GH concentrations may be attributable to a requirement for lipolysis, as high blood GH concentrations promote adipose tissue mobilisation and increase blood NEFA concentrations (Lucy, 2004) and is likely to underpin the preservation of metabolic homeostasis during limited feed supply (Bell, 1995; Breier, 1999; Sonntag et al., 1999). During subsequent re-alimentation, uncoupling of the somatotropic axis is typically reversed with systemic concentrations of IGF-I increasing rapidly, further suggesting a role in improved feed efficiency (Hornick et al., 1998b; Keogh et al., 2015d). Indeed, Li et al. (2012) reported greater systemic concentrations of GH following administration of its precursor GHRH in cattle undergoing compensatory growth and suggested a direct role in supporting accelerated growth through GH's well-defined functions in stimulating protein accretion and fat catabolism.

The majority of systemic IGF-I is produced in hepatocytes (Philippou et al., 2007), however, other peripheral tissues including skeletal muscle and adipose are also capable of synthesising the hormone (Duan et al., 2010). Indeed, using a porcine model, Novakofski and McCusker (1993) documented that approximately two-thirds of postnatal growth to normal adult size is attributed to local tissue production and utilisation of IGF-I. Signalling of IGF-I has been noted as a critical factor in the regulation of skeletal muscle growth (Duan et al., 2010), with the physiological actions of IGF-I signalling including stimulation of protein synthesis and inhibition of protein degradation (Oddy and Owens, 1996; Hill et al., 1999). The release of GH from the anterior pituitary gland and subsequent receptor binding in peripheral tissues is a primary activator of the somatotropic axis and stimulation of anabolic processes such as cell division and protein synthesis (Møller and Nørrelund, 2003; Jessen et al., 2005). At the molecular level Barendse et al. (2007) and Sherman et al. (2008) have both identified associations between a SNP located in intron 4 of the *GHR* gene and RFI. In hepatic tissue, Chen et al. (2011) also observed greater transcript abundance for *GHR* in high compared to low RFI cattle. Conversely though, the inverse was observed in *M. longissimus dorsi* in the work of Kelly et al. (2013), where greater expression of *GHR* was evident in low RFI cattle. Following a period of dietary restriction and associated poor feed efficiency, lower expression of the GHR, *GHR1A* and *IGF1* were apparent (Keogh et al., 2015d), further evidencing refractory signalling. Moreover, this was underpinned through up-regulation of *IGFBP1* and *IGFBP2* at the same time, both of which function in the inhibition of IGF-1 bioactivity and bioavailability (Maddux et al., 2006). The two main binding proteins thought to have the greatest roles in the growth and development of skeletal muscle are IGFBP3 and IGFBP5 (Kelly et al., 2013). Expression of *IGFBP5* was greater in high RFI cattle in the data of Welch et al. (2013). Chen et al. (2011) also observed up-regulation of *IGFBP3* in hepatic tissue of low RFI cattle. However, Kelly et al. (2013) did not detect any differences between RFI groups in the expression of either binding protein in skeletal muscle tissue. IGF-I can also function in an autocrine/paracrine manner through interaction with the IGF receptors. Activation of the IGF-I receptor initiates intracellular signal transduction cascades including the PI3-K pathway, which is the predominant pathway involved in stimulating muscle protein synthesis (Schiaffino and Mammucari, 2011). A negative association was evident between *IGF-IR* and RFI in the data of Kelly et al. (2013). This result was consistent with the lack of difference between contrasting RFI phenotypes in the size and rate of growth of *M. longissimus dorsi* tissue in that study (Kelly et al., 2013).

Overall, differences in systemic IGF-I in response to dietary restriction and subsequent re-alimentation suggest a clear role for the somatotropic axis in compensatory growth, however, the same is not apparent for RFI status, where the published relationship between that trait and systemic IGF-I response is variable. This inconsistency between published RFI based studies is likely to have been contributed to by factors including diet type and age, which, in their own right affect functionality of this mitogenic signalling pathway (Welch et al., 2013).

It must be noted that the insulin signalling pathway and the somatotropic axis should not be viewed in isolation but rather collectively, given their interactive nature. Despite differences in the role of these pathways in their contribution to explaining variation in feed efficiency, previous research suggests that further work on components of these pathways such as fatty acid synthesis may enhance our understanding of mechanisms underpinning feed efficiency as illustrated in Figure 3.

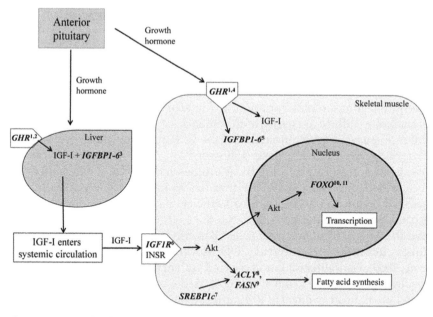

Figure 3: Proposed effect of feed efficiency on the somatotropic axis and insulin signalling pathway in beef cattle. Genes coding for components of the somatotropic axis and insulin signalling pathway were affected by feed efficiency state as follows (highlighted in bold italics in figure): 1: associations evident between a SNP in the *GHR* gene and RFI (Barendse et al., 2007; Sherman et al., 2008); 2: greater expression of *GHR* in high compared to low RFI (Chen et al., 2011); 3: *IGFBP3* up-regulated in low RFI (Chen et al., 2011); 4: lower expression of *GHR* in high RFI (Kelly et al., 2013); 5: *IGFBP5* expression greater in high RFI (Welch et al., 2013); 6: *IGF-1R* was negatively associated with RFI (Kelly et al., 2013); 7: *SREBP1c* was positively associated with RFI (Fitzsimons et al., 2014c); 8: *ACLY* up-regulated in cattle undergoing compensatory growth (Keogh et al., 2015c); 9: *FASN* expression was higher in feed efficient cattle (Welch et al., 2013); 10: *FOXO1, FOXO3* and *FOXO4* were up-regulated in cattle undergoing compensatory growth (Keogh et al., 2015c); 11: *FOXO1* expression was positively correlated with DMI (Fitzsimons et al., 2014c).

Metabolism and Maintenance

The proportion of total energy intake required solely for body maintenance is typically in excess of 50% in adult cattle thus representing the single most important factor that determines biological efficiency (Caton et al., 2000; Arango and Van Vleck, 2002). Similarly, in growing cattle the proportion of total ME intake used for body maintenance related processes is rarely less than 40%, even in situations where animals are consuming maximal feed intake (NRC, 2000). Such a significant underlying energetic requirement to maintain basic energetic homeostasis is contributed to by a number of physiological and biochemical processes and some of these are discussed below.

Size of and Metabolic Processes within the Visceral Organs

The GIT and liver are important energy sinks in cattle, and although only representing approximately 7% and 2.5% of body weight, these organs account for in the order of 18% and 25% of total oxygen consumption, respectively (McBride and Kelly, 1990). As a consequence, it is highly likely that inter-animal variation in the size and functionality of these organs may influence energy requirements for basal metabolism. Indeed the size and energy requirement of the GIT was found to increase with increasing level of feed intake (Johnson et al., 1990) and greater volume of digesta and supply of nutrients (Ortigues and Doreau, 1995). However, the limited published literature that has examined the contribution of visceral organs to variation in feed efficiency is equivocal. For example, some studies employing cattle offered a high concentrate diet, reported that the weight of stomach complex (Basarab et al., 2003; Fitzsimons et al., 2014a), intestines (Basarab et al., 2003; Meyer et al., 2014) and liver (Basarab et al., 2003) was lighter for low compared with high RFI steers, whereas other similar studies (Richardson et al., 2001; Mader et al., 2009; Cruz et al., 2010) failed to establish an effect of RFI status on the weight of any visceral organ measured. Where cattle are exposed to fluctuations in feed supply, studies have reported lower liver and GIT mass following a period of dietary restriction (Ryan et al., 1993; Yambayamba et al., 1996a; Keogh et al., 2015a) though this is quickly reversed upon realimentation (Ryan et al., 1993; Yambayamba et al., 1996a; Keogh et al., 2015a), with such splanchnic tissues apparently taking precedence for increasing energy supply. Our data also indicate greater transcriptional activity of rumen papillae during compensatory growth and replenishment of the GIT (Keogh et al., 2016b). Transcript abundance for genes including those involved in translation (*EIF4G2, EIF4G3, ELL2*), transcription (*EMG1, FOXN1, FOXP4, INTS3*) and histone functionality (*HIST1H2AC, HISTH2BO, HIST2H4A, KAT2A*) were particularly affected. During compensatory growth the initial enhanced rate of tissue gain is primarily comprised of protein deposition, consistent with the necessity to replenish metabolically active organs and further augment metabolic capacity and the demands of increased nutrient supply (Rompala et al., 1985; Ryan, 1990; Carstens et al., 1991; Ryan et al., 1993; Yambayamba et al., 1996a; Keogh et al., 2015a). There is also an increase in the efficiency of protein utilisation during the early phase of compensatory growth and associated improved feed efficiency. Indeed, Gonzaga Neto et al. (2011) and Keogh et al. (2015b) both observed a reduction in systemic concentrations of total protein and albumin at the start of compensatory growth which may be indicative of increased nitrogen utilisation efficiency and repletion of tissues, respectively. Additionally, Ellenberger et al. (1989) and Keogh et al. (2015b) reported an initial decrease in plasma urea concentrations in cattle undergoing compensatory growth, which may have reflected the high protein demand required for supporting increased visceral tissue replenishment. Furthermore, both Keogh et al. (2016a) and Connor et al. (2010b) observed up-regulation of ribosomal and protein synthesis related genes in hepatic tissue at the end of a period of dietary restriction. However, following 55 days of re-alimentation and compensatory growth, Keogh et al. (2016a) observed down-regulation of one ribosomal gene *RPS27* and up-regulation of hepatic genes involved in fatty acid biosynthesis including *FADS1*, *SCD* and *SREBF1*, thus suggesting a switch in tissue prioritisation from protein to lipid accretion as energy balance further improved during re-alimentation.

Although, as previously indicated, the size of visceral organs is responsive to the prevailing plane of nutrition (Johnson et al., 1990), oxygen consumption or energy expenditure of these organs increases after feeding and changes in accordance with level of feed intake (Seal and Reynolds, 1993). This suggests that the physical size of visceral organs alone may not be the sole contributory factor to variation in energetic efficiency. Indeed, McBride and Kelly (1990) have shown that, within tissues, different metabolic processes such as transport of sodium and potassium ions and protein synthesis and degradation in the GIT and liver, vary in their energetic efficiency. These discrepancies in energetic efficiency of metabolic processes within tissues may explain the absence of a consistent effect of RFI phenotype on the mass of visceral organs in cattle (Fitzsimons et al., 2014a).

Despite conflicting literature on mitochondrial function, which will be dealt with in greater detail in a subsequent section of this chapter, there is evidence for differential expression of genes involved in regulation of cellular energy status and other metabolic processes in hepatocytes and tissues of the GIT, between cattle of divergent feed efficiency potential. Chen et al. (2011) found differences in hepatic gene expression between high and low RFI animals for processes involved in carbohydrate metabolism, lipid metabolism and protein synthesis, in particular. Similarly, Tizioto et al. (2015), examining global liver gene expression profiles, suggest that high RFI Nellore steers have increased hepatic oxidative metabolism and melatonin degradation compared to their low RFI contemporaries. These authors (Tizioto et al., 2015) found that *EGR1* and *FOS* were upregulated in low RFI steers implicating a particular role for hepatic cellular growth, differentiation and response to oxidative stress in the efficiency of feed utilisation in cattle. Similarly, Keogh et al. (2016d) reported up-regulation of both of these genes during a state of improved feed efficiency during re-alimentation induced compensatory growth in skeletal muscle. Moreover, consistent with this, these genes were down-regulated in the same study during the preceding dietary restriction phase in accordance with a less feed efficient state (Keogh et al., 2016d). Additionally, in cattle undergoing compensatory growth, genes involved in lipid and carbohydrate metabolic processes were up-regulated in hepatic tissue (Keogh et al., 2016a). Similarly, Connor et al. (2010b) found that steers exhibiting compensatory growth and improved feed efficiency, had increased expression of hepatic genes involved in processes such as cellular metabolism, oxidative phosphorylation and glycolysis. Greater feed efficiency during compensatory growth was also associated with greater expression of genes involved in jejunal epithelial metabolism including *PFKB3, SDS, SDSL* (Keogh et al., 2016c). In agreement with these data, Serao et al. (2013), working at the genomic level, reported a SNP associated with feed efficiency that mapped to several genes with functions pertaining to digestion and absorption, transport of compounds from the intestinal epithelial cells to the intracellular space and other functions such as glycerophospholipid metabolism and immune-response related biological processes. Overall, these results imply that greater propensity for an upregulation of metabolic processes is consistent with an improved feed efficiency phenotype.

These findings further emphasise the need to examine cellular and molecular differences in organs that have high metabolic activity such as the GIT and liver between animals of differing feed efficiency potential. The conflicting findings of the gene expression studies using visceral organ tissues reflect the complex nature of the trait, however, consistency in processes related to oxidative stress, mitochondrial function and lipid and carbohydrate metabolism appear to be recurrent in the literature and further and more detailed exploration of the constituent biochemical pathways may identify the key underlying processes as mechanisms underpinning variation in RFI in beef cattle.

Maintenance and Metabolism of Muscle and Adipose Tissue

Despite protein accretion being more efficient than fat accretion (Herd et al., 2004), the energy cost associated with maintenance of lean tissue is greater than that of fat, due to the requirement to support the energetically expensive processes of protein degradation and turnover (Schiavon and Bittante, 2012). Authors have suggested that the observed variation of protein turnover and degradation between muscles of individual animals may be an important contributor to feed efficiency potential (Oddy, 1999). Indeed, Richardson and Herd (2004) suggested that high RFI steers have a higher rate of protein degradation in muscle and liver and a less efficient mechanism for protein deposition than their more efficient contemporaries. However, no relationship between RFI and systemic concentrations of creatinine, which is indicative of muscle mass and protein catabolism (Istasse et al., 1990), was observed by Fitzsimons et al. (2014a) or Nascimento et al. (2015) in bulls or heifers, respectively. In agreement, Lines et al. (2014) also found no differences in protein metabolism between efficient and inefficient heifers that were the result of three generations of divergent selection for RFI, though the actual feed efficiency of the animals reported was not measured in that study. Despite these findings however, Fitzsimons et al. (2013), Lawrence et al. (2012), and Richardson et al. (2004) all found that RFI was negatively associated with circulating concentrations of plasma creatinine. Other authors have observed increases in systemic creatinine as a consequence of dietary restriction with a reversal evident upon re-alimentation and improved feed efficiency (Hornick et al., 1998b; Cabaraux et al., 2003; Keogh et al., 2015b). McDonagh et al. (2001) found higher levels of calpastatin but a reduced rate of myofibril fragmentation suggesting reduced protein degradation in high efficiency steers. The variation between animals in the rate of protein turnover observed by some authors suggests a genetic origin for this trait, which may contribute to variation in energy requirements for maintenance and growth. These findings were further confirmed by Karisa et al. (2013) where SNPs in the calpastatin gene (among many others) were found to be both positively and negatively associated with RFI in beef cattle, depending on genotype. Additionally, greater expression of the calpastatin gene, *CAST* was apparent in skeletal muscle of animals displaying improved feed efficiency during re-alimentation induced compensatory growth (Keogh et al., 2016d).

In addition to muscle, variation in metabolism of different adipose tissue depots such as visceral, omental and subcutaneous depots may also contribute to differences in basal metabolism and energy utilisation (Welch et al., 2012). As previously discussed, circulating concentrations of leptin were found to have either a neutral (Kelly et al., 2010a, 2011a) or a positive association (Richardson and Herd, 2004) with RFI in cattle. The greater gene expression for leptin in adipose tissue of low RFI steers observed by Perkins et al. (2014) agrees with previous research where SNPs in the leptin promoter were associated with the quantity of feed consumed (Nkrumah et al., 2005). However, the associations between these SNPs and serum leptin concentrations and indeed feed efficiency measures such as RFI and FCR were both inconsistent and tenuous (Nkrumah et al., 2005). Karisa et al. (2014), through analysis of gene networks found that a hub associated with insulin induced gene 1 (*INSIG1*) was affiliated with RFI phenotype. The genes found within the *INSIG1* hub were all involved in biological processes related to energy, steroid and lipid metabolism. Indeed, differential expression of *INSIG1* during lower and improved states of feed efficiency has been reported in the context of compensatory growth, across tissues and studies. These included down regulation of *INSIG1* during lower feed efficiency associated with dietary restriction in skeletal muscle (Doran, 2013; Keogh et al., 2016d) and jejunum tissues (Keogh et al., 2016c). Furthermore, *INSIG1* expression was found to be up-regulated in hepatic tissue during compensatory growth and improved feed efficiency in the data of both Connor et al. (2010b) and Keogh et al. (2016a). Up-regulation of genes involved in lipid, amino acid and carbohydrate metabolism was apparent in *M. longissimus dorsi* in addition to greater expression of genes involved in protein synthesis in the same tissue in cattle undergoing compensatory growth and displaying greater feed efficiency (Keogh et al., 2016d). Greater metabolic function in muscle tissue was further evidenced through up-regulation of genes involved in nutrient transport in *M. longissimus dorsi* during greater efficiency of compensatory growth (Keogh et al., 2016d).

In addition to investigating metabolic pathways within muscle tissue that contribute to differences in feed efficiency, Welch et al. (2013) also examined muscle fibre type composition of cattle divergent for feed efficiency. Energy expenditure is dependent on muscle fibre type which potentially may be contributing to feed efficiency phenotype. Welch et al. (2013) found that type IIb fibres, which are glycolytic and have greater energy expenditure than type I fibres (oxidative), were more abundant in muscle tissue of high RFI steers at slaughter. Concurrently, type I fibres tended to be more abundant in low RFI steers. Nevertheless, fibre type composition was not associated with feed efficiency at the end of the post-weaning RFI evaluation period.

Further research exploring differences in the metabolic regulation, function and morphology of adipocytes and myocytes would assist in the understanding of the potential role these processes have in explaining variation in feed efficiency of beef cattle.

Mitochondrial Function

Mitochondria are cellular organelles, responsible for approximately 90% of oxygen consumption and the bulk of ATP synthesis (Lehninger et al., 1993; Kolath et al., 2006a). Bottje and Carstens (2012) showed that feed efficiency affects the coupling of the enzyme reactions of the electron transport chain, with inefficient animals exhibiting leakage of protons out of the inner mitochondrial membrane, thus reducing the amount of ATP produced. The reverse was evident in mitochondria of feed efficient animals. Although there is indirect evidence to suggest that feed efficient mice have enhanced mitochondrial biogenesis, there is little information for beef cattle (Bottje and Carstens, 2012). In order to address this void in the literature, McKenna et al. (unpublished), using citrate synthase activity as a biomarker of mitochondrial content, has recently established no difference in mitochondrial number in either muscle or hepatic tissue between bulls divergent for feed efficiency, using the animal models of Fitzsimons et al. (2014a) (see Figure 4) and Keogh et al. (2015a).

Moreover, studies have shown, that compared with inefficient animals, ADP-control of oxidative phosphorylation (Lancaster et al., 2014) and

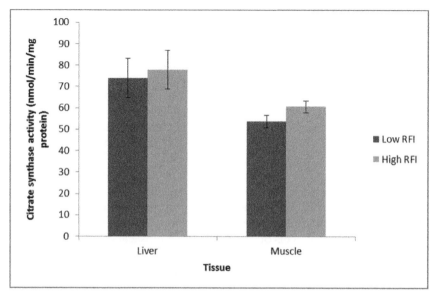

Figure 4: Citrate synthase activity (proxy for mitochondrial abundance) in muscle and liver tissue of beef bulls divergent for residual feed intake (RFI) (McKenna et al., unpublished).

mitochondrial complex protein concentration (Davis, 2009) are all higher, while mitochondrial derived reactive oxygen species (ROS) production (Bottje and Carstens, 2009) is lower in energetically efficient animals. Lancaster et al. (2014) explored hepatic mitochondrial function in cattle divergent for RFI and found no differences between efficiency groups for respiration rate of complex 2, 3 and 4 and indices of proton leakage rates but that acceptor control ratio (which is an indicator or respiratory rate within the mitochondrion) was greater in low RFI cattle. However, Acetoze et al. (2015) also reported no differences in hepatic mitochondrial respiration rate in steer progeny of bulls divergent for RFI. Greater mitochondrial complex I protein concentration (mitochondrial NADH) was found in lymphocytes of steers differing in RFI (Ramos and Kerley, 2013). The results of Lancaster et al. (2014) and Acetoze et al. (2015) disagree with an earlier study by Kolath et al. (2006a) where it was found that relative to high RFI, mitochondrial respiration in *M. longissimus dorsi* was increased in low RFI steers. In agreement, Connor et al. (2010b) observed that genes associated with oxidative phosphorylation and mitochondrial efficiency were up-regulated in hepatic tissue during compensatory growth, suggesting improved cellular energetic efficiency in animals of an improved feed efficiency phenotype. Conversely, the opposite effect was reported by Keogh et al. (2016d) for *M. longissimus dorsi*, where, during a period of restricted feed intake and lower feed efficiency, greater expression of component genes of the tricarboxylic cycle and oxidative phosphorylation was evident. However, upon re-alimentation and improved feed efficiency, these genes were subsequently down-regulated (Keogh et al., 2016d). A similar effect was observed in hepatic tissue, with up-regulation of genes encoding oxidative phosphorylation proteins evident following a period of dietary restriction (Keogh et al., 2016a). However, functional assays revealed no difference in mitochondrial abundance or complex rate in these animals in either muscle or liver tissues (McKenna et al., 2016). Consistent with this, the findings of Lancaster et al. (2014) and Acetoze et al. (2015) suggest no effect of RFI status on mitochondrial proton leak in beef cattle.

Several studies have examined differential expression of genes pertaining to mitochondrial oxidative phosphorylation in beef cattle to determine the contribution of this process to variation in feed efficiency. In an earlier study, Kolath et al. (2006b) reported no effect of feed efficiency on mRNA expression levels of uncoupling proteins 2 and 3 (*UCP2, UCP3*) in muscle tissue of steers. In agreement, Fonseca et al. (2015) and Marks et al. (2014) also observed no difference in expression of *UCP2* and *UCP3* in muscle tissue of beef cattle. However, Kelly et al. (2011b) reported that *UCP3* gene expression tended to be upregulated in high RFI heifers which agrees with the findings of Sherman et al. (2008) where a SNP in the *UCP3* gene was found to be positively and negatively associated with FCR and PEG, respectively, implicating the role of this gene in feed efficiency. Nevertheless, in hepatic tissue Fonseca et al. (2015) found that *UCP2* was upregulated in high RFI steers suggesting a reduction in ATP production which is contradictory to the hypothesis of enhanced, or more efficient, mitochondrial energy production in feed efficient animals (Bottje and Carstens, 2012). During the lower feed efficient state associated with dietary restriction, *UCP2* expression was found to be up-regulated, whilst expression of *UCP3* was down-regulated in *M. longissimus dorsi* tissue (Keogh et al., 2016d).

Transcription factors play a key role in oxidative phosphorylation via the precise regulation of gene expression. Such transcription factors include the peroxisome proliferator activated-receptors *PPARGC1α, PPARα* and *PPARy*, nuclear respiratory factor 1 (*NRF1*) and mitochondrial transcription factor A (*TFAM*). *PPARGC1α* controls mitochondrial proliferation, lipid metabolism and energy homeostasis (Lin et al., 2005) and mRNA expression of this gene was found to be upregulated in the *M. longissimus dorsi* of feed efficient heifers (Kelly et al., 2011b) suggesting enhanced mitochondrial biogenesis and respiration. Similarly, expression of *PPARα* was down-regulated in hepatic tissue of cattle offered a restricted dietary allowance (Keogh et al., 2016a). Conversely though, down-regulation of *NRF1* and *TFAM* was evident in skeletal muscle of cattle undergoing compensatory growth and displaying improved feed efficiency (Keogh et al., 2016d). In subsequent studies, however, using muscle (Welch et al., 2013; Fitzsimons et al., 2014c; Fonseca et al., 2015; Marks et al., 2014) and liver (Fonseca et al., 2015; Marks et al., 2014) tissue, no differences were found in levels of expression of *PPARGC1α* between feed efficiency phenotypes. Similarly, mRNA expression levels of *PPARα* and *PPARγ*, and *NRF1* were not affected by feed efficiency phenotype in either muscle (Kelly et al., 2011b; Welch et al., 2013; Fitzsimons et al., 2014c; Marks et al., 2014) or liver tissue in relation to *PPARα* and *NRF1* (Marks et al., 2014). However, Marks et al. (2014) did report upregulation of *PPARγ* in liver tissue of high RFI bulls suggesting that differential gene expression patterns may not be consistent across tissues. In support of this premise, while mRNA expression of *TFAM* in muscle was upregulated in low RFI cattle in both the studies of Fonseca et al. (2015) and Marks et al. (2014) contrasting results were observed between the two studies for liver, with Fonseca et al. (2015) reporting upregulation and Marks et al. (2014) reporting no differences in mRNA expression of this gene.

In the study of Kelly et al. (2011b), mRNA expression of a range of other genes involved in oxidative phosphorylation was also examined. It was found that feed efficiency had no effect on expression levels of protein kinase, AMP-activated, alpha 1 catalytic subunit (*PRKAA1*), NADH-Coenzyme Q reductase (*NDUFS6*), Succinate Dehydrogenase Complex, Subunit D, Integral Membrane Protein (*SDHD*), Ubiquinol-Cytochrome C Reductase Core Protein II (*UQCRC2*), ATPase (*ATP6D*), Coenzyme Q4 (*COQ4*) and cytochrome c (*CSCS*). Nevertheless, there was a feed efficiency x diet type interaction observed for transcript abundance of adenine nucleotide translocator (*ANT1*) and mitochondrially encoded cytochrome c oxidase II (*COX2*). It was found that mRNA expression of *ANT1* was greater for low RFI heifers consuming a high forage diet but that this difference disappeared when the heifers were offered a low forage diet. Given its involvement in the control of ATP supply to the cell, Kelly et al. (2011b) suggest that the production and rate of exchange of ATP may be greater in feed efficient cattle but that this effect may be mediated by diet type. Conversely, when heifers in the study of Kelly et al. (2011b) were offered a low forage diet, expression of *COX2* was upregulated in low RFI heifers while when the same animals were subsequently offered a high forage diet, these differences were not apparent. Elevated *COX2* gene expression in feed efficient animals further supports the hypothesis of enhanced efficiency of the cellular oxidative phosphorylation process in beef cattle, however, the effect may be diet dependent.

The foregoing discussion suggests that of the transcription factors regulating mitochondrial function, *TFAM* and the genes *ANT1* and *COX2* appear to be genes that warrant further investigation. However, additional studies of multiple tissue types may provide a more insightful means of determining the magnitude of the role of oxidative phosphorylation in contributing to the variation in feed efficiency between beef cattle.

Stress Physiology

The stress response of an animal triggers increased metabolic rate, energy consumption and catabolic processes such as lipolysis and protein degradation, via activation of the hypothalamic pituitary adrenal (HPA) axis (Minton, 1994; Knott et al., 2010). The release of glucocorticoids and catecholamines in response to a stress stimulus in cattle increases the mobilisation of energy rich substances for metabolism, e.g., increasing gluconeogenesis (Brockman and Laarveld, 1986). Animals that are stressed have a higher demand for nutrients and divert these nutrients away from growth towards the stress response (Colditz, 2004; Knott et al., 2010). Differences in the stress response in high and low RFI animals may account for some of the observed variation in RFI, as indications of higher stress levels in high RFI cattle (Richardson et al., 2002, 2004; Montanholi et al., 2010) may account for a significant source of energy wastage (Richardson and Herd, 2004; Herd and Arthur, 2009; Welch et al., 2012). Results from an adrenocorticotropic hormone (ACTH) challenge on pregnant and non-pregnant beef heifers previously ranked on RFI (Lawrence et al., 2011), found that low RFI heifers had a reduced sensitivity to exogenous ACTH, suggesting differing HPA axis function in high and low RFI animals. In contrast, Kelly et al. (2016) found no differences in the stress response in growing beef heifers previously ranked on RFI following a corticotropin-releasing hormone (bCRH) challenge. The conflicting results in the literature suggest that further research on the role that an animals responsiveness to stress may have in determining feed efficiency phenotype in beef cattle is needed.

It can be deduced from the preceding discussion that acquiring a greater knowledge of the underlying variation in metabolism and maintenance related processes may increase our understanding of the factors that contribute to variation in feed efficiency. Within tissues of the GIT, there are numerous biological processes, the efficiency of which, may contribute to overall feed utilisation. Additionally, there are many metabolic processes that span several organs with their own varying efficiencies suggesting that processes within individual organs alone may not necessarily be the only mechanisms worth investigating. Maintenance and metabolism of muscle and adipose tissue may also be significant contributors to variation in feed efficiency however, further research in many aspects of both of these tissues, from metabolic regulation to morphology, is required. Processes pertaining to mitochondrial function are common across all organs and tissues within the body. From the published literature to-date, it is apparent that variation in processes revolving around mitochondrial function and associated biochemical pathways may contribute to variation in feed efficiency potential. Although not involved with metabolism and maintenance *per se*, response to a stressor affects metabolic rate, energy consumption and catabolism, therefore implying that this process may in turn, affect feed efficiency status. Thus, mechanisms related to metabolism and maintenance provide a vast reservoir of information that may potentially shed further light on the biological regulation of feed efficiency in beef cattle.

Genomic Selection for Feed Efficiency

It is now well established that feed efficiency has a strong underlying genetic basis and is, for the most part, not antagonistically correlated with other economically relevant traits (Crews, 2005; Crowley et al., 2010). Indeed, the foregoing analysis of the published literature provides a compelling argument towards significant differences in several key biochemical processes across a number of tissues in cattle divergent for energetic efficiency status. However, continued genetic progress for the trait, using traditional quantitative genetic selection approaches will be hampered by the on-going limited availability of appropriate phenotypes. Thus given the cost and logistical difficulties in procuring phenotypic information together with its obvious molecular basis, suggests that the trait is a prime candidate for the application of genomic technologies (Rolf et al., 2011). Genomic selection refers to selection decisions that are based on breeding values predicted using genome wide marker data such as SNP (Meuwissen et al., 2001). The approach is aimed at increasing selection accuracy and accelerating genetic improvement by focusing on the SNP most strongly correlated to phenotype, although the genes and sequence variants directly affecting the phenotype remain largely unknown (Snelling et al., 2013). Unlike phenotype based genetic evaluation, genomic prediction has the capacity to estimate genetic merit of selection candidates at birth before phenotypes become available, which offers great promise in the prediction of genetic potential of selection candidates for traits that are difficult and expensive to measure such as feed efficiency. The success of genomic selection largely depends on the accuracy of the predicted genomic breeding values. However, genomic prediction accuracy in beef cattle is still not sufficiently high to allow selection of candidates without appropriate phenotypic measurement (Bolormaa et al., 2013). Accurate genomic estimated breeding values (GEBV) for feed efficiency would improve genetic gain and shorten the generation interval for the trait. The calculation of GEBV depends on the generation of a reference population where feed efficiency has already been measured and animals genotyped for appropriate genomic markers (Hayes et al., 2009; VanRaden et al., 2009).

Assembling sufficiently large reference populations for accurate GEBV prediction is a major challenge for a trait like feed efficiency, where factors such as contrasting breeds, age and nutritional management (i.e., potential for G x E) will affect the utility of amalgamating datasets.

Whether a breed-specific reference population should be used for each breed requiring GEBV or whether a common reference population should be used for all breeds is a major issue for consideration. Chen et al. (2013) pooled data from different breeds to form the reference population which improved the accuracy of across breed genomic prediction for RFI in beef cattle. Indeed, combining breeds into a common reference population increases its size, but linkage disequilibrium exists only over shorter genomic distances in a multi-breed population and this tends to reduce the accuracy of GEBV and may counter the benefits of a larger reference population. In essence, the accuracy of GEBV for feed efficiency could decline as a result of conducting across breed evaluations. Another method of improving the portability and accuracy of genomic predictions across breeds and into crossbred populations involves shifting focus from the assayed SNPs to variants more likely to have functional effects (Snelling et al., 2013). Integrating information about gene function and regulation with SNP genotyping assays is the most promising approach to enhance knowledge of genomic mechanisms affecting complex traits (Snelling et al., 2012). Using gene transcript expression and pathway analysis will be critical for the characterisation of sequence variation in expressed genes enabling the future development of relatively small, inexpensive marker sets that are sufficiently robust to describe phenotypic variation in feed efficiency in cattle. A study of beef tenderness measured in crosses of different breeds illustrates that genomic selection based on functional SNPs, incorporating existing functional annotation, can be more accurate across populations than genomic predictions based only on genotype-phenotype associations assessed in one population (Snelling et al., 2013). In addition, a recent Canadian study identified SNPs in genes involved in digestive and metabolic processes and showed these SNPs to be associated with feed efficiency in beef cattle (Abo-Ismail et al., 2013). Similarly, Onteru et al. (2013) identified functionally important genomic regions associated with RFI in pigs using whole genome association studies which are the focus of causative SNP identification. Therefore, enhanced knowledge of genomic control of biological mechanisms affecting feed efficiency is required to enable reliable development of genomic selection approaches.

Current and future research in the area of genomic selection for feed efficiency will focus on the identification of panels of genetic variants of biological significance to feed efficiency. However these polymorphisms need to be sufficiently robust across breed, phase of development and dietary regimen. Strategies are underway worldwide to build small, more accurate low-cost panels for individual genotyping which will include functional SNP for selection of a particular trait or suite of traits such as feed efficiency. Future success in breeding for feed efficiency in beef cattle will depend on the discovery of functional SNP regulating this trait which will be incorporated into national and international multi-trait genomically assisted breeding programmes.

Conclusion

The literature discussed in this review of the molecular physiology of feed efficiency in beef cattle highlights significant variation across a variety of metabolic organs and constituent biochemical processes amongst animals of contrasting feed efficiency phenotype. The main likely physiological processes involved are summarised in Figure 5. Responses to fluctuations in nutrient supply, which induce differential coping mechanisms, may provide valuable insights into the inherent biological regulation of the feed efficiency trait. Nevertheless, it is apparent that much equivocation exists in the literature in relation to inter-animal responses in appetite regulation, digestion along the entirety of the GIT, host-ruminal microbiome interactions, nutrient partitioning, myocyte and adipocyte metabolism, metabolic processes, including mitochondrial function, as well as other

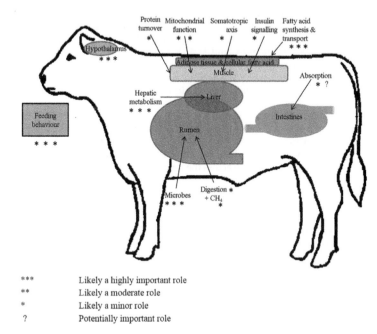

***	Likely a highly important role
**	Likely a moderate role
*	Likely a minor role
?	Potentially important role

Figure 5: Suggested relative contribution of various biological processes underlying variation in feed efficiency of beef cattle.

physiological processes such as response to stressors. Current and emerging technology including transcriptomic and proteomic platforms will further enlighten our understanding of and focus on the major regulators of energetic efficiency. The advent of genomically assisted selection technology has the potential to provide a powerful vehicle to implement the outcomes of these fundamental analyses within the context of multi-trait, profit based selection indices for sustained genetic progress. Notwithstanding this, continued comprehensive investigation into and a thorough understanding of, the biochemical regulation of feed efficiency and related economically important traits will be critical to future progress as well as the potential to obviate the on-going necessity to collect individual animal feed intake measurements.

Keywords: Appetite, feed intake, digestion, nutrient partitioning, composition of growth, metabolism, maintenance, muscle, adipose, mitochondria, stress physiology

References

Abo-Ismail, M.K., M.J. Kelly, E.J. Squires, K.C. Swanson, S. Bauck, and S.P. Miller. 2013. Identification of single nucleotide polymorphisms in genes involved in digestive and metabolic processes associated with feed efficiency and performance traits in beef cattle. J. Anim. Sci. 91: 2512–2529.

Abo-Ismail, M.K., G. Vander Voort, J.J. Squires, K.C. Swanson, I.B. Mandell, X. Liao, P. Stothard, S. Moore, G. Plastow, and S.P. Miller. 2014. Single nucleotide polymorphisms for feed efficiency and performance in crossbred beef cattle. BMC Genet. 15: 14.

Acetoze, G., K.L. Weber, J.J. Ramsey, and H.A. Rossow. 2015. Relationship between liver mitochondrial respiration and proton leak in low and high RFI steers from two lineages of RFI Angus bulls. Int. Scholarly Res. Notices 1–5.

Ahima, R.S. 2005. Central actions of adipocyte hormones. Trends Endocrinol. Metab. 16: 307–313.

Alam, T., D.A. Kenny, T. Sweeney, F. Buckley, R. Prendiville, M. McGee, and S.M. Waters. 2012. Expression of genes involved in energy homeostasis in the duodenum and liver of Holstein-Friesian and Jersey cows and their F1 hybrid. Physiol. Genomics 44: 198–209.

Albornoz, R.I., J.R. Aschenbach, D.R. Barreda, and G.B. Penner. 2013. Feed restriction reduces short-chain fatty acid absorption across the reticulorumen of beef cattle independent of diet. J. Anim. Sci. 91: 4730–4738.

Allen, M.S. 2014. Drives and limits to feed intake in ruminants. Anim. Prod. Sci. 54: 1513–1524.

Arango, J., and L.D. Van Vleck. 2002. Size of beef cows: Early ideas, new developments. Genet. Mol. Res. 1: 51–63.

Arthur, P.F., G. Renand, and D. Krauss. 2001. Genetic and phenotypic relationships among different measures of growth and feed efficiency in young Charolais bulls. Livest. Prod. Sci. 68: 131–139.

Ashfield, A., M. Wallace, M. McGee, and P. Crosson. 2014. Bioeconomic modelling of compensatory growth for grass-based dairy calf-to-beef production systems. J. Agric. Sci. 152: 805–816.

Barendse, W., A. Reverter, R.J. Bunch, B.E. Harrison, W. Barris, and M.B. Thomas. 2007. A validated whole-genome association study of efficient food conversion in cattle. Genetics 176: 1893–1905.

Basarab, J.A., M.A. Price, J.L. Aalhus, E.K. Okine, W.M. Snelling, and K.L. Lyle. 2003. Residual feed intake and body composition in young growing cattle. Can. J. Anim. Sci. 83: 189–204.

Basarab, J.A., D. McCartney, E.K. Okine, and V.S. Baron. 2007. Relationships between progeny residual feed intake and dam productivity traits. Can. J. Anim. Sci. 87: 489–502.

Bell, A.W. 1995. Regulation of organic nutrient metabolism during transition from late pregnancy to early lactation. J. Anim. Sci. 73: 2804–2819.

Berry, D.P., and J.J. Crowley. 2013. Cell biology symposium: Genetics of feed efficiency in dairy and beef cattle. J. Anim. Sci. 91: 1594–1613.

Bolormaa, S., J.E. Pryce, K. Kemper, K. Savin, B.J. Hayes, W. Barendse, Y. Zhang, C.M. Reich, B.A. Mason, R.J. Bunch, B.E. Harrison, A. Reverter, R.M. Herd, B. Tier, H.U. Graser, and M.E. Goddard. 2013. Accuracy of prediction of genomic breeding values for residual feed intake and carcass and meat quality traits in Bos taurus, Bos indicus, and composite beef cattle. J. Anim. Sci. 91: 3088–3104.

Bottje, W.G., and G.E. Carstens. 2009. Association of mitochondrial function and feed efficiency in poultry and livestock species. J. Anim. Sci. 87(14_suppl): E48–E63.

Bottje, W.G., and G.E. Carstens. 2012. Variation in metabolism: Biological efficiency of energy production and utilization that affects feed efficiency. pp. 251–274. *In*: R.A. Hill (ed.). Feed Efficiency in the beef industry. John Wiley & Sons, Inc. Ames, Iowa.

Bouquet, A., M.N. Fouilloux, G. Renand, and F. Phocas. 2010. Genetic parameters for growth, muscularity, feed efficiency and carcass traits of young beef bulls. Livest. Sci. 129: 38–48.

Breier, B.H. 1999. Regulation of protein and energy metabolism by the somatotropic axis. Domest. Anim. Endocrinol. 17: 209–218.

Brockman, R.P., and B. Laarveld. 1986. Hormonal regulation of metabolism in ruminants; A review. Livest. Prod. Sci. 14: 313–334.

Brown, E.G. 2005. Sources of Biological Variation in Residual Feed Intake in Growing and Finishing Steers. PhD Diss. Texas A&M Uni. College Station.

Cabaraux, J.F., M. Kerrour, C. Van Eenaeme, I. Dufrasne, L. Istasse, and J.L. Hornick. 2003. Different modes of food restriction and compensatory growth in double-muscled Belgian Blue bulls: Plasma metabolites and hormones. Anim. Sci. 77: 205–214.

Carberry, C.A., D.A. Kenny, S. Han, M.S. McCabe, and S.M. Waters. 2012. Effect of phenotypic residual feed intake and dietary forage content on the rumen microbial community of beef cattle. Appl. Environ. Microbiol. 78: 4949–4958.

Carberry, C.A., D.A. Kenny, A.K. Kelly, and S.M. Waters. 2014a. Quantitative analysis of ruminal methanogenic microbial populations in beef cattle divergent in phenotypic residual feed intake (RFI) offered contrasting diets. J. Anim. Sci. Biotechnol. 5: 41.

Carberry, C.A., S.M. Waters, D.A. Kenny, and C.J. Creevey. 2014b. Rumen methanogenic genotypes differ in abundance according to host residual feed intake phenotype and diet type. Appl. Environ. Microbiol. 80: 586–594.

Carstens, G.E., D.E. Johnson, M.A. Ellenberger, and J.D. Tatum. 1991. Physical and chemical components of the empty body during compensatory growth in beef steers. J. Anim. Sci. 69: 3251–3264.

Castro Bulle, F.C.P., P.V. Paulino, A.C. Sanches, and R.D. Sainz. 2007. Growth, carcass quality, and protein and energy metabolism in beef cattle with different growth potentials and residual feed intakes. J. Anim. Sci. 85: 928–936.

Caton, J.S., M.L. Bauer, and H. Hidari. 2000. Metabolic components of energy expenditure in growing beef cattle—Review. Asian-Aust. J. Anim. Sci. 13: 702–710.

Channon, A.F., J.B. Rowe, and R.M. Herd. 2004. Genetic variation in starch digestion in feedlot cattle and its association with residual feed intake. Aust. J. Exper. Agric. 44: 469–474.

Chen, L., F. Schenkel, M. Vinsky, D.H. Crews, and C. Li. 2013. Accuracy of predicting genomic breeding values for residual feed intake in Angus and Charolais beef cattle. J. Anim. Sci. 91: 4669–4678.

Chen, Y., C. Gondro, K. Quinn, R.M. Herd, P.F. Parnell, and B. Vanselow. 2011. Global gene expression profiling reveals genes expressed differentially in cattle with high and low residual feed intake. Anim. Genet. 42: 475–490.

Cheng, Z., Y. Tseng, and M.F. White. 2010. Insulin signaling meets mitochondria in metabolism. Trends Endocrinol. Metab. 21: 589–598.

Chilliard, Y., M. Doreau, F. Bocquier, and G.E. Lobley. 1995. Digestive and metabolic adaptations of ruminants to variations in food supply. pp. 329–360. *In*: M. Journet, E. Grenet, M.-H. Farce, M. Theriez, and C. Demarquilly (eds.). Recent Developments in the Nutrition of Herbivores. Proc. IVth Int. Symp. Nutr. Herbivores. INRA Editions, Clermont-Ferrand, France.

Clemmons, D.R. 2009. Role of IGF-I in skeletal muscle mass maintenance. Trends Endocrinol. Metab. 20: 349–356.

Colditz, I.G. 2004. Some mechanisms regulating nutrient utilisation in livestock during immune activation: An overview. Aust. J. Exper. Agric. 44: 453–457.

Connor, E.E., R.W. Li, R.L. Baldwin, and C. Li. 2010a. Gene expression in the digestive tissues of ruminants and their relationships with feeding and digestive processes. Animal 4: 993–1007.

Connor, E.E., S. Kahl, T.H. Elsasser, J.S. Parker, R.W. Li, C.P. Van Tassell, R.L. Baldwin, and S.M. Barao. 2010b. Enhanced mitochondrial complex gene function and reduced liver size may mediate improved feed efficiency of beef cattle during compensatory growth. Funct. Integr. Genomics 10: 39–51.

Conroy, S.B., M.J. Drennan, M. McGee, M.G. Keane, D.A. Kenny, and D.P. Berry. 2010. Predicting beef carcass meat, fat and bone proportions from carcass conformation and fat scores or hindquarter dissection. Animal 4: 234–241.

Crews, D.H. 2005. Genetics of efficient feed utilization and national cattle evaluation: A review. Genet. Mol. Res. 4: 152–165.

Crowley, J.J., M. McGee, D.A. Kenny, D.H. Crews, R.D. Evans, and D.P. Berry. 2010. Phenotypic and genetic parameters for different measures of feed efficiency in different breeds of Irish performance tested beef bulls. J. Anim. Sci. 88: 885–894.

Crowley, J.J., R.D. Evans, N. McHugh, T. Pabiou, D.A. Kenny, M. McGee, D.H. Crews, and D.P. Berry. 2011. Genetic associations between feed efficiency measured in a performance test station and performance of growing cattle in commercial beef herds. J. Anim. Sci. 89: 3382–3393.

Cruz, G.D., J.A. Rodriguez-Sanchez, J.W. Oltjen, and R.D. Sainz. 2010. Performance, residual feed intake, digestibility, carcass traits, and profitability of Angus-Hereford steers housed in individual or group pens. J. Anim. Sci. 88: 324–329.

Davis, M. 2009. Influence of diet, production traits, blood hormones and metabolites and mitochondrial complex protein concentrations on residual feed intake in beef cattle. PhD Diss. Univ. Missouri-Columbia.

de Menezes, A.B., E. Lewis, M. O'Donovan, B.F. O'Neill, N. Clipson, and E.M. Doyle. 2011. Microbiome analysis of dairy cows fed pasture or total mixed ration diets. FEMS Microbiol. Ecol. 78: 256–265.

Delavaud, C., F. Bocquier, Y. Chilliard, D.H. Keisler, A. Gertler, and G. Kann. 2000. Plasma leptin determination in ruminants: Effect of nutritional status and body fatness on plasma leptin concentrations assessed by a specific RIA in sheep. J. Endocrinol. 165: 519–526.

Dijkstra, J., E. Kebreab, J.A.N. Mills, W.F. Pellikaan, S. Lopez, A. Bannink, and J. France. 2007. Predicting the profile of nutrients available for absorption: From nutrient requirement to animal response and environmental impact. Animal 1: 99–111.

Dmitriew, C.M. 2011. The evolution of growth trajectories: What limits growth rate? Biol. Rev. 86: 97–116.

Doran, A.G. 2013. Systems level investigation of the genetic basis of bovine muscle growth and development. PhD thesis, National University of Ireland Maynooth, Ireland.

Duan, C., H. Ren, and S. Gao. 2010. Insulin-like growth factors (IGFs), IGF receptors, and IGF-binding proteins: Roles in skeletal muscle growth and differentiation. Gen. Comp. Endocrinol. 167: 344–351.

Durunna, O.N., M.G. Colazo, D.J. Ambrose, D. McCartney, V.S. Baron, and J.A. Basarab. 2012. Evidence of residual feed intake reranking in crossbred replacement heifers. J. Anim. Sci. 90: 734–741.

Ellenberger, M.A., D.E. Johnson, G.E. Carstens, K.L. Hossner, M.D. Holland, T.M. Nett, and C.F. Nockels. 1989. Endocrine and metabolic changes during altered growth rates in beef cattle. J. Anim. Sci. 67: 1446–1454.

Finneran, E., P. Crosson, P. O'Kiely, L. Shalloo, D. Forristal, and M. Wallace. 2010. Simulation modeling of the cost of producing and utilising feeds for ruminants on Irish farms. J. Farm. Manag. 14: 95–116.

Fitzsimons, C., D.A. Kenny, A.G. Fahey, and M. McGee. 2013. Methane emissions, body composition and rumen fermentation traits of beef heifers differing in phenotypic residual feed intake. J. Anim. Sci. 91: 5789–5800.

Fitzsimons, C., D.A. Kenny, and M. McGee. 2014a. Visceral organ weights, digestion and carcass characteristics of beef bulls differing in residual feed intake offered a high concentrate diet. Animal 8: 949–959.

Fitzsimons, C., D.A. Kenny, A.G. Fahey, and M. McGee. 2014b. Feeding behavior, rumen fermentation variables and body composition traits of pregnant beef cows differing in phenotypic residual feed intake. J. Anim. Sci. 92: 2170–2181.

Fitzsimons, C., D.A. Kenny, S.M. Waters, B. Earley, and M. McGee. 2014c. Effects of phenotypic residual feed intake on response to an *in-vivo* glucose tolerance test and gene expression in the insulin signalling pathway in beef cattle. J. Anim. Sci. 92: 4616–4631.

Florini, J.R., D.Z. Ewton, and S.A. Coolocan. 1996. Growth hormone and the insulin-like growth factor system in myogenesis. Endocr. Rev. 17: 481–517.

Fonseca, L., D. Gimenez, M. Mercadante, S. Bonilha, J. Ferro, F. Baldi, F. De Souza, and L. De Albuquerque. 2015. Expression of genes related to mitochondrial function in Nellore cattle divergently ranked on residual feed intake. Mol. Biol. Rep. 42: 1–7.

Foote, A.P., K.E. Hales, C.A. Lents, and H.C. Freetly. 2014. Association of circulating active and total ghrelin concentrations with dry matter intake, growth, and carcass characteristics of finishing beef cattle. J. Anim. Sci. 92: 5651–5658.

Foote, A.P., K.E. Hales, R.G. Tait, E.D. Berry, C.A. Lents, J.E. Wells, A.K. Lindholm-Perry, and H.C. Freetly. 2016. Relationship of glucocorticoids and hematological measures with feed intake, growth, and efficiency of finishing beef cattle. J. Anim. Sci. 94: 275–283.

Foster, D.L., and S. Nagatani. 1999. Physiological perspectives on leptin as a regulator of reproduction: Role in timing puberty. Biol. Reprod. 60: 205–215.

Gill, M. 2013. Converting feed into human food: the multiple dimensions of efficiency. pp. 1–13. *In*: H.P.S. Makkar, and D. Beever (eds.). Optimization of feed use efficiency in ruminant production systems. Proc. FAO Symposium, 27 November 2012, Bangkok, Thailand. FAO Anim. Prod. Health Proc. No. 16. Rome, FAO and Asian-Aust. Assoc. Anim. Prod. Soc.

Gomes, R.C., R.D. Sainz, S.L. Silva, M.C. César, M.N. Bonin, and P.R. Leme. 2012. Feedlot performance, feed efficiency reranking, carcass traits, body composition, energy requirements, meat quality and calpain system activity in Nellore steers with low and high residual feed intake. Livest. Sci. 150: 265–273.

Gomes, R.D.C., R.D. Sainz, and P.R. Leme. 2013. Protein metabolism, feed energy partitioning, behavior patterns and plasma cortisol in Nellore steers with high and low residual feed intake. R. Bras. Zootec. 42: 44–50.

Gonzaga Neto, S., L.R. Bezerra, A.N. Medeiros, M.A. Ferreira, E.C. Pimenta Filho, E.P. Candido, and R.L. Oliveira. 2011. Feed restriction and compensatory growth in Guzera females. Asian-Aust. J. Anim. Sci. 24: 791–799.

Grimaud, P., D. Richard, A. Kanwe, C. Durier, and M. Doreau. 1998. Effect of undernutrition and refeeding on digestion in *Bos Taurus* and *Bos Indicus* in a tropical environment. Anim. Sci. 67: 49–58.

Guan, L.L., J.D. Nkrumah, J.A. Basarab, and S.S. Moore. 2008. Linkage of microbial ecology to phenotype: Correlation of rumen microbial ecology to cattle's feed efficiency. FEMS Microbiol. Letters 288: 85–91.

Han, C., M. Vinsky, N. Aldai, M.E.R. Dugan, T.A. McAllister, and C. Li. 2013. Association analyses of DNA polymorphisms in bovine SREBP-1, LXRα, FADS1 genes with fatty acid composition in Canadian commercial crossbred beef steers. Meat Sci. 93: 429–436.

Hayashida, T., K. Murakami, K. Mogi, M. Nishihara, M. Nakazato, M.S. Mondal, Y. Horii, M. Kojima, K. Kangawa, and N. Murakami. 2001. Ghrelin in domestic animals: Distribution in stomach and its possible role. Domest. Anim. Endocrinol. 21: 17–24.

Hayes, B.J., P.J. Bowman, A. Chamberlain, and M.E. Goddard. 2009. Invited Review: Genomic selection in dairy cattle: Progress and challenges. J. Dairy Sci. 92: 433–443.

Henderson, G.F., S. Cox, A. Ganesh, A. Jonker, W. Young, and P.H. Janssen. 2015. Rumen microbial community composition varies with diet and host, but a core microbiome is found across a wide geographical range. Nature Scientific Reports 5: 14567.

Herd, R.M., V.H. Oddy, and E.C. Richardson. 2004. Biological basis for variation in residual feed intake in beef cattle. 1. Review of potential mechanisms. Aust. J. Exper. Agric. 44: 423–430.

Herd, R.M., and P.F. Arthur. 2009. Physiological basis for residual feed intake. J. Anim. Sci. 87: E64–71.

Hernandez-Sanabria, E., L.A. Goonewardene, Z. Wang, O.N. Durunna, S. Moore, and L.L. Guan. 2012. Impact of feed efficiency and diet on adaptive variations in the bacterial community in the rumen fluid of cattle. Appl. Environ. Microbiol. 78: 1203.

Hill, R.A., R.A. Hunter, D.B. Lindsay, and P.C. Owens. 1999. Action of Long (R3)-IGF-1 on protein metabolism in beef heifers. Domest. Anim. Endocrinol. 16: 219–229.

Hills, J.L., W.J. Wales, F.R. Dunshea, S.C. Garcia, and J.R. Roche. 2015. Invited review: An evaluation of the likely effects of individualized feeding of concentrate supplements to pasture-based dairy cows. J. Dairy Sci. 98: 1363–1401.

Hocquette, J.F., I. Ortigues-Marty, D. Pethick, P. Herpin, and X. Fernandez. 1998. Nutritional and hormonal regulation of energy metabolism in skeletal muscles of meat-producing animals. Livest. Prod. Sci. 56: 115–143.

Hocquette, J.F., P. Bas, D. Bauchart, M. Vermorel, and Y. Geay. 1999. Fat partitioning and biochemical characteristics of fatty tissues in relation to plasma metabolites and hormones in normal and double-muscled young growing bulls. Comp. Biochem. Physiol. Part A: Mol. Integr. Physiol. 122: 127–138.

Hornick, J.L., C. Van Eenaeme, A. Clinquart, M. Diez, and L. Istasse. 1998a. Different periods of feed restriction before compensatory growth in Belgian Blue bulls: I. animal performance, nitrogen balance, meat characteristics, and fat composition. J. Anim. Sci. 76: 249–259.

Hornick, J.L., C. Van Eenaeme, M. Diez, V. Minet, and L. Istasse. 1998b. Different periods of feed restriction before compensatory growth in Belgian Blue bulls: II. Plasma metabolites and hormones. J. Anim. Sci. 76: 260–271.

Hornick, J.L., C. Van Eenaeme, O. Gerard, I. Dufrasne, and L. Istasse. 2000. Mechanisms of reduced and compensatory growth. Domest. Anim. Endocrinol. 19: 121–132.

Horton, J.D., J.L. Goldstein, and M.S. Brown. 2002. SREBPs: activators of the complete program of cholesterol and fatty acid synthesis in the liver. J. Clin. Invest. 109: 1125–1131.

Huhtanen, P., S. Ahvenjärvi, M.R. Weisbjerg, and P.N. Ørgaard. 2006. Digestion and passage of fibre in ruminants. pp. 87–138. *In*: K. Sejrsen, T. Hvelplund, and M.O. Nielson (eds.). Ruminant Physiology: Digestion, Metabolism and Impact of Nutrition on Gene Expression, Immunology and Stress. Wageningen Acad. Publ., Wageningen, The Netherlands.

Istasse, L., C. Van Eenaeme, P. Evrard, A. Gabriel, P. Baldwin, G. Maghuin-Rogister, and J.M. Bienfait. 1990. Animal performance, plasma hormones and metabolites in Holstein and Belgian Blue growing-fattening bulls. J. Anim. Sci. 68: 2666–2673.

Janssen, P.H. 2010. Influence of hydrogen on rumen methane formation and fermentation balances through microbial growth kinetics and fermentation thermodynamics. Anim. Feed Sci. Technol. 160: 1–22.

Jessen, N., C.B. Djurhuus, J.O.L. Jørgensen, L.S. Jensen, N. Møller, S. Lund, and O. Schmitz. 2005. Evidence against a role for insulin-signaling proteins PI 3-kinase and Akt in insulin resistance in human skeletal muscle induced by short-term GH infusion. Am. J. Physiol. Endocrinol. Met. 288: E194–E199.

Johnson, D.E., K.A. Johnson, and R.L. Baldwin. 1990. Changes in liver and gastrointestinal tract energy demands in response to physiological workload in ruminants. J. Nutr. 120: 649–655.

Johnston, D.J., R.M. Herd, M.J. Kadel, H.U. Graser, P.F. Arthur, and J.A. Archer. 2002. Evidence of IGF-I as a genetic predictor of feed efficiency traits in beef cattle. pp. 0–4. Proc. 7th World Congress Genetics Applied to Livestock Production, Montpellier, France.

Karisa, B.K., J. Thomson, Z. Wang, P. Stothard, S.S. Moore, and G.S. Plastow. 2013. Candidate genes and single nucleotide polymorphisms associated with variation in residual feed intake in beef cattle. J. Anim. Sci. 91: 3502–3513.

Karisa, B., S. Moore, and G. Plastow. 2014. Analysis of biological networks and biological pathways associated with residual feed intake in beef cattle. Anim. Sci. J. 85: 374–387.

Keady, S.M. 2011. Examination of the expression of genes and proteins controlling M. longissimus thorais et lumborum growth in steers. PhD Diss. Natl. Univ. Ireland, Maynooth.

Kelly, A.K., M. McGee, D.H. Crews, A.G. Fahey, A.R. Wylie, and D.A. Kenny. 2010a. Effect of divergence in residual feed intake on feeding behavior, blood metabolic variables, and body composition traits in growing beef heifers. J. Anim. Sci. 88: 109–123.

Kelly, A.K., M. McGee, D.H. Crews, T. Sweeney, T.M. Boland, and D.A. Kenny. 2010b. Repeatability of feed efficiency, carcass ultrasound, feeding behavior, and blood metabolic variables in finishing heifers divergently selected for residual feed intake. J. Anim. Sci. 88: 3214–3225.

Kelly, A.K., M. McGee, D.H. Crews, C.O. Lynch, A.R. Wylie, R.D. Evans, and D.A. Kenny. 2011a. Relationship between body measurements, metabolic hormones, metabolites and residual feed intake in performance tested pedigree beef bulls. Livest. Sci. 135: 8–16.

Kelly, A.K., S.M. Waters, M. McGee, R.G. Fonseca, C. Carberry, and D.A. Kenny. 2011b. mRNA expression of genes regulating oxidative phosphorylation in the muscle of beef cattle divergently ranked on residual feed intake. Physiol. Genomics 43: 12–23.

Kelly, A.K., S.M. Waters, M. McGee, J.A. Browne, D.A. Magee, and D.A. Kenny. 2013. Expression of key genes of the somatotorpic axis in *longissimus dorsi* muscle of beef heifers phenotypically divergent for residual feed intake. J. Anim. Sci. 91: 159–167.

Kelly, A.K., B. Earley, M. McGee, A.G. Fahey, and D.A. Kenny. 2016. Endocrine and hematological responses of beef heifers divergently ranked for residual feed intake following a bovine corticotrophin-releasing hormone challenge. J. Anim. Sci. 94: 1703–1711.

Keogh, K., S.M. Waters, A.K. Kelly, and D.A. Kenny. 2015a. Feed restriction and subsequent re-alimentation in Holstein Friesian bulls: I. Effect on animal performance; muscle, fat and linear body measurements; and slaughter characteristics. J. Anim. Sci. 93: 3578–3589.

Keogh, K., S.M. Waters, A.K. Kelly, A.R.G. Wylie, H. Sauerwein, T. Sweeney, and D.A. Kenny. 2015b. Feed restriction and subsequent re-alimentation in Holstein Friesian bulls: II. Effect on blood pressure and systemic concentrations of metabolites and metabolic hormones. J. Anim. Sci. 93: 3590–3601.

Keogh, K., D.A. Kenny, A.K. Kelly, and S.M. Waters. 2015c. Insulin secretion and signalling in response to dietary restriction and subsequent re-alimentation in cattle. Physiol. Genomics 47: 344–354.

Keogh, K., S.M. Waters, A.K. Kelly, A.R.G. Wylie, and D.A. Kenny. 2015d. Effect of feed restriction and subsequent re-alimentation on hormones and genes of the somatotropic axis in cattle. Physiol. Genomics 47: 264–273.

Keogh, K., D.A. Kenny, P. Cormican, A.K. Kelly, and S.M. Waters. 2016a. Effect of dietary restriction and subsequent re-alimentation on the transcriptional profile of hepatic tissue in cattle. BMC Genomics. 17: 244.

Keogh, K., S.M. Waters, P. Cormican, A.K. Kelly, E. O'Shea and D.A. Kenny. 2016b. Transcriptional profile of bovine rumen papillae in response to diet restriction and re-alimentation. In proceedings of the European Association of Animal Production meeting, 29th of August to 2nd of September, Belfast, United Kingdom. p 550.

Keogh, K., S.M. Waters, P. Cormican, A.K. Kelly and D.A. Kenny. 2016c. Response of bovine jejunal transcriptome to dietary restriction and subsequent compensatory growth. In proceedings of the European Association of Animal Production meeting, 29th of August to 2nd of September, Belfast, United Kingdom. p 185.

Keogh, K., D.A. Kenny, P. Cormican, M. McCabe, A.K. Kelly, and S.M. Waters. 2016d. Effect of dietary restriction and subsequent re-alimentation on the transcriptional profile of bovine skeletal muscle. PLoS One 11: e0149373.

Kittelmann, S., C.S. Pinares-Patino, H. Seedorf, M.R. Kirk, G. Ganesh, J.C. McEwan, and P.H. Janssen. 2014. Two different bacterial community types are linked with the low-methane emission trait in sheep. PLoS One 9: e103171.

Knott, S.A., L.J. Cummins, F.R. Dunshea, and B.J. Leury. 2010. Feed efficiency and body composition are related to cortisol response to adrenocorticotropin hormone and insulin-induced hypoglycemia in rams. Domest. Anim. Endocrinol. 39: 137–146.

Kolath, W.H., M.S. Kerley, J.W. Golden, and D.H. Keisler. 2006a. The relationship between mitochondrial function and residual feed intake in Angus steers. J. Anim. Sci. 84: 861–865.

Kolath, W.H., M.S. Kerley, J.W. Golden, S.A. Shahid, and G.S. Johnson. 2006b. The relationships among mitochondrial uncoupling protein 2 and 3 expression, mitochondrial deoxyribonucleic acid single nucleotide polymorphisms, and residual feed intake in Angus steers. J. Anim. Sci. 84: 1761–1766.

Krehbiel, C.R. 2014. Invited Review: Applied nutrition of ruminants: Fermentation and digestive physiology. Prof. Anim. Scientist 30: 129–139.

Krueger, W.K., G.E. Carstens, R.R. Gomez, B.M. Bourg, P.A. Lancaster, L.J. Slay, J.C. Miller, R.C. Anderson, S.M. Horrocks, N.A. Kreuger, and T.D.A. Forbes. 2009. Relationships between residual feed intake and apparent nutrient digestibility, *in vitro* methane producing activity and VFA concentrations in growing Brangus heifers. J. Anim. Sci. 87(E-Suppl. 2): 129 (Abstr.).

Lancaster, P.A., G.E. Carstons, F.R.B. Ribero, M.E. Davis, J.G. Lyons, and T.H. Welsh. 2008. Effects of divergent selection for serum IGF-1 concentration on performance, feed efficiency and ultrasound measures of carcass composition traits in Angus bulls and heifers. J. Anim. Sci. 83: 2862–2871.

Lancaster, P.A., G.E. Carstens, F.R.B. Ribeiro, L.O. Tedeschi, and D.H. Crews. 2009. Characterization of feed efficiency traits and relationships with feeding behavior and ultrasound carcass traits in growing bulls. J. Anim. Sci. 87: 1528–1539.

Lancaster, P.A., G.E. Carstens, J.J. Michal, K.M. Brennan, K.A. Johnson, and M.E. Davis. 2014. Relationships between residual feed intake and hepatic mitochondrial function in growing beef cattle. J. Anim. Sci. 92: 3134–3141.

Lawrence, P., D.A. Kenny, B. Earley, D.H. Crews, and M. McGee. 2011. Grass silage intake, rumen and blood variables, ultrasonic and body measurements, feeding behavior and activity in pregnant beef heifers differing in phenotypic residual feed intake. J. Anim. Sci. 89: 3248–3261.

Lawrence, P., D.A. Kenny, B. Earley, and M. McGee. 2012. Grazed grass herbage intake and performance of beef heifers with predetermined phenotypic residual feed intake classification. Animal 6: 1648–1661.

Lehninger, A.L., D.L. Nelson, and M.M. Cox. 1993. Principles of Biochemistry. Worth Publishers, New York, NY.

Lenard, N.R., and H.R. Berthoud. 2008. Central and peripheral regulation of food intake and physical activity: Pathways and genes. Obesity (Silver Spring, Md). 16: S11–S22.

Levesque, C.L., L. Skinner, J. Zhu, and C.F.M. De Lange. 2015. Dynamic changes in digestive capability may contribute to compensatory growth following a nutritional insult in newly weaned pigs. J. Anim. Sci. 90: 236–238.

Li, Z.H., S.K. Kang, Y.C. Jin, Q.K. Zhang, Z.S. Hong, H. ThidarMyint, H. Kuwayama, J.S. Lee, S.B. Lee, T. Wang, E.J. Kim, Y.J. Choi, and H.G. Lee. 2012. Responses to administration of growth hormone releasing hormone and glucose in steers receiving stair-step and extended restriction on feeding. Livest. Sci. 150: 229–235.

Lin, J., C. Handschin, and B.M. Spiegelman. 2005. Metabolic control through the PGC-1 family of transcription coactivators. Cell Metabolism 1: 361–370.

Lines, D.S., W.S. Pitchford, C.D.K. Bottema, R.M. Herd, and V.H. Oddy. 2014. Selection for residual feed intake affects appetite and body composition rather than energetic efficiency. Anim. Prod. Sci. http://dx.doi.org/10.1071/AN13321.

Lucy, M.C. 2004. Mechanisms linking the somatotropic axis with insulin: Lessons from the postpartum dairy cow. Proc. NZ Soc. Anim. Prod. 64: 19–23.

Maddux, B.A., A. Chan, E.A. De Filippis, L.J. Mandarino, and I.D. Goldfine. 2006. IGF-binding protein-1 levels are related to insulin-mediated glucose disposal and are a potential serum marker of insulin resistance. Diabetes Care 29: 1535–1537.

Mader, C.J., Y.R. Montanholi, Y.J. Wang, S.P. Miller, I.B. Mandell, B.W. McBride, and K.C. Swanson. 2009. Relationships among measures of growth performance and efficiency with carcass traits, visceral organ mass, and pancreatic digestive enzymes in feedlot cattle. J. Anim. Sci. 87: 1548–1557.

Mao, F., L. Chen, M. Vinsky, E. Okine, Z. Wang, J. Basarab, D.H. Crews, and C. Li. 2013. Phenotypic and genetic relationships of feed efficiency with growth performance, ultrasound, and carcass merit traits in Angus and Charolais steers. J. Anim. Sci. 91: 2067–2076.

Marks, H., D.A. Kenny, M. McGee, C. Fitzsimons and S. Waters. 2014. Effect of residual feed intake phenotype on the expression of genes regulating oxidative phosphorylation in the liver and muscle of beef cattle. In proceedings of the British Society of Animal Science meeting, 29th to 30th of April, Nottingham, United Kingdom. p 94.

McBride, B.W., and J.M. Kelly. 1990. Energy cost of absorption and metabolism in the ruminant gastrointestinal tract and liver: A review. J. Anim. Sci. 68: 2997–3010.

McCabe, M.S., P. Cormican, K. Keogh, A. O'Connor, E. O'Hara, A.R. Palladino, D.A. Kenny, and S.M. Waters. 2015. Illumina MiSeq phylogenetic amplicon sequencing shows a large reduction of an uncharacterised *Succinivibrionaceae* and an increase of the *Methanobrevibacter gottschalkii* clade in feed restricted cattle. PLoS One. 10: e0133234.

McCann, J.C., L.M. Wiley, T.D. Forbes, F.M. Rouquette, and L.O. Tedeschi. 2014. Relationship between the rumen microbiome and residual feed intake-efficiency of Brahman bulls stocked on Bermudagrass pastures. PLoS One 9: 3.

McDonagh, M.B., R.M. Herd, E.C. Richardson, V.H. Oddy, J.A. Archer, and P.F. Arthur. 2001. Meat quality and the calpain system of feedlot steers following a single generation of divergent selection for residual feed intake. Aust. J. Exper. Agric. 41: 1013–1021.

McDonnell, R.P., K. Hart, T.M. Boland, A.K. Kelly, M. McGee, and D.A. Kenny. 2016. Effect of divergence in phenotypic residual feed intake on methane emissions, ruminal fermentation and apparent whole-tract digestibility of beef heifers across three contrasting diets. J. Anim. Sci. 94: 1179–1193.

McGee, M., M.J. Drennan, and P.J. Caffrey. 2005. Effect of suckler cow genotype on energy requirements and performance in winter and subsequently at pasture. Ir. J. Agric. Food Res. 44: 157–171.

McGee, M., C.M. Welch, J.B. Hall, W. Small, and R.A. Hill. 2013. Evaluation of Wagyu for residual feed intake: Optimizing feed efficiency, growth, and marbling in Wagyu cattle. Prof. Anim. Scientist 29: 51–56.

McGee, M., J.A. Ramirez, G.E. Carstens, W.J. Price, J.B. Hall, and R.A. Hill. 2014. Relationships of feeding behaviors with efficiency in RFI-divergent Japanese Black cattle. J. Anim. Sci. 92: 3580–3590.

McKenna, C.E., K. Keogh, S. Waters, D. Kenny, R. Porter and A. Kelly. 2016. Relationship between mitochondria and compensatory growth in Holstein Friesian bulls. In proceedings of the European Association of Animal Production meeting, 29th of August to 2nd of September, Belfast, United Kingdom. p 244.

McNeill, D.M. 2013. Forages for ruminants, cereals for human food and fuel. pp. 15–32. In: H.P.S. Makkar, and D. Beever (eds.). Optimization of Feed use Efficiency in Ruminant Production Systems. Proc. FAO Symposium, 27 November 2012, Bangkok, Thailand. FAO Anim. Prod. Health Proc., No. 16. Rome, FAO and Asian-Aust. Assoc. Anim. Prod. Soc.

Meuwissen, T.H., B.J. Hayes, and M.E. Goddard. 2001. Prediction of total genetic value using genome-wide dense marker maps. Genetics 157: 1819–1829.

Meyer, A.M., B.W., Hess, S.I. Paisley, M. Du, and J.S. Caton. 2014. Small intestinal growth measures are correlated with feed efficiency in market weight cattle, despite minimal effects of maternal nutrition during early to midgestation. J. Anim. Sci. 92: 3855–3867.

Minton, J.E. 1994. Function of the hypothalamic-pituitary-adrenal axis and the sympathetic nervous system in models of acute stress in domestic farm animals. J. Anim. Sci. 72: 1891–1898.

Møller, N., and H. Nørrelund. 2003. The role of growth hormone in the regulation of protein metabolism with particular reference to conditions of fasting. Horm. Res. Paediatr. 59: 62–68.

Montanholi, Y.R., K.C. Swanson, R. Palme, F.S. Schenkel, B.W. McBride, D. Lu, and S.P. Miller. 2010. Assessing feed efficiency in beef steers through feeding behavior, infrared thermography and glucocorticoids. Animal 4: 692–701.

Montanholi, Y., A. Fontoura, K. Swanson, B. Coomber, S. Yamashiro, and S. Miller. 2013. Small intestine histomorphometry of beef cattle with divergent feed efficiency. Acta Veterinaria Scandinavica 55: 9.

Moore, K.L., D.J. Johnston, H. Graser, and R. Herd. 2005. Genetic and phenotypic relationships between insulin-like growth factor-I (IGF-I) and net feed intake, fat, and growth traits in Angus beef cattle. Aust. J. Agric. Res. 56: 211–218.

Moss, A.R., J.P. Jouany, and J. Newbold. 2000. Methane production by ruminants: Its contribution to global warming. Ann. Zootech. 49: 231–253.

Nascimento, C.F., R.H. Branco, S.F.M. Bonilha, J.N.S. Cyrillo, J.A. Negrão, and M.E.Z. Mercadante. 2015. Residual feed intake and blood variables in young Nellore cattle. J. Anim. Sci. 93: 1318–1326.

National Research Council (NRC). 2000. Energy. pp. 3–15. *In*: Nutrient Requirements for Beef Cattle. National Academy Press, Washington DC.

Nielsen, M.K., M.D. MacNeil, J.C.M. Dekkers, D.H. Crews, T.A. Rathje, R.M. Enns, and R.L. Weaber. 2013. Review: Life-cycle, total-industry genetic improvement of feed efficiency in beef cattle: Blueprint for the Beef Improvement Federation. Professional Anim. Scientist 29: 559–565.

Niemann, H., B. Kuhla, and G. Flachowsky. 2011. Perspectives for feed-efficient animal production. J. Anim. Sci. 89: 4344–4363.

Nkrumah, J.D., C. Li, J. Yu, C. Hansen, D.H. Keisler, and S.S. Moore. 2005. Polymorphisms in the bovine leptin promoter associated with serum leptin concentration, growth, feed intake, feeding behavior, and measures of carcass merit. J. Anim. Sci. 83: 20–28.

Nkrumah, J.D., E.K. Okine, G.W. Mathison, K. Schmid, C. Li, J.A. Basarab, M.A. Price, Z. Wang, and S.S. Moore. 2006. Relationships of feedlot feed efficiency, performance, and feeding behavior with metabolic rate, methane production, and energy partitioning in beef cattle. J. Anim. Sci. 84: 145–153.

Novakofski, J., and R.H. McCusker. 1993. Physiology and principles of muscle growth. pp. 33–48. *In*: G.R. Hollis (ed.). Growth of the Pig. CAB International, Wallingford, UK.

Nozière, P., I. Ortigues-Marty, C. Loncke, and D. Sauvant. 2010. Carbohydrate quantitative digestion and absorption in ruminants: From feed starch and fibre to nutrients available for tissues. Animal 4: 1057–1074.

Oddy, H., and P.C. Owens. 1996. Insulin-like growth factor I inhibits degradation and improves retention of protein in hindlimb muscle of lambs. Am. J. Physiol. 271: 973–982.

Oddy, V.H. 1999. Genetic variation in protein metabolism and implications for variation in efficiency of growth. Recent Advances in Animal Nutrition in Australia. 12: 23–29.

Oksbjerg, N., F. Gondret, and M. Vestergaard. 2004. Basic principles of muscle development and growth in meat-producing mammals as affected by the insulin-like growth factor (IGF) system. Domes. Anim. Endocrinol. 27: 219–240.

Onteru, S.K., D.M. Gorbach, J.M. Young, D.J. Garrick, J.C. Dekkers, and M.F. Rothschild. 2013. Whole genome association studies of residual feed intake and related traits in the pig. PLoS One 26: e61756.

O'Shea, E., S.M. Waters, A.K. Kelly, K. Keogh, and D.A. Kenny. 2016. Examination of the molecular control of ruminal epithelial function in response to dietary restriction and subsequent compensatory growth in cattle. J. Anim. Sci. Biotech. 7: 53.

Ortigues, I., and M. Doreau. 1995. Responses of the splanchnic tissues of ruminants to changes in intake: Absorption of digestion end products, tissue mass, metabolic activity and implications to whole animal energy metabolism. Ann. Zootech. 44: 321–346.

Owens, F.N., S. Qi, and D.A. Sapienza. 2014. Invited Review: Applied protein nutrition of ruminants—Current status and future directions. Prof. Anim. Scientist 30: 150–179.

Paradis, F., S. Yue, J.R. Grant, P. Stothard, J.A. Basarab, and C. Fitzsimmons. 2015. Transcriptomic analysis by RNA sequencing reveals that hepatic interferon-induced genes may be associated with feed efficiency in beef heifers. J. Anim. Sci. 93: 3331–3341.

Perkins, S.D., C.N. Key, C.F. Garrett, C.D. Foradori, C.L. Bratcher, L.A. Kriese-Anderson, and T.D. Brandebourg. 2014. Residual feed intake studies in Angus-sired cattle reveal a potential role for hypothalamic gene expression in regulating feed efficiency. J. Anim. Sci. 92: 549–560.

Philippou, A., M. Maridaki, A. Halapas, and M. Koutsilieris. 2007. The role of the insulin-like growth factor 1 (IGF-1) in skeletal muscle physiology. *In Vivo* 21: 45–54.

Pitta, D., W. Pinchak, S. Dowd, J. Osterstock, V. Gontcharova, E. Youn, K. Dorton, I. Yoon, B.R. Min, J.D. Fulford, T.A. Wickersham, and D.P. Malinowski. 2010. Rumen bacterial diversity dynamics associated with changing from bermudagrass hay to grazed winter wheat diets. Microb. Ecol. 59: 511–522.

Pliquett, R.U., D. Fuhrer, S. Falk, S. Zysset, D.Y. Von Cramon, and M. Stumvoll. 2006. The effects of insulin on the central nervous system—focus on appetite regulation. Horm. Metab. Res. 38: 442–446.

Ramos, M.H. 2011. Mitochondrial complex I protein is related to residual feed intake in beef cattle. PhD Diss. Univ. Missouri-Columbia.

Ramos, M.H., and M.S. Kerley. 2013. Mitochondrial complex I protein differs among residual feed intake phenotype in beef cattle. J. Anim. Sci. 91: 3299–3304.

Randel, R.D. and T.H. Welsh. 2013. Joint alpharma-beef species symposium: Interactions of feed efficiency with beef heifer reproductive development. J. Anim. Sci. 91: 1323–1328.

Redden, R.R., R.B. McCosh, R.W. Kott, and J.G. Berardinelli. 2011. Effect of residual feed intake on temporal patterns of glucose, insulin, and NEFA concentrations after a glucose challenge in Targhee ewes. Proc. West. Sect. Am. Soc. Anim. Sci. 62: 237–240.

Reynolds, C.K., L.A. Crompton, and J.A.N. Mills. 2011. Improving the efficiency of energy utilisation in cattle. Anim. Prod. Sci. 51: 6–12.

Richardson, E.C., R.M. Herd, P.F. Arthur, J. Wright, G. Xu, K. Dibley, and H. Oddy. 1996. Possible physiological indicators for net feed conversion efficiency. Proc. Aust. Soc. Anim. Prod. 21: 103–106.

Richardson, E.C., R.M. Herd, V.H. Oddy, J.M. Thompson, J.A. Archer, and P.F. Arthur. 2001. Body composition and implications for heat production of Angus steer progeny of parents selected for and against residual feed intake. Aust. J. Exper. Agric. 41: 1065–1072.

Richardson, E.C., R.M. Herd, I.G. Colditz, J.A. Archer, and P.F. Arthur. 2002. Blood cell profiles of steer progeny from parents selected for and against residual feed intake. Aust. J. Exper. Agric. 42: 901–908.

Richardson, E.C., and R.M. Herd. 2004. Biological basis for variation in residual feed intake in beef cattle. 2. Synthesis of results following divergent selection. Aust. J. Exper. Agric. 44: 431–440.

Richardson, E.C., R.M. Herd, J.A. Archer, and P.F. Arthur. 2004. Metabolic differences in Angus steers divergently selected for residual feed intake. Aust. J. Exp. Agric. 44: 441–452.

Rius, A.G., S. Kittelmann, K.A. Macdonald, G.C. Waghorn, P.H. Janssen, and E. Sikkema. 2012. Nitrogen metabolism and rumen microbial enumeration in lactating cows with divergent residual feed intake fed high-digestibility pasture. J. Dairy Sci. 95: 5024–5034.

Roche, J.R., D. Blache, J.K. Kay, D.R. Miller, A.J. Sheahan, and D.W. Miller. 2008. Neuroendocrine and physiological regulation of intake with particular reference to domesticated ruminant animals. Nutr. Res. Rev. 21: 207–234.

Rolf, M.M., J.F. Taylor, R.D. Schnabel, S.D. McKay, M.C. McClure, S.L. Northcutt, M.S. Kerley, and R.L. Weaber. 2011. Genome-wide association analysis for feed efficiency in Angus cattle. Anim. Genet. 43: 367–374.

Rompala, R.E., S.D.M. Jones, J.G. Buchanan-Smith, and H.S. Bayley. 1985. Feedlot performance and composition of gain in late-maturing steers exhibiting normal and compensatory growth. J. Anim. Sci. 61: 637–646.

Ryan, W.J. 1990. Compensatory growth in cattle and sheep. Nutr. Abstr. Rev. 60: 653–664.

Ryan, W.J., I.H. Williams, and R.J. Moir. 1993. Compensatory growth in sheep and cattle II. Changes in body composition and tissue weights. Aust. J. Agric. Res. 44: 1623–1633.

Savietto, D., D.P. Berry, and N.C. Friggens. 2014. Towards an improved estimation of the biological components of residual feed intake in growing cattle. J. Anim. Sci. 467–476.

Schiaffino, S., and C. Mammucari. 2011. Regulation of skeletal muscle growth by the IGF1-Akt/PKB pathway: Insights from genetic models. Skeletal Muscle 1: 4.

Schiavon, S., and G. Bittante. 2012. Double-muscled and conventional cattle have the same net energy requirements if these are related to mature and current body protein mass, and to gain composition. J. Anim. Sci. 90: 3973–3987.

Seal, C.J., and C.K. Reynolds. 1993. Nutritional implications of gastrointestinal and liver metabolism in ruminants. Nutr. Res. Rev. 6: 185–208.

Serao, N., D. Gonzalez-Pena, J. Beever, D. Faulkner, B. Southey, and S. Rodriguez-Zas. 2013. Single nucleotide polymorphisms and haplotypes associated with feed efficiency in beef cattle. BMC Genet. 14: 94.

Shafer, G.L. 2011. Insulin sensitivity in tropically adapted cattle with divergent residual feed intake. Master's Diss. Texas A&M Univ.

Sherman, E.L., J.D. Nkrumah, B.M. Murdoch, C. Li, Z. Wang, A. Fu, and S.S. Moore. 2008. Polymorphisms and haplotypes in the bovine neuropeptide Y, growth hormone receptor, ghrelin, insulin-like growth factor 2, and uncoupling proteins 2 and 3 genes and their associations with measures of growth, performance, feed efficiency, and carcass merit in beef cattle. J. Anim. Sci. 86: 1–16.

Smuts, M., H.H. Meissner, and P.B. Cronje. 1995. Retention time of digesta in the rumen: Its repeatability and relationship with wool production of Merino rams. J. Anim. Sci. 73: 206–210.

Snelling, W.M., R.A. Cushman, M.R. Fortes, A. Reverter, G.L. Bennett, J.W. Keele, L.A. Kuehn, T.G. McDaneld, R.M. Thallman, and M.G. Thomas. 2012. Physiology and endocrinology symposium: How single nucleotide polymorphism chips will advance our knowledge of factors controlling puberty and aid in selecting replacement beef females. J. Anim. Sci. 90: 1152–1165.

Snelling, W.M., R.A. Cushman, J.W. Keele, C. Maltecca, M.G. Thomas, M.R. Fortes, and A. Reverter. 2013. Networks and pathways to guide genomic selection. J. Anim. Sci. 91: 537–552.

Sonntag, W.E., C.D. Lynch, W.T. Cefalu, R.L. Ingram, S.A. Bennett, P.L. Thornton, and A.S. Khan. 1999. Pleiotropic effects of growth hormone and insulin-like growth factor (IGF)-1 on biological aging: Inferences from moderate caloric-restricted animals. J. Gerontol. A Biol. Sci. Med. Sci. 54: B521–538.

Steinert, R.E., C. Feinle-Bisset, N. Geary, and C. Beglinger. 2013. Digestive physiology of the pig symposium: Secretion of gastrointestinal hormones and eating control. J. Anim. Sci. 91: 1963–1973.

Strat, A.L., T.A. Kokta, M.V. Dodson, A. Gertler, Z. Wu, and R.A. Hill. 2005. Early signaling interactions between the insulin and leptin pathways in bovine myogenic cells. Biochimica et Biophysica Acta 1744: 165–174.

Sun, Z.H., Z.X. He, Q.L. Zhang, Z.L. Tan, X.F. Han, S.X. Tang, C.S. Zhou, M. Wang, and Q.X. Yan. 2013. Effects of energy and protein restriction, followed by nutritional recovery on morphological development of the gastrointestinal tract of weaned kids. J. Anim. Sci. 91: 4336–4344.

Susenbeth, A., R. Mayer, B. Koehler, and O. Neumann. 1998. Energy requirement for eating in cattle. J. Anim. Sci. 76: 2701–2705.

Susenbeth, A., T. Dickel, K.H. Südekum, W. Drochner, and H. Steingaß. 2004. Energy requirements of cattle for standing and for ingestion, estimated by a ruminal emptying technique. J. Anim. Sci. 82: 129–136.

Taniguchi, C.M., K. Ueki, and R. Kahn. 2005. Complementary roles of IRS-1 and IRS-2 in the hepatic regulation of metabolism. J. Clin. Invest. 115: 718–727.

Thornhill, J.B., L.C. Marett, M.J. Auldist, J.S. Greenwood, J.E. Pryce, B.J. Hayes, and W.J. Wales. 2014. Whole-tract dry matter and nitrogen digestibility of lactating dairy cows selected for phenotypic divergence in residual feed intake. Anim. Prod. Sci. 54: 1460–1464.

Tizioto, P.C., L.L. Coutinho, J.E. Decker, R.D. Schnabel, K.O. Rosa, P.S. Oliveira, M.M. Souza, G.B. Mourão, R.R. Tullio, A.S. Chaves, D.P.D. Lanna, A. Zerlotini-Neto, M.A. Mudadu, J.F. Taylor, and L.C.A. Regitano. 2015. Global liver gene expression differences in Nelore steers with divergent residual feed intake phenotypes. BMC Genomics 16: 1–14.

Van Nevel, C.J., and D.I. Demeyer. 1996. Influence of antibiotics and a deaminase inhibitor on volatile fatty acids and methane production from detergent washed hay and soluble starch by rumen microbes *in vitro*. Anim. Feed Sci. Technol. 37: 21–31.

VanRaden, P.M., C.P. Van Tassell, G.R. Wiggans, T.S. Sonstegard, R.D. Schnabel, J.F. Taylor, and F.S. Schenkel. 2009. Invited review: Reliability of genomic predictions for North American Holstein bulls. J. Dairy Sci. 92: 16–24.

Vega, R.S.A., H. Lee, H. Hidari, and H. Kuwayama. 2009. Changes in plasma hormones and metabolites during compensatory growth in Holstein steers. Vet. Anim. Sci. 35: 1–14.

Welch, C.M., M. McGee, T.A. Kokta, and R.A. Hill. 2012. Muscle and adipose tissue: Potential roles in driving variation in feed efficiency. pp. 175–198. *In*: R.A. Hill (ed.). Feed Efficiency in the Beef Industry. John Wiley & Sons, Inc. Ames, Iowa.

Welch, C.M., K.J. Thornton, G.K. Murdoch, K.C. Chapalamadugu, C.S. Schneider, J.K. Ahola, J.B. Hall, W.J. Price, and R.A. Hill. 2013. An examination of the association of serum IGF-I concentration, potential candidate genes, and fiber type composition with variation in residual feed intake in progeny of Red Angus sires divergent for maintenance energy EPD. J. Anim. Sci. 91: 5626–5636.

Wertz-Lutz, A.E., J.A. Daniel, J.A. Clapper, A. Trenkle, and D.C. Beitz. 2008. Prolonged, moderate nutrient restriction in beef cattle results in persistently elevated circulating ghrelin concentrations. J. Anim. Sci. 86: 564–575.

Yambayamba, E.S.K., M.A. Price, and S.D.M. Jones. 1996a. Compensatory growth of carcass tissues and visceral organs in beef heifers. Livest. Prod. Sci. 46: 19–32.

Yambayamba, E.S., M.A. Price, and G.R. Foxcroft. 1996b. Hormonal status, metabolic changes, and resting metabolic rate in beef heifers undergoing compensatory growth. J. Anim. Sci. 74: 57–69.

Zhang, S., R.I. Albornoz, J.R. Aschenbach, D.R. Barreda, and G.B. Penner. 2013. Short-term feed restriction impairs the absorptive function of the reticulo-rumen and total tract barrier function in beef cattle. J. Anim. Sci. 91: 1685–1695.

CHAPTER-7

Hormonal Control of Energy Substrate Utilization and Energy Metabolism in Domestic Animals

Colin G. Scanes[1] and *Rodney A. Hill*[2,*]

IIIIIIIIIIIIIIIIIIIIIIIIIIIIIIIIIIIIIIIIIIIIIIIIIIIIIIIIIIIIIIIIIIIIIIIIIIIIIIIIIIIIIIIIIIIIIIIIIIIIIIIIIIIIIIIIIIIIIIIIIIIIIIIIIIIIIIIIIIIIIIIIIIIIIIIIIIIIIIIIIIIIIIIIIIIIIIIIIIIIIIII

INTRODUCTION

In this chapter, we explore the key molecules involved in energy storage and energy utilization in domestic animals. We also describe the hormonal regulatory pathways that modulate the processing, shuttling and oxidation of these molecules in order to maintain life processes. In domestic animals that are kept for production purposes or for recreation, the most efficient use of energy substrates will have effects on the economics of animal production and on the general well-being of the animal.

As may be expected, these key pathways and the roles of the energy storage-utilization molecules are highly conserved across species. However, differences across species mean that optimizing production and animal well-being requires knowledge of nutrient utilization and also the biological processes within animals that regulate them.

Much is known about these molecules and processes. To address this topic comprehensively, several volumes would be required. The present chapter takes a snap-shot of a selection of these molecules and processes. We describe some of the recent discoveries applicable to domestic animals and take the reader through discussion of elements that we believe are

[1] University of Wisconsin, Milwaukee, Wisconsin, USA.
[2] School of Biomedical Science, Charles Sturt University, Wagga Wagga, New South Wales, Australia.
* Corresponding author

interesting, especially at the interface of the pathways we discuss. As more becomes understood about these molecules and processes, it is at the interface of energy storage and utilization pathways that deeper understanding and new discoveries are uncovered.

We begin the story in this chapter with glucose, perhaps the most described molecule involved in energy storage-utilization. This story evolves to include glycogen (a glucose polymer) and then we move on to discuss lipids and fatty acids. The third broad class of molecules in our discussion, proteins are also described in terms of their roles in these energy pathways. Along the way we also explore the interactions and shuttling between these major classes of energy molecules. Depending on the animal's energetic state and needs, each of these classes of molecules can be utilized for energy storage and for oxidation to release energy. The variation between domestic animal species is intertwined through discussion of these molecules and pathways, and aims to provide the reader with knowledge and a depth of understanding to underpin reader insights into what is presently known and to point to glimpses of new directions and discoveries in the field.

Glucose

Glucose is a major source of energy. Glucose is absorbed in the small intestine in non-ruminants mammals and birds. It is metabolized by glycolysis and the citric acid (or tricarboxylic acid cycle or Krebs cycle) to generate ATP. At times when glucose is in short supply, glucose is synthesized from glycogenic precursors, such as lactate, pyruvate and alanine, in the liver and kidneys by the process of gluconeogenesis (reviewed, e.g., Scanes, 2015).

Circulating concentrations of glucose are maintained under close homeostatic control. Without this in diabetes mellitus, circulating concentrations of glucose are very high following a glucose load (hyperglycemia) but can drop to low levels between meals (hypoglycemia). There is diabetes in domestic animals. Diabetes is found in dogs and cats with incidence rates of 0.34% in dogs (Mattin et al., 2014) and 11.6 cases of diabetes per 10,000 cat-years at risk (CYAR) (Öhlund et al., 2015). Moreover, diabetes can be induced by pancreatectomy. This approach led to the discovery of insulin (Banting et al., 1922; also see review: Best, 1945).

In ruminants, volatile fatty acids (acetate, propionate and butyrate) are the major energy sources instead of glucose. Volatile fatty acids are absorbed from the rumen (e.g., Kristensen and Harmon, 2004). In ruminants, glucose is synthesized from propionate in the liver by gluconeogenesis (e.g., Aiello et al., 1989; Kristensen and Harmon, 2004) and fatty acids from acetate.

Glucose in the circulation is critically important particularly to brain function. Brain metabolism requires glucose with glucose accounting for

Table 1: Comparison of energy sources used by different tissues in anesthetized sheep (calculated from Lindsay and Setchell 1976).

Metabolite	Utilization in metabolites (nmole g^{-1} minute^{-1})		
	Brain	Heart	Skeletal muscle
Glucose	275	266	21
Acetoacetate (ketone body)	33	371	28
3-Hydroxybutyrate	9	308	16
Acetate	10	233	21.5

over 60% of energy needs in non-ruminants (Reviewed: Berg, 2002) and over 95% of energy needs in ruminants (Table 1) (Lindsay and Setchell, 1976).

In conscious sheep, brain utilization of glucose was markedly higher (508 nmole g^{-1} minute^{-1}) (Lindsay and Setchell, 1976). Glucose is transported into neurons by the glucose transporter, GLUT3, into glial cells by GLUT5 and through the blood-brain barrier by GLUT1 (Vannucci et al., 1997; also see review: Berg, 2002). In contrast, GLUT4 transports glucose into skeletal muscle and adipose tissue.

In birds, plasma concentrations of glucose are markedly elevated compared to the situation in mammals (reviewed: Scanes, 2015). The basis for this is unclear but may be related to resistance to insulin (reviewed: Scanes, 2015).

Glucose is stored as glycogen in the liver and skeletal muscles. Glycogenesis increases at times of surplus. There are large increase in the glycogen content in the liver following a meal (e.g., chicken: Ekmay et al., 2010). Similarly, very large increases in hepatic concentrations of glycogen are reported on the day of feeding in broiler breeder chickens fed alternate days during the growing period (de Beer et al., 2007). Hocquette et al. (1998) reported that in well-fed sheep, liver glycogen contributes around 18% of glucose to the glucose pool at rest. Thus, as in humans and other well-studied species (e.g., rodents), poultry and ruminants, during periods of a higher plane of nutrition, readily store energy in the form of glycogen.

During the post absorptive phase or during fasting, glycogenolysis increases markedly yielding glucose-6-phosphate as an energy source (skeletal muscles) or in the liver to generate glucose via glucose-6-phosphatase. The glucose, in turn, is moved via glucose transporters into the blood. Liver glycogen, for example, decreases following the peak after a meal in chickens (Ekmay et al., 2010). Moreover, there is an 80% decline during the non-feeding day in broiler breeder chickens fed alternate days (de Beer et al., 2007). Similarly, in young turkeys, hepatic concentrations of glycogen decline by over 99% after 30 hours of fasting (Kurima et al., 1994a). In sheep (Hocquette et al., 1998) during exercise at or above 50–60% VO$_2$ max, there is an increase in the levels of blood glucose released from liver glycogen stores mobilized for energy. During sustained exercise, blood glucose concentration increases, and lactate and pyruvate levels increase during the first fifteen minutes, but then fall below pre-exercise values, indicating low Cori cycle activity (Pethick et al., 1987).

The Cori-cycle is a cycle in which lactate produced by muscle glycolysis is transported via the blood to the liver and used for gluconeogenesis, providing glucose that can be utilized by the heart or returns to the muscle (Voet et al., 1999). Both glycogenesis and glycogenolysis are controlled by hormones—see the following section.

Glucose-6-phosphatase is a critically important hepatic enzyme allowing the formation of glucose. It is not found in muscle. The enzyme catalyzes the conversion of glucose-6-phosphate to glucose. Glucocorticoids increase the hepatic expression of glucose-6-phosphatase and other gluconeogenic enzymes and the liver concentration of glycogen in late gestation (reviewed: Fowden et al., 1995). Cortisol increases hepatic glucose-6-phosphatase activity in fetal pigs (Fowden et al., 1995). There is elevated glucose-6-phosphase activity and expression in the liver of neonatal pigs from dams on a low protein diet (Jia et al., 2012). This is presumably due to a glucocorticoid effect. Neonatal pigs from dams on a low protein diet also have increased hepatic glucocorticosteroid receptor expression (Jia et al., 2012). Administration of either dexamethasone or cortisol increased hepatic glucose-6-phosphatase activity in chickens (Joseph and Ramachandran, 1992).

Hormones and Glucose Metabolism

Insulin and Glucose Metabolism

Circulating concentrations of glucose are decreased by injections of exogenous insulin (chickens: Heald et al., 1965; Langslow et al., 1970). Elimination of insulin from the circulation of an animal can be achieved by pancreatectomy or streptozotocin administration or somatostatin administration. These approaches have disadvantages with pancreatectomy also eliminating glucagon and other pancreatic hormones, streptozotocin influencing release of pancreatic hormones and somatostatin influences rerelease of pancreatic hormones and potentially and directly metabolism. Following pancreatectomy, plasma concentrations of glucose are increased in the fasted (basal) state and/or after a glucose load (e.g., chicken: Danby et al., 1982; Simon and Dubois, 1983; pigs: Sells et al., 1972; fetal sheep: Fowden and Comline, 1984). There is evidence that insulin plays a physiological, perhaps the major, hormone controlling circulating concentrations of glucose in chickens. Administration of antisera against insulin greatly increases the circulating concentrations of glucose in fed chickens (Table 2) (Simon et al., 2000; Dupont et al., 2008).

The changes in global transcription and metabolomics in chickens receiving antisera to insulin (a treatment that prevents insulin from functioning) have been characterized (Ji et al., 2012). There is increased expression of pyruvate dehydrogenase kinase, isozyme 4, glucagon, and angiotensin II receptor and reduced expression of glucagon receptor (Ji et al., 2012). There are increased concentrations of glutamine but decreased concentrations of glycerol-3-phosphate and D-glucono-1,5-lactone-6-phosphate.

Moreover, circulating concentrations of insulin exhibit changes that would be expected if insulin were playing a major role in glucose homeostasis. For instance, circulating concentrations of insulin are depressed in fasted chickens (Christensen et al., 2013). Moreover, circulating concentrations of insulin are elevated following feeding in meal fed broiler breeder chickens during the growing period with an even larger increase when fed alternate days (de Beer et al., 2008).

Insulin increases the uptake of glucose by the liver, skeletal muscle and adipose tissue. In the absence of insulin, glucose uptake by the liver does not increase when a glucose load is delivered via the hepatic portal vein (dogs: Pagliassotti et al., 1992). *In vivo*, insulin increases the uptake of glucose by the hind limb, and predominantly skeletal muscle, of cattle, pigs and sheep as demonstrated by atrial-venous differences (Table 3) (Vernon et al., 1990; Dunshea et al., 1995; Wray-Cahen et al., 1995).

In vivo administration of insulin is followed by large increases in the uptake of glucose [2-deoxy-D-[1-H^3]glucose] by the liver of chickens fasted for 12 hours (Tokushima et al., 2005). There were also moderate increases in glucose uptake by skeletal muscles but no changes in adipose tissue (Tokushima et al., 2005; Nishiki et al., 2008; Zhao et al., 2009). Similarly, glucose uptake by hind limbs is increased in fetal sheep infused with insulin from 33.5 to 45.4 μmol·min^{-1}·kg tissue^{-1} (Anderson et al., 2005). In addition, insulin administration to fetal sheep increases fetal glucose

Table 2: Effect of antisera to insulin on plasma concentrations of glucose in chickens.

Time (hours)	Plasma concentrations of glucose		
	Control		Antisera to insulin (mg dl^{-1})
1	264 ± 2	**	434 ± 21
5	279 ± 6	**	747 ± 28

(data from Dupont et al., 2008).

Table 3: Effect of *in vivo* insulin infusion on uptake of glucose by the hind limb in mammals.

Treatment	Glucose (A-V difference)		
	Sheep	Cattle	Pig
	(umol/ml)		(mmoles min^{-1})
Control	0.15	0.23	0.4
Insulin	0.95	0.56	1.26

(based on Vernon et al., 1990; Dunshea et al., 1995; Wray-Cahen et al., 1995).

uptake from the dam (Simmons et al., 1978). *In vitro* studies also support insulin increasing glucose uptake by skeletal muscle. Insulin stimulates uptake of glucose by the *musculus semitendinosus*, as indicated by uptake of 3-O-methylglucose (MG), in pigs and cattle (Duehlmeier et al., 2005). *In vitro* insulin increases glucose (2-deoxy-D-[1-H^3]glucose) uptake by chicken skeletal muscle (*M. fibularis longus*) *in vitro* (Zhao et al., 2009) and chick embryonic myoblasts (Zhao et al., 2012). In *in vitro* studies, insulin increases glucose uptake by adipose tissue (pig: Akanbi et al., 1990; Gardan et al., 2006). Moreover, insulin increases glucose [14C orthomethyl glucose (OMG)] uptake in chicken adipose tissue from fasted but not fed chickens (Rudas and Scanes, 1983).

Insulin and Glucose Transporters

Glucose moves into hepatic, skeletal muscle and adipose tissue via glucose transporters (GLUT1-12) in the plasma membrane. Hormones, particularly insulin, increase the number of GLUT, and particularly GLUT4, in the plasma membrane by translocation from intracellular vesicles and by new synthesis. GLUT4 has been characterized in domestic animals (e.g., cattle: Abe et al., 1997; pig: Chiu et al., 1994). The glucose transporters, GLUT 1 and GLUT4, are expressed in skeletal muscles in cattle, goats and pigs (Duehlmeier et al., 2005, 2007). Insulin acts by induction of translocation of the GLUT4 glucose transporters to the cell membrane (Duehlmeier et al., 2005). Consistent with this observation, GLUT4 glucose transporters in the plasma membrane are increased in fetal sheep infused with insulin (Anderson et al., 2005).

GLUT4 appears to be absent in chickens (Seki et al., 2003) and other birds (hummingbird: Welch et al., 2013; sparrow: Sweazea and Braun, 2006). Despite this, insulin stimulates glucose uptake (see above). Moreover, insulin increases both the levels of expression and amount of GLUT1 protein in chick embryonic myoblasts *in vitro* (Table 4) (Zhao et al., 2012).

Table 4: Effect of insulin and/or glucocorticoid on expression of glucose transporters in chick embryonic myoblasts *in vitro*.

Treatment	GLUT1	GLUT 3	GLUT8
Control	100[b]	100[c]	100[b]
Insulin (100 nM)	142[c]	76[b]	66[a]
DEX (200 nM)	72[a]	84[b]	114[b]
Insulin and DEX	74[a]	61[a]	78[a]

(adapted from Zhao et al., 2012).

Different superscript letter indicate difference (p < 0.05).

Insulin and Glycogenesis/Glycogenolysis and Gluconeogenesis

Insulin increases glycogen synthase in sheep hepatocytes *in vitro* (Morand et al., 1990), and glycogen synthase is also activated in the presence of fructose and propionate in this model. Counter-regulatory pathways via either glucagon or alpha 1-adrenergic agents decrease glycogen synthase with the effects mediated by increase intracellular cAMP and Ca^{++} (Morand et al., 1988). Insulin also increases the concentrations of glycogen in hepatocytes from pre-weaning and non-ruminant calves but was without effect in ruminating calves (Donkin and Armentano, 1995), suggesting that in ruminants, metabolic regulation of glucose in the pre-ruminating phase is similar to that of other non-ruminant animals. Although some elements of glucose metabolism in mature ruminants remain similarly regulated by insulin. For example, in cattle infused with insulin, there is a decreased rate of endogenous glucose production (Dunshea et al., 1995).

The evidence for a role of insulin in the control of gluconeogenesis in domestic animals is equivocal. The stimulation of glucose formation in lamb hepatocytes by glucagon was partially inhibited by insulin (Clark et al., 1976). In contrast, insulin infusion did not influence the rate of gluconeogenesis from propionate in sheep but decreased endogenous glucose synthesis (Brockman, 1990).

Glucagon and Glucose Metabolism

In vivo, exogenous glucagon administration is followed by increased circulating concentrations of glucose, for instance, in neonatal pigs (Boyd et al., 1985), adult miniature pigs (Müller et al., 1988), fetal or neonatal sheep (Philipps et al., 1983), in pancreatectomized dogs (Muller et al., 1978) and in poultry (chickens: Heald et al., 1965; Braganza et al., 1973; turkeys: McMurtry et al., 1996).

Plasma concentrations of glucose in fed chickens are not affected by the administration of the glucagon antagonist, des-His1(Glu9) glucagon amide (Simon et al., 2000). This suggests that while glucagon can increase circulating concentrations of glucose, it is not playing a critical physiological role in the control of circulating concentrations of glucose at least in the fed state. The increase in circulating concentrations of glucose, may be due to some combination of: increased hepatic glycogenolysis, increased hepatic gluconeogenesis and/or reduced glucose utilization.

Glucagon increases glycogenolysis. *In vivo* infusion of glucagon increases hepatic glucose production in miniature pigs (Müller et al., 1988). In dogs in which glucagon and insulin secretion is suppressed by somatostatin, administration of glucagon is followed initially by a large but only short term (< 4 hours) increase in glycogenolysis and release of glucose from the liver (Cherrington et al., 1978). Glucagon administration also depressed hepatic glycogen concentrations in chickens (Palokangas et al., 1973). Glucagon also reduced the concentrations of glycogen in hepatocytes from either pre-weaning and non-ruminant calves or ruminating calves *in vitro* (Donkin and Armentano, 1994, 1995). *In vitro* glucagon stimulates glucose release/glycogenolysis by hepatocytes (chickens: McCumbee and Hazelwood, 1978; Onoagbe, 1993).

In non-ruminants, the prevailing view is that glucagon increases the rate of gluconeogenesis. For instance, gluconeogenesis from lactate is reported to be increased in neonatal pigs during infusion of glucagon (Helmrath and Bieber, 1975). Glucagon administration to dogs, with suppressed glucagon secretion, is followed by increased gluconeogenesis (Cherrington et al., 1978; Brockman and Greer, 1980). Moreover, glucagon stimulated the formation of glucose from propionate or lactate in hepatocytes from pre-weaning and non-ruminant calves but was ineffective in cells derived from ruminating calves (Donkin and Armentano, 1994, 1995). Further evidence that glucagon does not have a role in the control of gluconeogenesis in ruminants comes from studies in which somatostatin is infused to suppress glucagon secretion and there is no effect on gluconeogenesis in adult sheep (Brockman and Greer, 1980).

Glucagon also increased gluconeogenesis from either propionate or lactate in sheep hepatocytes *in vitro* (Faulkner and Pollock, 1990). However, it had no effect on gluconeogenesis from alanine or lactate in pig liver slices after glucagon infusion (Boyd et al., 1985). Similarly, glucagon increased the rate of glucose formation in lamb hepatocytes with media including galactose or propionate or lactate (Clark et al., 1976). Glucagon has also been shown to stimulate gluconeogenesis in chicken (Dickson and Langslow, 1978) and in rabbit (Yorek et al., 1980) hepatocytes *in vitro*.

Glucagon reduced glucose utilization by adipose tissue depressing labelled glucose incorporation in glycerol/glyceryl, glycogen and carbon dioxide (Goodridge, 1968b). There was no effect of glucagon on fatty acid oxidation or glucose synthesis in hepatocytes from neonatal pigs (Lepine et al., 1993). In pigs infused with glucagon, there was increased blood flow to the liver in the hepatic artery (Gelman et al., 1987), suggesting increased delivery of glucose for hepatic uptake.

Catecholamines: Glucose and Carbohydrate Metabolism

Beta-adrenergic agonists increased gluconeogenesis in both sheep and rabbit hepatocytes *in vitro* (from either propionate or lactate in sheep: Faulkner and Pollock, 1990; and from either lactate or dihydroxyacetone or D-fructose in rabbit: Yorek et al., 1980).

Glucocorticoids and Carbohydrate Metabolism

Infusion of the glucocorticoid, dexamethasone, increased hepatic gluconeogenesis in the perfused chicken liver (Kobayashi et al., 1989). *In vivo*, administration of a glucocorticoid (cortisol) decreased hepatic hexokinase (O'Neill and Langslow, 1978). *In vivo* cortisol treatment increased *in vitro* uptake of glucose by chicken skeletal muscle (Zhao et al., 2009). Glucocorticoids also induced increases in both circulating concentrations of insulin and expression of GLUT1 compared to a pair fed control in chickens (Zhao et al., 2012).

Many actions of cortisol on carbohydrate metabolism in chickens appear to occur via increasing insulin resistance (Dupont et al., 1999; Yuan et al., 2008). *In vivo* cortisol treatment prevented insulin stimulated glucose (2-deoxy-D-[1-H^3]glucose) uptake by chicken skeletal muscle *in vitro* (Zhao et al., 2009). Similarly, in the presence of dexamethasone, insulin did not stimulate glucose uptake in chick embryonic myoblasts *in vitro* (Zhao et al., 2012). In addition, dexamethasone also decreased hepatic insulin binding, and intracellular insulin receptor substrate-1 levels (Dupont et al., 1999) and increased the glucose infusion rate required for euglycemia with insulin administration (Hamano, 2006). Thus, these regulators (or their synthetic analogs), those that are increased during stress, appear to have similar effects in domestic animals to those in other species.

Other Hormones: Glucose and Carbohydrate Metabolism

In young pigs, plasma concentrations of glucose are increased with growth hormone (GH) treatment (Chung et al., 1985). *In vivo* administration of GH is followed by decreased porcine hepatic expression of GLUT4 (Donkin et al., 1996). Chronic infusion of GH to pigs also decreased glucose uptake in the presence of insulin by the hindlimb (see Table 5 below) (Wray-Cahen et al., 1995).

Table 5: Effect of Insulin and/or GH on uptake of Glucose or NEFA by a Hind-limb of Pigs (Adapted from Wray-Cahen et al., 1995).

Treatment	Glucose uptake (mmoles min^{-1})
Control	0.15
Insulin (low[x])	0.29
Insulin (high[y])	0.59
GH[z]	0.10
GH + Insulin (low)	0.12
GH + Insulin (high)	0.74

[x] 14 ng kg^{-1} min^{-1}, [y] 360 ng kg^{-1} min^{-1}, [z] 120 ug kg^{-1} min^{-1} for 7 days.

In contrast, plasma concentrations of glucose are depressed following administration of adiponectin (Hu et al., 2007). This cytokine has been characterized in pigs (Dai et al., 2006) and is expressed by adipose tissue with adiponectin receptors (adipoR1) expressed in multiple tissues including skeletal muscle and adipose tissue while adipoR2 in the liver, skeletal muscle and adipose tissue (pig: Dai et al., 2006; chicken: Ramachandran et al., 2007). Thus, it appears that GH and adiponectin have opposing roles in regulating plasma glucose.

Hormonal Interactions with Substrate Utilization

The utilization of energy substrates may be considered in the context of the interaction of muscle and adipose tissue. As well as being energetically important, these tissues are the focus of meat industries and thus, in livestock, understanding the interactions between muscle (principally an energy utilizing tissue, in this context) and adipose tissue (principally an energy storage tissue), provides a useful frame-work to think about substrate utilization and the interplay between the hormones that regulate these processes.

Muscle tissue and cells undergo regulated growth and differentiation processes, and those, as well as substrate utilization and energy partitioning, are also affected by a range of factors (Brooks, 1998; Hocquette et al., 1998). The (signaling) interactions between myogenic cells and adipocytes has been implicated as playing a significant role in the rate and extent of adipogenesis, myogenesis, and lipogenesis/lipolysis (Boone et al., 2000; Fruhbeck et al., 2001; Diamond, 2002; Welch et al., 2009; Welch et al., 2012). Key factors in these processes include leptin, insulin-like growth factors, and adiponectin (Fruhbeck et al., 2001). Leptin and leptin binding proteins, by direct actions and interactions with other hormones, are thought to play an important role in the communication between adipocytes and myogenic cells (Fruhbeck et al., 2001; Margetic et al., 2002).

Control of Lipid Metabolism

Following the theme of the section above, we now consider the multiple regulatory factors that control lipid metabolism. We begin with the process of adipogenesis, the fundamental process of lipid accumulation. It should be noted that energy storage in adipocytes provides a source to the animal in its most reduced form. Thus, per equivalent mass, lipid contains approximately double the energy content compared to carbohydrate. Another aspect of the evolutionary importance of lipids as energy storage molecules is that water is not required as a solute (as it is for carbohydrate). Thus, for the animal, storage of lipid also results in the saving of the energy needed to carry water associated with carbohydrate storage. In other words, storing energy as lipids is energetically efficient in comparison to energy storage in carbohydrates.

Thus, energy is principally stored as triglyceride in the adipose tissue together with some in the liver. For instance, there were large increases in liver weight and lipid contents in chickens, following meals (Boone et al., 1999; de Beer et al., 2007; Ekmay et al., 2010). Fatty acids are synthesized from glucose and other precursors including volatile fatty acids such as acetate (in ruminants) in the process of lipogenesis. The predominant site of lipogenesis is adipose tissue in livestock but the liver in chickens (O'Hea and Leveille, 1969). At times of energy surplus, fatty acids are esterified with glycerol-3 phosphate to form triglycerides (triacylglycols). At times of energy deficiency, triglycerides are hydrolyzed to free fatty acids or non-esterified fatty acids (NEFA) and glycerol. These processes occur simultaneously with the regulation of adipose cell number and thus are integral through the processes of adipogenesis (increase in adipose cell number) and lipolysis that is thought to reduce the number of lipid bearing cells. Although the evidence is not clear, lipid-free adipocytes are thought to persist (not undergoing apoptosis, but remaining in a dormant state). There are some reports that adipocytes may revert to some form of progenitor state (Fernyhough et al., 2005).

Control of Adipocyte Differentiation and the Key Regulators

During differentiation of preadipocytes to adipocytes (Boone et al., 2000), preadipocytes undergo changes in morphology as well as gene expression. The process is well-described and is modulated by nutritional factors through key signaling pathways. Peroxisome proliferator activated receptor γ (PPAR-γ), PPAR-α, and CCAAT/enhancer binding protein factor (C/EBP) α, β, δ and ζ are important transcription factors involved in the regulation of adipocyte differentiation (Lee et al., 1999; Kersten et al., 2000; Lacasa et al., 2001; Cheguru et al., 2010; Cheguru et al., 2012; Ji et al., 2013). The inhibitory effects of retinoids on differentiation of adipocytes is believed to be mediated through PPAR-γ and C/EBP-β (Boone et al., 2000). C/EBP-β transactivates C/EBP-α expression, promoting adipocyte differentiation. C/EBP-α binds the promoter region of the adipose-specific genes such as leptin and adipocyte lipid binding protein (aP2) (Lee et al., 1999; Boone et al., 2000; Kersten et al., 2000; Lacasa et al., 2001). Also expressed is GLUT 4, an important insulin-mediated glucose transporter in adipocytes, as is the case in muscle. aP2 (also known as A-FABP) is an intracellular fatty acid binding protein that is expressed during differentiation along with fatty acid binding protein (FABP) and fatty acid transferase (FAT), and these molecules are responsible in the transport of fatty acids into the adipocyte and subsequent lipid accumulation (Boone et al., 2000; Zimmerman and Veerkamp, 2002; Cheguru et al., 2012). C/EBP-α is necessary for triglyceride accumulation and adipocyte differentiation. C/EBP-ζ plays a role in the negative regulation of C/EBP-α, as does the growth factor c-*myc*, which blocks the induction of C/EBP-α, thereby inhibiting adipose conversion (Lacasa et al., 2001). Adipogenesis can also be inhibited by long chain polyunsaturated fatty acids, which, when taken up by the cell, act as transcription repressors, resulting in a reduction in C/EBP-α and PPAR-γ (Boone et al., 2000; Lacasa et al., 2001). Table 6 summarizes the range of markers expressed during adipocyte differentiation.

Clearly, energy substrates need to be present in excess in order for the process of adipocyte differentiation to occur. Thus when animals are well-fed and on a high plane of nutrition, adipogenesis and energy storage into lipids is an energetically favorable process. Interestingly, micronutrients such a vitamins may also play a role in the regulation of adipogenesis (Sato and Hiragun, 1988; Ji et al., 2013). Thus, these examples of the regulation of energy substrate storage provide an indication of the complexity of the processes and the many factors that interact in their modulation. Next we consider some of the endocrine factors that regulate adipogenesis.

Regulation of Adipose Differentiation by Glucocorticoids, Insulin, and Insulin-Like Growth Factors

Glucocorticoids, insulin, and insulin-like-growth factors (IGFs) are all involved in the regulation of adipocyte proliferation and differentiation (Hauner et al., 1987; Boone et al., 2000; Jia and Heersche, 2000; Bellows and Heersche, 2001). Glucocorticoids activate C/EBPs, showing another possible mechanism by which differentiation is regulated (Lee et al., 1999). Glucocorticoids act through the glucocorticoid receptor (GR), resulting in an allosteric change which enables the hormone receptor complex to bind the glucocorticoid response element (GRE), the classical glucocorticoid promoter, and modulate transcription (Floyd and Stephens, 2003). These authors also report that signal transducer and activator of transcription 5A (STAT 5A) interacts with the GR during adipogenesis, resulting in inhibition of adipocyte differentiation, indicating a potential regulatory role in adipocyte gene expression. The glucocorticoid, dexathasone, stimulates both differentiation of pre-adipocytes and subsequent fat filling of adipocytes *in vitro* in both pigs (Richardson et al., 1992) and chickens (Ramsay and Rosebrough, 2003).

Insulin-like-growth factors (IGFs) also regulate adipogenesis, providing an example of a paracrine interaction between skeletal muscle and adipose tissue. Insulin-like Growth Factor-1 (IGF-I) is essential for preadipocyte differentiation into adipocytes, although it is not clear if the mechanism

Table 6: Markers expressed during preadipocyte/adipofibroblast differentiation into an adipocyte (after (Kokta et al., 2004) reproduced with permission).

Factor	Time Expressed	Effect	Reference
AD3	Early	Preadipocyte recruitment	(Yu and Hausman, 1998) (Hausman and Richardson, 1998)
ADD1/SREBP1	Early	Stimulates PPAR-γ, transactivates leptin and FAS	(Kim et al., 1998)
CEBP-β	Early	Transactivates CEBP-α, promotes differentiation Activates PPAR-γ	(Lee et al., 1999) (Boone et al., 2000) (Tang et al., 2002) (Sorisky, 1999)
CEBP-δ	Early	Activates PPAR-γ and CEBP-α	(Boone et al., 2000) (Sorisky, 1999)
CEBP-ζ	Early	Negative regulation of CEBP-α	(Lee et al., 1999)
FAT	Early	Fatty acid transport, lipid accumulation	(Boone et al., 2000)
LPL	Early	Fatty acid metabolism	(Boone et al., 2000) (Lacasa et al., 2001) (Sorisky, 1999)
Pref-1	Early	Inhibits differentiation	(Lee et al., 2003), (Mei et al., 2002)
PPAR-γ	Mid	Preadipocyte differentiation, activates Glut 4	(Kersten et al., 2000) (Yamamoto et al., 2002) (Sorisky, 1999)
CEBP-α	Mid	Binds promoter region of leptin and AP2, inhibits proliferation	(Lee et al., 1999) (Sorisky, 1999)
Adipsin	Late	Terminal differentiation	(Diamond, 2002)
AP2	Late	Intracellular fatty acid binding protein, lipid shuttle	(Hansen et al., 1998) (Han et al., 2002)
GLUT-4	Late	Glucose transport	(Sorisky, 1999)
Leptin	Late	Terminal differentiation	(Diamond, 2002)
GPDH	Late	Triacylglycerol accumulation	(Ailhaud, 1997) (Sorisky, 1999)
A2-adrenoceptor	Late	Anti-lipolytic	(Saulnier-Blache et al., 1991)
HSL	Late	Triacylglycerol release	(Sorisky, 1999)

of action is mainly through the Type 1 receptor or the insulin receptor pathway (Smith et al., 1988). It was also demonstrated (Jia and Heersche, 2000) that dexamethasone stimulated the proliferation of preadipocytes in the presence of IGF, but dexamethasone or IGFs alone were unable to stimulate differentiation, suggesting that both glucocorticoids and IGFs may be required for differentiation *in vitro*, and the IGF responsiveness of adipocyte progenitors is a result of dexamethasone stimulation. Consistent with the data reported in muscle (Boney et al., 2000), IGF-1 stimulates both the proliferation and differentiation of preadipocytes via the IGF-1 receptor which ultimately leads to activation of mitogen activated protein kinase (MAP-K). MAP-K inhibition stimulates preadipocyte differentiation, and there is a decrease in MAP-K in cells in the latter stages of differentiation, similar to what is observed in muscle. The loss of MAP-K activity in differentiating cells was a result of the loss of Shc and not IRS-1. Consequently, it was concluded that the IGF-1 signaling switch from proliferation to differentiation is a result of the switch from Shc to IRS-1 mediated signaling. It could be noted, however, that the aforementioned studies were performed in 3T3-L1 cell lines which may behave differently from primary cell lines (Boone et al., 1999).

Catabolic Processes that Mobilize Lipid and Activate Lipolysis

Glucagon

Glucagon is the major stimulator of lipolysis in the chicken. Administration of glucagon is followed by large increases in circulating concentrations of free fatty acids (Heald et al., 1965) with a 3.4 fold increase reported (Braganza et al., 1973). Glucagon infusion to young turkeys is followed by increases in circulating concentrations of NEFA (Kurima et al., 1994b). Glucagon increased glycerol release from chicken adipose tissue *in vitro* (Goodridge, 1968a; Langslow and Hales, 1969) with an ED50 0.7 ng glucagon ml^{-1} (Oscar, 1991). There was also increased sensitivity to glucagon in late embryonic and neonatal development in chicks (Goodridge, 1968a). However, glucagon is without effect on lipolysis in domesticated mammals including cattle (Etherton et al., 1977), dogs (Prigge and Grande, 1971), pigs (Mersmann et al., 1976; Mersmann, 1986), or sheep (Etherton et al., 1977). Glucagon down regulated glucagon receptors with pre-incubation of chicken adipocytes with glucagon reducing both glucagon binding and lipolytic responsiveness to glucagon (Oscar, 1996a). There was, however, no relationship between circulating concentration of glucagon and adipose tissue weight (Sun et al., 2006).

Glucagon decreased lipogenesis or synthesis of fatty acid from ^{14}C Acetate by chick hepatocytes (Goodridge, 1973). Moreover in the presence of glucagon, there were lower activities of both fatty acid synthetase and malic enzyme in chick hepatocytes incubated in the presence of T_3 (Goodridge et al., 1974). *In vivo* glucagon decreased the expression of malic enzyme in chicken liver (Chendrimada et al., 2006). In addition, glucagon decreased stearoyl-CoA desaturase activity and expression in hepatocytes (Lefevre et al., 1999).

Norepinephrine and Epinephrine (Catecholamines)

Norepinephrine (NE) and/or epinephrine (E) are potent stimulator of lipolysis in multiple mammals including dogs (Grund et al., 1975; Connolly et al., 1991; Steiner et al., 1991), guinea pigs (Van den Bergh et al., 1992), horses (Breidenbach et al., 1999), pigs (Helmrath and Bieber, 1975; Mersmann et al., 1976; Hu et al., 1987) and sheep and cattle (Etherton et al., 1977; Smith and McNamara, 1989). The ability of epinephrine, albeit low, to increase lipolysis is not seen in the chicken embryo but is present following hatching (Langslow, 1972).

Insulin and Lipolysis

Insulin inhibits glucagon stimulated lipolysis in adipose tissue *in vitro* from rats but not rabbits (Prigge and Grande, 1971). In livestock mammals, lipolysis is inhibited by insulin. Circulating concentrations of NEFA are decreased with insulin infusion (cattle: Dunshea et al., 1995; pigs: Wray-Cahen et al., 1995; sheep: Vernon et al., 1990). Insulin decreases expression of triglyceride lipase (pig preadipocytes: Deiuliis et al., 2008). Insulin inhibits isoproterenol induced lipolysis by adipocytes *in vitro* (pig: Mills, 1999; Ramsay, 2001). Surprisingly, insulin increased basal and beta-adrenergic agonist stimulated lipolysis in sheep adipose explants after exposure for 48 hours (Watt et al., 1991).

In chickens, *in vitro* insulin has no effect on basal or glucagon stimulated lipolysis (Goodridge, 1968a; Langslow and Hales, 1969; Langslow, 1971; McCumbee and Hazelwood, 1978). Moreover, in chickens, administration of insulin is followed by increased circulating concentrations of fatty acids (Heald et al., 1965); suggesting that insulin has a lipolytic effect. Equally, this may be explicable by insulin increasing secretion of glucagon. Interestingly, administration of antisera to insulin is followed by increased plasma concentrations of NEFA. This may suggest that insulin has an anti-lipolytic role in chickens (Table 7).

However, the increase in circulating concentrations of free fatty acids may also be explicable to the increase in circulating concentrations of glucagon (Table 8) (Dupont et al., 2008).

Table 7: Effect of antisera to insulin on circulating concentrations of NEFA in chickens.

Time hours	Plasma concentrations of NEFA (mg dl⁻¹)		
	Control		Antisera to insulin
1	7.6 ± 0.3	*	9.6 ± 0.6
5	7.3 ± 0.3	*	11.6 ± 0.6

(Dupont et al., 2008).
* indicates difference $p < 0.05$.

Table 8: Effect of antisera to insulin on circulating concentrations of glucagon in chickens.

Time hours	Plasma concentrations of glucagon (ng ml¹)		
	Control		Antisera to insulin
1	0.12 ± 0.05	*	0.39 ± 0.20
5	0.05 ± 0.02	*	1.08 ± 0.18

(Dupont et al., 2008).
* indicates difference $p < 0.05$.

Insulin is anti-lipolytic in fish. This is supported by studies with the administration of insulin to rainbow trout leading to reductions in both circulating concentrations of NEFA and hepatic triacylglycerol lipase activity (Harmon and Sheridan, 1992).

Chronic infusion of insulin to pigs appeared to have little effect on NEFA uptake/utilization by the hind limb (pig: Wray-Cahen et al., 1995). However, it has more recently become clear that there are multiple interactions between the hormones leptin and insulin in regulation of substrate utilization, specifically regulation of lipids (fatty acids) and glucose.

Insulin-Leptin Interactions

In studies of energy partitioning in muscle, it appears that some actions of the hormones leptin and insulin may be antagonistic as insulin inhibits oxidation of free fatty acids and leptin appears to suppress this insulin effect (Figure 1; Welch et al., 2012) (Muoio et al., 1997; Bryson et al., 1999; Muoio et al., 1999; Ceddia et al., 2001).

Evidence for the intersection of the leptin and insulin pathways is also supported as an increase in fatty acid oxidation in muscle (associated with insulin resistance) appears to be linked to an increase in diacylglycerol synthesis and activation of protein kinase C (Griffin et al., 1999; Boden et al., 2001; Yu et al., 2001); a reduction in insulin-stimulated IRS-1-associated PI3-kinase activity, a blunting of insulin-stimulated IRS-1 tyrosine phosphorylation (Griffin et al., 1999); and inhibition of glucose transport and glucose phosphorylation (Roden et al., 1996). The effects of leptin treatment on muscle may be partially attenuated by a synthetic

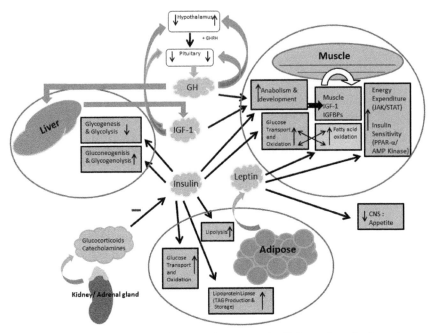

Figure 1: A brief overview of the pathways and processes that link muscle and adipose tissues across processes and pathways that regulate energy metabolism. Note: The pathways depicted are shown in the case of positive energy balance. The factors involved include growth hormone (GH), insulin-like growth factor-1 (IGF-1), insulin, leptin and glucocorticoids. Figure Key: the main tissues of focus are shown along with tissue-specific processes (grey boxes). Broad, grey arrows link tissues to endocrine factors and endocrine feed-back loops. Heavy, black arrows link endocrine signals to tissue-specific processes. Black arrows within process boxes indicate either up-regulation or down-regulation responses. Within muscle, some additional processes are depicted. The crossed arrows indicate the competing interactions of insulin and leptin that stimulate glucose oxidation and fatty acid oxidation, respectively. Interactions of these pathways can repartition oxidation between these two substrates. Each pathway also may inhibit the action of the other (see text for a more detailed description). The broad white arrow indicates that stimulation of locally produced IGF-1 and IGF binding proteins (IGFBPs) results in autocrine/paracrine signaling that also regulates anabolic processes in muscle. (Reproduced with permission from Welch et al., 2012.)

blockade of PI3-kinase activity (Muoio et al., 1999). Leptin also directly stimulates fatty-acid oxidation in muscle by activating the 5'-AMP-activated protein kinase (AMPK), an enzyme that phosphorylates and subsequently inactivates acetyl CoA carboxylase (Minokoshi et al., 2002). A paradigm for the role of leptin in the periphery and the interaction of the leptin and insulin signaling axes has been suggested. Solinas et al. (2004) propose that leptin may stimulate thermogenesis in skeletal muscle via a mechanism independent of decoupling of beta oxidation from ATP synthesis (through uncoupling proteins). In this scenario, leptin may induce a futile cycling between *de novo* lipogenesis and lipid oxidation, a process also requiring glucose and thus interdependence on insulin signaling. It is proposed that AMPK in response to leptin and PI3-kinase in response to insulin play pivotal roles in this process. Thus, this appears to be a potential mechanism that underlies individual variation in energetic efficiency.

The interactions between leptin and insulin signaling pathways are likely to be both tissue-specific and reflect physiological demand (Kim et al., 2000; Szanto and Kahn, 2000; Margetic et al., 2002; Kokta et al., 2004; McClelland et al., 2004). Studies in liver or hepatoma cells have demonstrated interaction between Janus kinases (JAKs) and IRS-1 with apparent downstream effects on signaling via the signal transducer and activator of transcription (STAT) (Carvalheira et al., 2003; Kuwahara et al., 2003) also observed in the hypothalamus (Carvalheira et al., 2001). Other studies have described effects of leptin on the IRS—PI3-kinase signaling pathway (Szanto and Kahn, 2000; Carvalheira et al., 2003), also apparent in hypothalamus (Niswender et al., 2001). Leptin may also interact with insulin signaling via the mitogen-activated protein kinase (MAPK) pathway (Kim et al., 2000; Hill et al., 2004; Strat et al., 2005). Thus, it appears that leptin and insulin signaling may intersect at multiple points in each pathway.

Leptin and Lipid Metabolism

Leptin is secreted by adipose tissue with its expression is influenced by hormones including GH (Houseknecht et al., 2000). Leptin has been demonstrated to influence lipid metabolism in pigs. For instance, leptin stimulates lipolysis by porcine adipocytes *in vitro* (Ramsay, 2001) and increases expression of triglyceride lipase in adipose tissue (Li et al., 2010). While, leptin has no effect on plasma concentrations of NEFA but there were increased plasma concentrations of NEFA compared to pair-fed controls (Ajuwon et al., 2003). Leptin has no effect on *de novo* lipogenesis (fatty acid synthesis) *in vitro* but decreased incorporation (esterification) of palmitate into fat (pigs: Ramsay, 2003, 2004). Leptin decreases glucose incorporation into triglyceride and glucose oxidation (pig: Ramsay, 2003, 2004). Leptin has been reported to increase proliferation of porcine pre-adipocytes (Ramsay, 2005). There is no discernible effect of leptin on adipocyte GPDH and lipoprotein lipase (LPL) activities (Ramsay, 2005).

In vivo exogenous leptin was demonstrated to increase hepatic expression of fatty acid synthase in chickens (Dridi et al., 2005). Earlier work on the effects of leptin in chickens were subject to doubt based on the concerns on the existence of leptin in birds (Sharp et al., 2008). However, leptin has now been definitively characterized in birds (Friedman-Einat et al., 2014).

Glucocorticoids and Lipid Metabolism

Administration of glucocorticoids increases adiposity in chickens (Bartov, 1985; Buyse et al., 1987). *In vivo*, administration of glucocorticoids increases circulating concentrations of NEFA (Hamano, 2006; O'Neill and Langslow, 1978). Dexamethasone also increases plasma concentrations of insulin together with elevated malic enzyme (ME) and fatty acid synthetase (FAS) activities and expression of acetyl-CoA carboxylase (ACC) and FAS (in the fasted state) in the liver and hence lipogenesis (Cai et al., 2009, 2011). *In vivo* administration of DEX depressed expression of an enzyme involved in fatty acid uptake, lipoprotein lipase (LPL) and oxidation, namely, carnitine palmitoyl transferase1 (L-CPT1), long-chain acyl-CoA dehydrogenase (LCAD) and AMP-activated protein kinase alpha 2 in one skeletal muscle, *M. biceps femoris,* but increases expression of LCAD in another skeletal muscle, *M. pectoralis major* (Wang et al., 2010).

Dexamethasone administration to chickens *in vivo* was followed by increased expression of adipose triglyceride lipase in abdominal and sub-cutaneous adipose tissue (Serr et al., 2011). Moreover, circulating concentrations of NEFA rose after administration of dexamethasone (Jiang et al., 2008; Serr et al., 2011).

ACTH and Lipid Metabolism

There is evidence that ACTH can directly stimulate lipolysis in some but not all species. For instance, ACTH stimulates *in vitro* lipolysis in chickens (Langslow and Hales, 1969) and some mammals: guinea pig, hamster, mouse, rat, and rabbit (Ng, 1990; Van den Bergh et al., 1992) and in the pig in the presence of theophylline (Mersmann, 1986). In contrast, ACTH does not affect lipolysis by primate adipocytes [marmoset (*Callithrix jacchus*), baboon (*Papio papio*), macaque (*Macaca fascicularis*) and human] (Bousquet-Mélou et al., 1995). Moreover, ACTH stimulated lipolysis is inhibited by somatostatin *in vitro* with chicken adipocytes (Strosser et al., 1983). Moreover, administration of ACTH is followed by increases in the circulating concentrations of fatty acids (Heald et al., 1965). This does not necessarily demonstrate that ACTH or ACTH of pituitary origin plays a physiological role in the control of lipolysis.

POMC expression has been reported in chicken adipose tissue (Takeuchi et al., 1999). There is expression of members of melanocortin receptor (MC-R) family namely MC4-R and MC5-R in chicken adipose tissue (Takeuchi and Takahashi, 1998). Moreover, the melanocortin receptor antagonist, agouti-related protein (AGRP) is expressed in chicken adipose tissue (Takeuchi et al., 2000).

Growth Hormone and Lipid Metabolism

Chronic infusion of GH alone to pigs elevated circulating concentrations of NEFA but when GH infusion was combined with insulin no such change was seen (Wray-Cahen et al., 1995).

Growth Hormone appears to stimulate lipolysis under some circumstances. GH exerts a lipolytic effect in the presence of theophylline in pig adipose tissue *in vitro* (Mersmann, 1986). Prolonged exposure (48 hours) of sheep adipose explants with insulin or GH increased basal and beta-adrenergic agonist stimulated lipolysis *in vitro* (Watt et al., 1991). The ability of adenosine to inhibit lipolysis is reduced by co-incubation with GH by cattle adipose tissue *in vitro* (Lanna and Bauman, 1999). GH stimulates lipolysis by adipose explants (chicken: Harvey et al., 1977; Campbell and Scanes, 1985) with the effect blocked by a GH antagonist (Campbell et al., 1993). While glucagon can stimulate lipolysis in gilthead seabream adipocytes, GH has a greater effect (Albalat et al., 2005a). The lipolytic effects of GH are illustrated by changes in NEFA uptake by pig hind-limbs (Table 9).

In contrast, pre-treatment of chicken adipocytes with GH decreases basal and glucagon-induced lipolysis (Harden and Oscar, 1993). Moreover, GH inhibits glucagon stimulated lipolysis (chicken: Campbell and Scanes, 1987).

Growth Hormone inhibits insulin stimulated lipogenesis in cattle (Etherton et al., 1987), and pigs (Walton and Etherton, 1986; Walton et al., 1986). *In vivo* in pigs administration of GH is followed by decreased hepatic expression of fatty acid synthase (Donkin et al., 1996).

Other Hormones and Lipid Metabolism

Adenosine

Adenosine can inhibit lipolysis as supported by the following. An adenosine agonist inhibited isoproterenol stimulated lipolysis by bovine adipose tissue

Table 9: Effect of Insulin and/or GH on uptake of NEFA by a Hind-limb of Pigs (Adapted from Wray-Cahen et al., 1995).

Treatment	NEFA uptake (umoles min^{-1})
Control	6.0
Insulin (low[x])	6.7
Insulin (high[y])	5.3
GH[z]	–2.1
GH + Insulin (low)	–2.1
GH + Insulin (high)	5.3

[x] 14 ng kg^{-1} min^{-1}, [y] 360 ng kg^{-1} min^{-1}, [z] 120 ug kg^{-1} min^{-1} for 7 days.

in vitro (Lanna and Bauman, 1999) and in pig adipose tissue *in vitro* (Mills, 1999). Adenosine increased and adenosine deaminase decreased lipogenesis in pig adipose tissue *in vitro* (Mills, 1999). Adenosine also inhibited lipolysis *in situ* (Sollevi and Fredholm, 1981). Moreover, administration of an adenosine agonist depressed lipolysis in dogs *in vivo* (Mittelman and Bergman, 2000). However, while administration of adenosine deaminase increased vascular resistance in adipose tissue, it failed to influence lipolysis (Martin and Bockman, 1986).

Atrial Natriuretic Peptide (ANP) or Brain Natriuretic Peptide (BNP)

These have been shown capable of stimulating lipolysis *in vitro* by primate (human and macaque) adipose tissue via a guanylate cyclase/cGMP mechanism. However, ANP is without effect in rodents (hamster, mouse and rat) and other mammals (dog and rabbit) (Sengenès et al., 2002). ANP also increases intracellular cGMP and lipolysis in rat adipocytes (Nishikimi et al., 2009) with cGMP inhibiting cAMP phosphodiesterase in rat adipocytes (Smith et al., 1991). There is no information on the effects of either ANP or BNP in birds or lower vertebrates.

Gut Hormones and Lipolysis

"Enteroglucagon"/gut glucagon (oxyntomodulin and glicentin) had no effects on lipolysis *in vitro* in chicken adipocytes (glycerol release—Langslow, 1973); or cAMP formation (Kitabgi et al., 1976). Similarly, vasoactive intestinal peptide (VIP) or secretin had little or no effect on lipolysis in these studies.

Ghrelin

There is a single report that hepatic expression of fatty acid synthase is depressed following administration of ghrelin (Buyse et al., 2009). Interestingly, chicken adipose tissue expresses both ghrelin and the ghrelin receptor, growth hormone secretagogue receptor 1a (GHS-R1a) (Nie et al., 2009) as does the liver (Chen et al., 2007). Evidence that the latter is related to the control of metabolism comes from the changes in GHS-R1a expression during fasting (Chen et al., 2007).

Neuropeptide Y Gene Family Peptides and Lipid Metabolism

The Neuropeptide Y (NPY) family of peptides consists of NPY, pancreatic polypeptide (PP) and peptide YY (PYY) (Cerdá-Reverter and Larhammar, 2000). These can influence metabolism. For instance, PP inhibits glucagon stimulated lipolysis and cAMP formation by chicken adipocytes

(McCumbee and Hazelwood, 1977, 1978). However, pre-treatment with PP greatly increases the basal rate of lipolysis (Oscar, 1993). Moreover, peptide YY and to a less extent NPY inhibited beta adrenergic stimulated lipolysis and lipolysis in the presence of ADA acting via a G_i protein and reduction in cAMP by dog adipocytes (Valet et al., 1990; Castan et al., 1992). In addition to effects on lipolysis, PP binds to the small intestine, spleen, bone marrow, proventriculus, liver and brain (chicken: Kimmel and Pollock, 1981; Adamo and Hazelwood, 1990). It is conceivable that PP influences hepatic metabolism. This is consistent with the observation that metabolic state influences PP secretion. For instance, fasting is accompanied by decreased plasma concentrations of PP (Johnson and Hazelwood, 1982). Alternatively, it is at least as persuasively argued that PP is acting in gastro-intestinal functioning.

Somatostatin and Lipid Metabolism

There is strong evidence from *in vitro* and *in vivo* studies that somatostatin (SRIF) inhibits lipolysis in poultry. For instance, SRIF depresses both glucagon and ACTH stimulated lipolysis by chicken adipocytes *in vitro* (Strosser et al., 1983). In contrast, prolonged exposure to somatostatin *in vitro* is followed by increased basal lipolysis (Oscar, 1996b). Infusion of somatostatin decreases circulating concentrations of free fatty acids (NEFA) in fed or fasted turkeys (Kurima et al., 1994a,b). Moreover, circulating concentrations of glucose/free fatty acids are increased in young chickens following the administration of antiserum to somatostatin (anti-SRIF) (Hall et al., 1986). It is not clear whether some of this anti-lipolytic effect is via increases in CORT secretion (Cheung et al., 1988) or shifts in the secretion of pancreatic hormones or of growth hormone.

In mammals, somatostatin has no effect on lipolysis in some mammals (guinea pig, hamster, mouse, rat and rabbit) adipocytes (Ng, 1990) but infusion of somatostatin decreased circulating concentrations of NEFA in dogs (Hendrick et al., 1987). In contrast, somatostatin stimulates fatty acid release from and triacylglycerol lipase activity in coho salmon liver slices *in vitro* (Sheridan and Bern, 1986).

Thyroid Hormones and Lipid Metabolism

Thyroid hormones influence both lipolysis and lipogenesis. There are reports that triiodothyronine T_3 influences lipolysis in chicken adipocytes incubated *in vitro*. Pre-incubation of chicken with T_3 increased both basal and glucagon stimulated lipolysis (Harden and Oscar, 1993; Suniga and Oscar, 1994) but somewhat attenuated somatostatin inhibition of lipolysis (Suniga and Oscar, 1994). Hepatocytes incubated with T_3 exhibit increases in fatty acid synthesis (chicken: Goodridge et al., 1974) with the malic enzyme having a T_3 response element (Hodnett et al., 1996).

Evolutionary Considerations of the Hormonal Control of Lipolysis

It is proposed that glucagon is the major ancestral lipolytic hormone in vertebrates but superseded by epinephrine and nor-epinephrine in mammals.

Glucagon

There are numerous examples of species with lipolytic responses to glucagon throughout most of the *Vertebrata*, namely representatives of the Superclass *Osteichthyes* and the *Tetrapoda* (classification based on the Tree of Life). Glucagon stimulates lipolysis *in vitro* in the following species:

Fish including gilthead seabream (*Sparus aurata*) (Albalat et al., 2005a) and rainbow trout (*Oncorhynchus mykiss*) (Harmon and Sheridan, 1992; Albalat et al., 2005b) but glucagon is without effect in one fish (*Hoplias malabaricus*) or a toad (*Bufo paracnemis*) (Migliorini et al., 1992), reptiles - snake (*Philodryas patagoniensis*) (Migliorini et al., 1992); birds including chickens (e.g., Goodridge, 1968a; Langslow and Hales, 1969), ducks, geese, and owls (Prigge and Grande, 1971); and in mammals (see below).

Generally, glucagon is without a direct effect on lipolysis in mammals with glucagon failing to influence lipolysis with adipose tissue *in vitro* from the following mammalian species (cattle: Etherton et al., 1977; dog: Prigge and Grande, 1971; pig: Mersmann et al., 1976; Mersmann, 1986; sheep: Etherton et al., 1977). However, it should be noted all the studies were from domestic animals. Glucagon stimulation of lipolysis by human adipose tissue has been reported in some studies (e.g., Richter et al., 1989) but not others (e.g., Bertin et al., 2001). In contrast, glucagon stimulates lipolysis with adipose tissue *in vitro* in some mammals (rabbit: Prigge and Grande, 1971; rat: Prigge and Grande, 1971). Moreover, glucagon administration is followed by increased plasma concentrations of fatty acids in northern elephant seals (Crocker et al., 2014) and also in humans when insulin secretion is suppressed (Gerich et al., 1976).

Norepinephrine and Epinephrine (Catecholamines)

Norepinephrine is a potent stimulator of lipolysis in multiple mammalian orders including: *In vitro, erissodactyla*—ponies but not horses (Breidenbach et al., 1999) and in horses in the presence of adenosine deaminase (ADA) or

8-phenyltheophylline stimulated lipolysis in adipose tissue (Breidenbach et al., 1998, 1999); in primates including marmosets (*Callithrix jacchus*), baboons (*Papio papio*), macaques (*Macaca fascicularis*) and in humans (Bousquet-Mélou et al., 1995); as well as in rodents and guinea pigs (*Cavia porcellus*) (Van den Bergh et al., 1992).

Epinephrine is a potent stimulator of lipolysis in multiple mammalian orders including:

In vitro, Artiodactyla—pigs (Mersmann et al., 1976), sheep and cattle adipose tissue *in vitro* (Etherton et al., 1977); in dogs (Prigge and Grande, 1971); and in primates including marmosets (*Callithrix jacchus*), baboons (*Papio papio*), macaques (*Macaca fascicularis*) with beta adrenergic agonists 1, 2 and 3 mimicking the effects of E or NE (Bousquet-Mélou et al., 1994); in rat (Prigge and Grande, 1971), but not in the *Lagomorpha*—rabbit (Prigge and Grande, 1971).

Moreover, epinephrine increases lipolysis *in vitro* in some birds (goose and owl) (Prigge and Grande, 1971) and to a very limited extent in chickens (Langslow and Hales, 1969) but not in the duck (Prigge and Grande, 1971). In lower vertebrates, NE or E are either without effect or inhibit lipolysis. Catecholamines did not influence lipolysis *in vitro* in a fish (*Hoplias malabaricus*), a toad (*Bufo paracnemis*) and a snake (*Philodryas patagoniensis*) (Migliorini et al., 1992). At low concentrations, epinephrine depressed lipolysis but at higher concentrations E increased lipolysis by sea turtle adipose tissue (Hamann et al., 2003). Surprisingly, in fish, NE decreases lipolysis. For instance, *in vitro* lipolysis in tilapia adipocytes is decreased by either NE or the β-adrenergic agonist, isoproterenol (Vianen et al., 2002) suggesting an effect via β-adrenergic receptors. Similarly, *in vivo* NE depresses circulating concentrations of glycerol and glycerol appearance rate in rainbow trout (Magnoni et al., 2008). In contrast, E elevates circulating concentrations of glycerol and glycerol appearance rate in rainbow trout (Magnoni et al., 2008).

An Overview of the Importance of Lipids in the Context of Animal Production

Lipids, Livestock and Meat Production

We now take a step back from the complex regulatory processes that modulate energy storage and utilization to introduce some concepts around broader animal and meat production that are of direct relevance to the regulation of energy metabolism. Of course, understanding of the above processes is fundamental to enable science to guide broader animal management scenarios.

As noted above, lipid storage is energetically favorable for the animal, but there is also a cost. We have explained that lipid storage occurs in times when animals are on a high plane of nutrition. High quality feeds are expensive and thus, deposition of fat in meat animals is costly to the producer. As a specific example of lipid storage and particular use of specific lipid storage depots, we will explore the metabolic costs and regulation of deposition of fat as marbling in beef. Marbling (due to intramuscular fat deposition) is highly prized and is an indicator of quality and specifically, flavor of beef. However, there is a high cost to depositing this lipid.

Energetic Costs of Marbled Beef

Considerable energy and investment in feed is expended to add adipose (fat) tissue to animals that are destined for slaughter. Feedlot cattle are often feed high-grain finishing diets that enhance marbling. However, there is a substantial cost in providing these diets on a large scale.

Role of Fat Depots in Fat Metabolism

Adipose tissue provides an efficient, long-term fuel store with the ability to mobilize during periods of nutritional deficiency. Thus, release of fatty acids from stored lipid provides an energy source for oxidation within organs in need (Trayhurn and Beattie, 2001). Each depot of adipose tissue differs in the fine regulation of energy storage and utilization. Of preadipocytes isolated from different fat depots, the omental depot has a higher lipid flux than the subcutaneous depot. The rate of triglyceride mobilization is also greater within visceral as opposed to omental fat depots. Furthermore, omental preadipocytes have a higher glucocorticoid receptor density, lipoprotein lipase activity, adenylyl cyclase activity, and apoptotic properties than subcutaneous adipose tissue (Trayhurn and Beattie, 2001).

Subcutaneous adipocytes have been shown to have higher levels of GLUT 4, glycogen synthase, and insulin receptors, as well as a higher rate of differentiation when compared to omental adipocytes (Niesler et al., 2001). These authors also reported that omental cells expressed higher levels of Cellular Inhibitor of Apoptosis Protein (cIAP), an anti-apoptotic protein, than subcutaneous adipocytes. This suggests that cells from different depots show inherent differences that account for the observed physiological variation between adipocytes isolated from different locations (Niesler et al., 2001).

Intramuscular fat deposition is regulated by different factors than those regulating deposition in other tissue depots. Thus there are metabolic differences between intramuscular and subcutaneous fat depots (Miller et al., 1991). Both this study and others (May et al., 1994; Eguinoa et al.,

2003) reported that in cattle, intramuscular adipocytes are smaller than subcutaneous fat cells. Additionally, activities of the glycolytic enzymes hexokinase and phosphofructokinase were higher in intramuscular depots. Subcutaneous fat depots had higher levels of the lipogenic enzymes NADP-malate dehydrogenase, 6-phosphogluconate dehydrogenase, and glucose 6-phosphate dehydrogenase. This indicates that different adipose depots have unique roles in and contributions to overall lipid metabolism. Furthermore, breed also appears to be a factor in the lipogenic activity of the tissue depots (Miller et al., 1991; May et al., 1994).

Intermuscular fat depots have similar characteristics to intramuscular fat depots. Eguinoa et al. (2003) reported that in cattle, the intermuscular fat depot had the smallest adipocyte size when compared to omental, peri-renal, and subcutaneous depots. The intermuscular depot additionally had a lower level of lipogenic enzyme activity than other depots, as was the case in intramuscular fat depots. However, when adjusting for adipocyte size, subcutaneous and intermuscular fat depots had higher enzyme activity than the other depots. This indicates a potential role of other factors such as blood flow and lipolytic activity as determinants of depot differences observed.

Vasculature development precedes adipose tissue growth and adipose tissue has the ability to grow throughout the lifetime of the animal, therefore Hutley et al. (2001) hypothesized that microvascular endothelial cells may secrete location specific factors that regulate adipose tissue growth. These factors may play a role in depot differences seen in adipose tissue. These researchers also found that endothelial cells stimulated preadipocyte proliferation, but there were negligible differences in differentiation observed between different depots (Hutley et al., 2001).

A comparison of adipogenic factors from cells isolated from subcutaneous abdominal, omental, and mesenteric fat depots (Montague et al., 1997; Van Harmelen et al., 2002) showed that subcutaneous preadipocytes developed the most, mesenteric cells were intermediate, and omental cells had the lowest levels of lipid, GDPH activity, and adipocyte fatty acid binding protein (aP2). Levels of the transcription factors PPAR-γ and C/EBP-α followed the same order of expression. There appears to be higher leptin mRNA levels in subcutaneous than omental fat depots, but this may be due to adipocyte volume, as it has been reported that leptin is correlated to adipocyte volume, and subcutaneous adipocytes are larger than omental adipocytes (Zhang et al., 2001).

This variation in fat depot metabolism and adipocyte size and physiological dynamics indicate that as animals vary in the relative proportions of fat that is accumulated across different depots, the energetic dynamics of their overall metabolism will vary. Thus, variation in fat depots is a potential difference at tissue level that can contribute to variation in maintenance requirements and energetic efficiency.

In addition, fat deposition is not uniform; therefore there is variation of adipocyte growth throughout animal development. During normal growth of beef cattle, omental fat reaches its maximal growth rate first, followed by intermuscular and then subcutaneous fat. Intramuscular fat depots grow during the later maturation stage of animals and are responsible for the characteristic marbling associated in beef carcasses. Intramuscular fat is highly valued by producers and the wider industry as it provides an excellent indicator for meat palatability. Thus, high intramuscular fat levels indicate a high quality product. As noted above, there is a considerable energetic price associated with producing a high quality product with high intramuscular fat level.

An Overview of the Energetics of Protein Metabolism

In the above sections, we have considered metabolism of molecules that are primarily utilized as energy substrates: carbohydrates and lipids. The next category of molecules we will consider in the context of metabolism are proteins. Proteins form the structural basis of organization at the cellular and molecular levels. They can be used as energy substrates, more so in times when animals are not well-fed. However, protein turnover is a fundamental element of growth and replacement of microstructure, and is a strong contributor to basal metabolism. Thus, although the net contribution of protein to energy metabolism can be small (especially when animals have available abundant and high quality feeds), we note that protein turnover is a continuous process of protein degradation and protein accretion, that can be tuned and directed to shuttle protein towards energy metabolism when needed.

In the following discussion, we have presented the relative contribution of protein to energy metabolism in the context of its efficient use as an energy substrate.

The energetic cost of deposition of a unit of fat is much greater than the cost of an equivalent unit of muscle. Deposition of similar weight lean tissue and fat has different energy expenditure values (Herd and Arthur, 2009). However, there is a greater amount of variation of protein turnover in lean muscle gain when compared to adipose, and during normal growth and development. Over time, muscle will use a larger amount of energy than will an equivalent unit of fat. Thus, there are both energy storage and energy turnover issues to consider in terms of efficient use of energy inputs, be they sourced from proteins, carbohydrates or lipids.

The feed efficiency literature provides some insights into relative energetic efficiency of fat versus protein. We use the example of the efficiency measure termed residual feed intake (RFI). It has been shown by several studies that RFI-inefficient cattle are fatter and that more RFI-efficient cattle are leaner (Carstens et al., 2002; Nkrumah et al., 2007). In fact, some authors who study RFI also include a correction for body fat in the model that predicts feed intake (Schenkel et al., 2004; Baker et al., 2006; Ahola et al., 2011), as it is widely understood that at least 5% of the variation in RFI is due to differences in body composition as described by Richardson and Herd (2004). Relative muscle mass and total proportion of body protein is relatively increased RFI-efficient animals. Basarab et al. (2003), found a modest correlation with dissectible carcass lean (r = –0.17).

A Molecular Basis for Regulation of Protein Metabolism— Context of Efficiency Structural Organization and Fiber type Composition of Muscle

Within the mammalian body, there are three types of muscle tissue: cardiac, skeletal, and smooth. When identifying underlying mechanisms associated with energetics of protein utilization, anatomical structure and physiological function can provide insight.

Great variation in protein composition of skeletal muscle tissue exists both within an individual muscle tissue and among all muscle tissues of the mammalian body. Muscle function and size are important factors involved in the variation of composition; however, muscle fiber type is considered a major contributor to variation. There are three major types of muscle fibers: type I, type IIa, and type IIb.

Ultimately it is this composition that dictates overall performance and energy utilization characteristics of the muscle. For example, muscles used primarily for posture (i.e., *longissiumus dorsi*) will contain a different combination of fibers than those used for force (i.e., *biceps femoris*). Ono et al. (1995) reported that posture is maintained by deep muscles expressing a greater proportion of type I fibers than those located superficially and involved in rapid movements. Furthermore, each fiber type displays characteristic biochemical and mechanical properties, allowing the muscle to perform certain tasks. Type I fibers (known as red) are slow-twitch, perform oxidative metabolism, and contract slowly. Type II fibers (known as white) are fast-twitch, perform anaerobic glycolysis, and contract quickly. Type II fibers are further categorized into two groups: type IIa and type IIb. Type IIa fibers have an intermediate capacity for oxidative metabolism when compared to type I and type IIb fibers, with type I fibers being primarily oxidative and type IIb fibers being primarily glycolytic (Bailey, 2004).

The variation in muscle composition is influenced by several environmental and genetic factors, such as species, breed, sex, nutrition, and gene expression. Muscle weight is related to the total number of muscle fibers, cross-sectional area of fibers, and fiber length (Lefaucheur and Gerrard, 2000). The total number of fibers is a determining factor of muscle growth capacity, which is a characteristic predetermined before birth in most meat-producing farm animals (see review in Picard et al. (2002)).

Glycolytic fibers (type IIb) often exhibit a greater cross-sectional area than that of oxidative fibers (type I). Theoretically, an increase in glycolytic fibers could result in an increase in muscle weight due to the increase in muscle mass. However, this is a relationship that is yet to be scientifically understood (Lefaucheur and Gerrard, 2000). According to Harrison et al. (1996), energy expenditure per unit of tension developed is lower in type I fibers (oxidative) than in type IIb fibers (glycolytic). Even though type I fibers are energetically more efficient, an increase in the proportion of type IIb fibers has the potential to improve protein mass (muscle weight). Thus, beef cattle exhibiting improved energetic efficiency could potentially have similar weight gains due to a greater proportion of type IIb fibers (glycolytic) being expressed, while consuming less energy. In this scenario, efficient animals are able to produce weight gain without increased feed intake. For individual animals whose muscles contain a relatively greater proportion of type IIb fibers compared to type I fibers, especially in those muscles used for movement instead of posture, it is this difference in muscle fiber type that may contribute to the variation in metabolic and energetic efficiency.

A Molecular Basis for Regulation of Protein Metabolism— Context of Efficiency: Energy Consumption and Metabolic Pathways of Muscle

On average skeletal muscle accounts for approximately 40–45% of the total body mass of vertebrates, regardless of their body size (Blaxter, 1989). When considering the whole animal, skeletal muscle can contribute approximately 60% or more to systemic metabolism. In comparing metabolic activity of other bodily tissues, skeletal muscle is one of the most energy-consuming tissues. When active, muscle must be supplied with energy-rich substrates that can accommodate the needs of the resulting increased metabolism. We provide here a brief summary of muscle substrates and their utilization by muscle.

Muscle as an Energetically Dynamic Tissue. Skeletal muscle is a unique tissue in that it has adaptive capabilities which allow it to alter composition and function in response to different physiological conditions. There are two main adaptations or changes that will be discussed in this section: (1) metabolic plasticity and (2) muscle fiber plasticity (anabolic growth mainly regulated via the insulin-like growth factor-1 (IGF-I) axis). Each of these changes affects protein turnover and energy substrate utilization and can dramatically affect processes occurring within skeletal muscle. As a result, function and structure of the muscle can become altered.

More importantly, due to the role of skeletal muscle in terms of the product (that becomes meat), it is critical that we understand how each of these conditions affect not only skeletal muscle at the tissue level, but also how it contributes to the overall growth of an animal. In addition, understanding how it might affect energetic (and feed) inputs required to grow the muscle or sustain it in its remodeled form is equally as important.

One of the most important alterations concerning metabolic plasticity is that of mitochondrial biogenesis. Mitochondrial biogenesis consists of two types of inclusive alterations within the muscle cell. First, there is a change in mitochondrial content per gram of tissue, and/or, second, there is a change in the mitochondrial composition (see review in Hood et al. (2006)). These alterations are highly specific and occur in response to particular types of exercise (i.e., resistance vs. endurance), with changes exhibited most evidently in low-oxidative, white muscle (type IIb) fibers (Hoppeler, 1986). In order to initiate mitochondrial biogenesis in skeletal muscle, a series of signaling events must occur, which include elevation of intracellular calcium (Ca^{2+}) and activation of Ca^{2+}-sensitive signaling molecules, activation of gene transcription that encodes mitochondrial proteins, messenger ribonucleic acid (mRNA) translation into protein, and protein structure assembly (Hood, 2001). The consequences of mitochondrial biogenesis are metabolically beneficial for skeletal muscle at the cellular level. Since skeletal muscle is such an energy-demanding tissue, an increase in the number of mitochondria will allow for cellular metabolic preference to utilize high-energy lipid substrates instead of carbohydrates (i.e., glucose and glycogen). Ultimately, this preference will sustain glycogen stores within muscle, reduce the formation of lactic acid (due to an increase in aerobic capabilities), and reduce muscle fatigue (Hood et al., 2006). Skeletal muscles with a high capacity for lipid oxidation will ultimately exhibit greater efficiency for the mobilization of adipose (fat) storage and have an increased endurance rate. In terms of fiber type, type I fibers are oxidative, allowing for sustained endurance over longer periods of time in comparison to that of type II fibers.

Conclusions

In this chapter, we have provided an overview of the hormonal control of energy substrate utilization and energy metabolism in domestic animals. In considering the metabolism of energy substrates, beginning with glucose, we have described species similarities and differences in the regulation and utilization of glucose (and other carbohydrate energy sources) via the hormones insulin, the catecholamines, glucocorticoids, growth hormone and adiponectin. We have considered the hormones that regulate partitioning of energy substrate utilization between carbohydrates and lipids; the implications for substrate switching on energetic efficiency and then moved to a discussion of the hormones that regulate lipid metabolism across domestic animal species; including a discussion of the evolution of regulation of lipid metabolism. Interactions between regulators such as leptin and insulin and the roles of glucocorticoids, ACTH and growth hormone have been considered in the context of lipid utilization. Finally, we have considered proteins in the context of energy substrate utilization, including a discussion of how differential utilization of protein for energy substrates may effect overall animal efficiency.

Thus, we have addressed the utilization of the three major classes of macromolecules as energy substrates across a range of species of domestic animals. Much new knowledge of variation across species is being facilitated by both specialist study of regulatory pathways, and broader approaches under the banners of metabolomics, nutrigenomics, proteomics, etc. (Brameld and Parr, 2016; Carrillo et al., 2016). As new "omics" tools are evolving, our ability to integrate knowledge from multiple animal species will also be enhanced. Detailed knowledge of specific pathway—substrate interaction presented in this chapter can inform design of studies that utilize the new technologies. The combination of elements from each of these approaches sets the scene for exciting new discoveries in the years ahead.

Keywords: Hormonal regulation, metabolism, adipocytes, adipogenesis, lipogenesis, lipolysis, muscle, anabolism, catabolism

References

Abe, H., M. Morimatsu, H. Nikami, T. Miyashige, and M. Saito. 1997. Molecular cloning and mRNA expression of the bovine insulin-responsive glucose transporter (GLUT4). J. Anim. Sci. 75: 182–188.

Adamo, M.L., and R.L. Hazelwood. 1990. Characterization of liver and cerebellar binding sites for avian pancreatic polypeptide. Endocrinology 126: 434–440.

Ahola, J.K., T.A. Skow, C.W. Hunt, and R.A. Hill. 2011. Relationship between residual feed intake and end product palatability in longissimus steaks from steers sired by Angus bulls divergent for intramuscular fat expected progeny difference. Professional Animal Scientist 27: 109–115.

Aiello, R.J., L.E. Armentano, S.J. Bertics, and A.T. Murphy. 1989. Volatile fatty acid uptake and propionate metabolism in ruminant hepatocytes. J. Dairy Sci. 72: 942–949.

Ailhaud, G. 1997. Molecular mechanisms of adipocyte differentiation. J. Endocrinol. 155: 201–202.

Ajuwon, K.M., J.L. Kuske, D.B. Anderson, D.L. Hancock, K.L. Houseknecht, O. Adeola, and M.E. Spurlock. 2003. Chronic leptin administration increases serum NEFA in the pig and differentially regulates PPAR expression in adipose tissue. J. Nutr. Biochem. 14: 576–583.

Akanbi, K.A., D.C. England, and C.Y. Hu. 1990. Effect of insulin and adrenergic agonists on glucose transport of porcine adipocytes. Comp. Biochem. Physiol. C. 97: 133–138.

Albalat, A., P. Gómez-Requeni, P. Rojas, F. Médale, S. Kaushik, G.J. Vianen, G. Van den Thillart, J. Gutiérrez, J. Pérez-Sánchez, and I. Navarro. 2005a. Nutritional and hormonal control of lipolysis in isolated gilthead seabream (*Sparus aurata*) adipocytes. Am. J. Physiol. 289: R259–265.

Albalat, A., J. Gutiérrez, and I. Navarro. 2005b. Regulation of lipolysis in isolated adipocytes of rainbow trout (*Oncorhynchus mykiss*): The role of insulin and glucagon. Comp. Biochem. Physiol. A 42: 347–354.

Anderson, M.S., M. Thamotharan, D. Kao, S.U. Devaskar, L. Qiao, J.E. Friedman, and W.W. Hay. 2005. Effects of acute hyperinsulinemia on insulin signal transduction and glucose transporters in ovine fetal skeletal muscle. Am. J. Physiol. 288: R473–481.

Bailey, J.G. 2004. Muscle physiology. pp. 871–902. *In*: W.O. Reece (ed.). Dukes' Physiology of Domestic Animals. Cornell University Press, Ithaca, NY.

Baker, S.D., J.I. Szasz, T.A. Klein, P.S. Kuber, C.W. Hunt, J.B. Glaze, Jr., D. Falk, R. Richard, J.C. Miller, R.A. Battaglia, and R.A. Hill. 2006. Residual feed intake of purebred Angus steers: Effects on meat quality and palatability. J. Anim. Sci. 84: 938–945.

Banting, F.G., C.H. Best, J.B. Collip, W.R. Campbell, and A.A. Fletcher. 1922. Pancreatic extracts in the treatment of diabetes mellitus. Can. Med. Assoc. J. 12: 141–146.

Bartov, I. 1985. Effects of dietary protein concentration and corticosterone injections on energy and nitrogen balances and fat deposition in broiler chicks. Br. Poult. Sci. 26: 311–324.

Basarab, J.A., M.A. Price, J.L. Aalhus, E.K. Okine, W.M. Snelling, and K.L. Lyle. 2003. Residual feed intake and body composition in young growing cattle. Can. J. Anim. Sci. 83: 189–204.

Bellows, C.G., and J.N. Heersche. 2001. The frequency of common progenitors for adipocytes and osteoblasts and of committed and restricted adipocyte and osteoblast progenitors in fetal rat calvaria cell populations. J. Bone Miner. Res. 16: 1983–1993.

Berg, J.M. 2002. Section 30.2 Each Organ Has a Unique Metabolic Profile. Biochemistry. 5th edition. http://www.ncbi.nlm.nih.gov/books/NBK22436/accessed November 2, 2015, 2015.

Bertin, E., P. Arner, J. Bolinder, and E. Hagström-Toft. 2001. Action of glucagon and glucagon-like peptide-1-(7-36) amide on lipolysis in human subcutaneous adipose tissue and skeletal muscle *in vivo*. J. Clin. Endocrinol. Metab. 86: 1229–1234.

Best, C.H. 1945. Insulin and diabetes-in retrospect and in prospect: The banting memorial lecture, 1945. Can. Med. Assoc. J. 53: 204–212.

Blaxter, K. 1989. Muscular work. pp. 147–179. *In*: Energy Metabolism in Animals and Man University Press, Cambridge, UK.

Boden, G., B. Lebed, M. Schatz, C. Homko, and S. Lemieux. 2001. Effects of acute changes of plasma free fatty acids on intramyocellular fat content and insulin resistance in healthy subjects. Diabetes 50: 1612–1617.

Boney, C.M., P.A. Gruppuso, R.A. Faris, and A.R.J. Frackelton. 2000. The critical role of Shc in insulin-like growth factor-I-mediated mitogenesis and differentiation in 3T3-L1 preadipocytes. Mol. Endocrinol. 14: 805–813.

Boone, C., F. Gregoire, and C. Remacle. 1999. Regulation of porcine adipogenesis *in vitro*, as compared with other species. Domest. Endocrinol. 17: 257–267.

Boone, C., J. Mourot, F. Gregiore, and C. Remacle. 2000. The adipose conversion process: Regulation by extracellular and intracellular factors. Reprod. Nutr. Dev. 40: 325–358.

Bousquet-Mélou, A., J. Galitzky, C. Carpéné, M. Lafontan, and M. Berlan. 1994. Beta-Adrenergic control of lipolysis in primate white fat cells: A comparative study with nonprimate mammals. Am. J. Physiol. 267: R115–123.

Bousquet-Mélou, A., J. Galitzky, M. Lafontan, and M. Berlan. 1995. Control of lipolysis in intra-abdominal fat cells of nonhuman primates: comparison with humans. J. Lipid Res. 36: 451–461.

Boyd, R.D., D.M. Whitehead, and W.R. Butler. 1985. Effect of exogenous glucagon and free fatty acids on gluconeogenesis in fasting neonatal pigs. J. Anim. Sci. 60: 659–665.

Braganza, A.F., R.A. Peterson, and R.J. Cenedella. 1973. The effects of heat and glucagon on the plasma glucose and free fatty acids of the domestic fowl. Poult. Sci. 52: 58–63.

Brameld, J.M., and T. Parr. 2016. Improving efficiency in meat production. Proc. Nutr. Soc. 75: 242–246.

Breidenbach, A., H. Fuhrmann, R. Busche, and H.P. Sallmann. 1998. Studies on equine lipid metabolism. 1. A fluorometric method for the measurement of lipolytic activity in isolated adipocytes of rats and horses. Zentralbl Veterinarmed A 45: 635–643.

Breidenbach, A., H. Fuhrmann, E. Deegen, A. Lindholm, and H.P. Sallmann. 1999. Studies on equine lipid metabolism. 2. Lipolytic activities of plasma and tissue lipases in large horses and ponies. Zentralbl Veterinarmed A 46: 39–48.

Brockman, R.P., and C. Greer. 1980. Effects of somatostatin and glucagon on the utilization of [2-(14)C]propionate in glucose production *in vivo* in sheep. Aust. J. Biol. Sci. 33: 457–464.

Brockman, R.P. 1990. Effect of insulin on the utilization of propionate in gluconeogenesis in sheep. Br. J. Nutr. 64: 95–101.

Brooks, G. 1998. Mammalian fuel utilization during sustained exercise. Comp. Biochem. Physiol. B 120: 89–107.

Bryson, J.M., J.L. Phuyal, V. Swan, and I.D. Caterson. 1999. Leptin has acute effects on glucose and lipid metabolism in both lean and gold thioglucose-obese mice. Am. J. Physiol. 277: E417–422.

Buyse, J., E. Decuypere, P.J. Sharp, L.M. Huybrechts, E.R. Kühn, and C. Whitehead. 1987. Effect of corticosterone on circulating concentrations of corticosterone, prolactin, thyroid hormones and somatomedin C and on fattening in broilers selected for high or low fat content. J. Endocrinol. 112: 229–237.

Buyse, J., S. Janssen, S. Geelissen, Q. Swennen, H. Kaiya, V.M. Darras, and S. Dridi. 2009. Ghrelin modulates fatty acid synthase and related transcription factor mRNA levels in a tissue-specific manner in neonatal broiler chicks. Peptides 30: 1342–1347.

Cai, Y., Z. Song, X. Wang, H. Jiao, and H. Lin. 2011. Dexamethasone-induced hepatic lipogenesis is insulin dependent in chickens (*Gallus gallus domesticus*). Stress 14: 273–281.

Cai, Y.L., Z.G. Song, X.H. Zhang, X.J. Wang, H.C. Jiao, and H. Lin. 2009. Increased *de novo* lipogenesis in liver contributes to the augmented fat deposition in dexamethasone exposed broiler chickens (*Gallus gallus domesticus*). Comp. Biochem. Physiol. A 150: 164–169.

Campbell, R.M., and C.G. Scanes. 1985. Lipolytic activity of purified pituitary and bacterially derived growth hormone on chicken adipose tissue *in vitro*. Proc. Soc. Exp. Biol. Med. 180: 513–517.

Campbell, R.M., and C.G. Scanes. 1987. Growth hormone inhibition of glucagon and cAMP-induced lipolysis by chicken adipose tissue *in vitro*. Proc. Soc. Exp. Biol. Med. 184: 456–460.

Campbell, R.M., W.Y. Chen, P. Wiehl, B. Kelder, J.J. Kopchick, and C.G. Scanes. 1993. A growth hormone (GH) analog that antagonizes the lipolytic effect but retains full insulin-like (antilipolytic) activity of GH. Proc. Soc. Exp. Biol. Med. 203: 311–316.

Carrillo, J.A., Y. He, Y. Li, J. Liu, R.A. Erdman, T.S. Sonstegard, and J. Songa. 2016. Integrated metabolomic and transcriptome analyses reveal finishing forage affects metabolic pathways related to beef quality and animal welfare. Sci. Rep. 6: 25948.

Carstens, G.E., C.M. Theis, M.B. White, T.H. Welsh, Jr., B.G. Warrington, R.D. Randel, T.D.A. Forbes, H. Lippke, L.W. Greene, and D.K. Lunt. 2002. Residual feed intake in beef steers: I. Correlations with performance traits and ultrasound measures of body composition. Proceedings of the Western Section Meeting, American Society of Animal Science 53: 552–555.

Carvalheira, J.B., R.M. Siloto, I. Ignacchitti, S.L. Brenelli, C.R. Carvalho, A. Leite, L.A.Velloso, J.A. Gontijo, and M.J. Saad. 2001. Insulin modulates leptin-induced STAT3 activation in rat hypothalamus. FEBS Lett. 500: 119–124.

Carvalheira, J.B., E.B. Ribeiro, F. Folli, L.A. Velloso, and M.J. Saad. 2003. Interaction between leptin and insulin signaling pathways differentially affects JAK-STAT and PI 3-kinase-mediated signaling in rat liver. Biol. Chem. 384: 151–159.

Castan, I., P. Valet, T. Voisin, N. Quideau, M. Laburthe, and M. Lafontan. 1992. Identification and functional studies of a specific peptide YY-preferring receptor in dog adipocytes. Endocrinology 131: 1970–1976.

Ceddia, R.B., W.N. William, Jr., and R. Curi. 2001. The response of skeletal muscle to leptin. Front Biosci. 6: D90–97.

Cerdá-Reverter, J.M., and D. Larhammar. 2000. Neuropeptide Y family of peptides: Structure, anatomical expression, function, and molecular evolution. Biochem. Cell. Biol. 78: 371–392.

Cheguru, P., M.E. Doumit, G.K. Murdoch, and H.R.A. 2010. Effect of fatty acids on adipocyte differentiation specific gene expression in 3T3-L1 cells. No. 88 (E Suppl. 2). Journal of Animal Science Denver CO.

Cheguru, P., K.C. Chapalamadugu, M.E. Doumit, G.K. Murdoch, and R.A. Hill. 2012. Adipocyte differentiation-specific gene transcriptional response to C18 unsaturated fatty acids plus insulin. Pflugers Archiv-EJP 463: 429–447.

Cheguru, P., M.E. Doumit, G. Murdoch, and R.A. Hill. 2010. Effect of fatty acids on adipocyte differentiation specific genes expression. J. Anim. Sci. 88 E Suppl. 2: 758.

Chen, L.L., Q.Y. Jiang, X.T. Zhu, G. Shu, Y.F. Bin, X.Q. Wang, P. Gao, and Y.L. Zhang. 2007. Ghrelin ligand-receptor mRNA expression in hypothalamus, proventriculus and liver of chicken (*Gallus gallus domesticus*): studies on ontogeny and feeding condition. Comp. Biochem. Physiol. A 147: 893–902.

Chendrimada, T., K. Adams, M. Freeman, and A.J. Davis. 2006. The role of glucagon in regulating chicken hepatic malic enzyme and histidase messenger ribonucleic acid expression in response to an increase in dietary protein intake. Poult. Sci. 85: 753–760.

Cherrington, A.D., J.L. Chiasson, J.E. Liljenquist, W.W. Lacy, and C.R. Park. 1978. Control of hepatic glucose output by glucagon and insulin in the intact dog. Biochem. Soc. Symp. 43: 31–45.

Cheung, A., S. Harvey, T.R. Hall, S.K. Lam, and G.S. Spencer. 1988. Effects of passive immunization with antisomatostatin serum on plasma corticosterone concentrations in young domestic cockerels. J. Endocrinol. 116: 179–183.

Chiu, P.Y., S. Chaudhuri, P.A. Harding, J.J. Kopchick, S. Donkin, and T.D. Etherton. 1994. Cloning of a pig glucose transporter 4 cDNA fragment: use in developing a sensitive ribonuclease protection assay for quantifying low-abundance glucose transporter 4 mRNA in porcine adipose tissue. J. Anim. Sci. 72: 1196–1203.

Christensen, K., J.P. McMurtry, Y.V. Thaxton, J.P. Thaxton, A. Corzo, C. McDaniel, and C.G. Scanes. 2013. Metabolic and hormonal responses of growing modern meat type chickens to fasting. Br. Poult. Sci. 54: 199–205.

Chung, C.S., T.D. Etherton, and J.P. Wiggins. 1985. Stimulation of swine growth by porcine growth hormone. J. Anim. Sci. 60: 118–130.

Clark, M.G., O.H. Filsell, and I.G. Jarrett. 1976. Gluconeogenesis in isolated intact lamb liver cells. Effects of glucagon and butyrate. Biochem. J. 156: 671–680.

Connolly, C.C., K.E. Steiner, R.W. Stevenson, D.W. Neal, P.E. Williams, K.G. Alberti, and A.D. Cherrington. 1991. Regulation of lipolysis and ketogenesis by norepinephrine in conscious dogs. Am. J. Physiol. 261: E466–742.

Crocker, D.E., M.A. Fowler, C.D. Champagne, A.L. Vanderlugt, and D.S. Houser. 2014. Metabolic response to a glucagon challenge varies with adiposity and life-history stage in fasting northern elephant seals. Gen. Comp. Endocrinol. 195: 99–106.

Dai, M.H., T. Xia, G.D. Zhang, X.D. Chen, L. Gan, S.Q. Feng, H. Qiu, Y. Peng, and Z.Q. Yang. 2006. Cloning, expression and chromosome localization of porcine adiponectin and adiponectin receptors genes. Domest. Anim. Endocrinol. 30: 117–125.

Danby, R.W., I.K. Martin, and W.R. Gibson. 1982. Effects of pancreatectomy, tolbutamide and insulin on glucose fluxes in chickens. J. Endocrinol. 94: 429–441.

de Beer, M., R.W. Rosebrough, B.A. Russell, S.M. Poch, M.P. Richards, and C.N. Coon. 2007. An examination of the role of feeding regimens in regulating metabolism during the broiler breeder grower period. 1. Hepatic lipid metabolism. Poult. Sci. 86: 1726–1738.

de Beer, M., J.P. McMurtry, D.M. Brocht, and C.N. Coon. 2008. An examination of the role of feeding regimens in regulating metabolism during the broiler breeder grower period. 2. Plasma hormones and metabolites. Poult. Sci. 87: 264–275.

Deiuliis, J.A., J. Shin, D. Bae, M.J. Azain, R. Barb, and K. Lee. 2008. Developmental, hormonal, and nutritional regulation of porcine adipose triglyceride lipase (ATGL). Lipids 43: 215–225.

Diamond, F. 2002. The endocrine function of adipose tissue. Growth Genet. Horm. 18: 17–23.

Dickson, A.J., and D.R. Langslow. 1978. Hepatic gluconeogenesis in chickens. Mol. Cell. Biochem. 22: 167–181.

Donkin, S.S., and L.E. Armentano. 1994. Regulation of gluconeogenesis by insulin and glucagon in the neonatal bovine. Am. J. Physiol. 266: R1229–1237.

Donkin, S.S., and L.E. Armentano. 1995. Insulin and glucagon regulation of gluconeogenesis in preruminating and ruminating bovine. J. Anim. Sci. 73: 546–551.

Donkin, S.S., P.Y. Chiu, D. Yin, I. Louveau, B. Swencki, J. Vockroth, C.M. Evock-Clover, J.L. Peters, and T.D. Etherton. 1996. Porcine somatotrophin differentially down-regulates expression of the GLUT4 and fatty acid synthase genes in pig adipose tissue. J. Nutr. 126: 2568–2577.

Dridi, S., J. Buyse, E. Decuypere, and M. Taouis. 2005. Potential role of leptin in increase of fatty acid synthase gene expression in chicken liver. Domest. Anim. Endocrinol. 29: 646–660.

Duehlmeier, R., K. Sammet, A. Widdel, W. von Engelhardt, U. Wernery, J. Kinne, and H.P. Sallmann. 2007. Distribution patterns of the glucose transporters GLUT4 and GLUT1 in skeletal muscles of rats (*Rattus norvegicus*), pigs (*Sus scrofa*), cows (*Bos taurus*), adult goats, goat kids (*Capra hircus*), and camels (*Camelus dromedarius*). Comp. Biochem. Physiol. A 146: 274–282.

Duehlmeier, R., A. Hacker, A. Widdel, W. von Engelhardt, and H.P. Sallmann. 2005. Mechanisms of insulin-dependent glucose transport into porcine and bovine skeletal muscle. Am. J. Physiol. 289: R187–197.

Dunshea, F.R., Y.R. Boisclair, D.E. Bauman, and A.W. Bell. 1995. Effects of bovine somatotropin and insulin on whole-body and hindlimb glucose metabolism in growing steers. J. Anim. Sci. 73: 2263–2271.

Dupont, J., M. Derouet, J. Simon, and M. Taouis. 1999. Corticosterone alters insulin signaling in chicken muscle and liver at different steps. J. Endocrinol. 162: 67–76.

Dupont, J., S. Tesseraud, M. Derouet, A. Collin, N. Rideau, S. Crochet, E. Godet, E. Cailleau-Audouin, S. Métayer-Coustard, M.J. Duclos, C. Gespach, T.E. Porter, L.A. Cogburn, and J. Simon. 2008. Insulin immuno-neutralization in chicken: effects on insulin signaling and gene expression in liver and muscle. J. Endocrinol. 197: 531–542.

Eguinoa, P., S. Brocklehurst, A. Arana, J.A. Mendizabal, R.G. Vernon, and A. Purroy. 2003. Lipogenic enzyme activities in different adipose depots of Pirenaican and Holstein bulls and heifers taking into account adipocyte size. J. Anim. Sci. 81: 432–440.

Ekmay, R.D., M. de Beer, R.W. Rosebrough, M.P. Richards, J.P. McMurtry, and C.N. Coon. 2010. The role of feeding regimens in regulating metabolism of sexually mature broiler breeders. Poult Sci. 89: 1171–1181.

Elmadhun, N.Y., A.D. Lassaletta, L.M. Chu, and F.W. Sellke. 2013. Metformin alters the insulin signaling pathway in ischemic cardiac tissue in a swine model of metabolic syndrome. J. Thorac. Cardiovasc. Surg. 145: 258–265.

Etherton, T.D., D.E. Bauman, and J.R. Romans. 1977. Lipolysis in subcutaneous and perirenal adipose tissue from sheep and dairy steers. J. Anim. Sci. 44: 1100–1106.

Etherton, T.D., C.M. Evock, and R.S. Kensinger. 1987. Native and recombinant bovine growth hormone antagonize insulin action in cultured bovine adipose tissue. Endocrinology 121: 699–703.

Faulkner, A., and H.T. Pollock. 1990. Effects of glucagon and alpha- and beta-agonists on glycogenolysis and gluconeogenesis in isolated ovine hepatocytes. Biochim. Biophys. Acta 1052: 229–234.

Fernyhough, M.E., D.L. Helterline, J.L. Vierck, G.J. Hausman, R.A. Hill, and M.V. Dodson. 2005. Dedifferentiation of mature adipocytes to form adipofibroblasts: 1. More than just a possibility. Adipocytes 1: 17–24.

Floyd, Z.E., and J.M. Stephens. 2003. STAT5A Promotes adipogenesis in nonprecursor cells and associates with glucocorticoid receptor during adipocyte differentiation. Diabetes 52: 308–314.

Fowden, A.L., and R.S. Comline. 1984. The effects of pancreatectomy on the sheep fetus *in utero*. Q. J. Exp. Physiol. 69: 319–330.

Fowden, A.L., R.S. Apatu, and M. Silver. 1995. The glucogenic capacity of the fetal pig: Developmental regulation by cortisol. Exp. Physiol. 80: 457–467.

Friedman-Einat, M., L.A. Cogburn, S. Yosefi, G. Hen, D. Shinder, A. Shirak, and E. Seroussi. 2014. Discovery and characterization of the first genuine avian leptin gene in the rock dove (*Columba livia*). Endocrinology 155: 3376–3384.

Fruhbeck, G., J. Gomez-Ambrosi, F.J. Muruzabal, and M.A. Burrell. 2001. The adipocyte: A model for integration of endocrine and metabolic signaling in energy metabolism regulation. Am. J. Physiol. 280: E827–847.

Gardan, D., F. Gondret, and I. Louveau. 2006. Lipid metabolism and secretory function of porcine intramuscular adipocytes compared with subcutaneous and perirenal adipocytes. Am. J. Physiol. 291: E372–380.

Gelman, S., E. Dillard, and D.A. Parks. 1987. Glucagon increases hepatic oxygen supply-demand ratio in pigs. Am. J. Physiol. 252: G648–653.

Gerich, J.E., M. Lorenzi, D.M. Bier, E. Tsalikian, V. Schneider, J.H. Karam, and P.H. Forsham. 1976. Effects of physiologic levels of glucagon and growth hormone on human carbohydrate and lipid metabolism. Studies involving administration of exogenous hormone during suppression of endogenous hormone secretion with somatostatin. J. Clin. Invest. 57: 875–884.

Goodridge, A.G. 1968a. Lipolysis *in vitro* in adipose tissue from embryonic and growing chicks. Am. J. Physiol. 214: 902–907.

Goodridge, A.G. 1968b. Metabolism of glucose-U-14C *in vitro* in adipose tissue from embryonic and growing chicks. Am. J. Physiol. 214: 897–901.

Goodridge, A.G. 1973. Regulation of fatty acid synthesis in isolated hepatocytes prepared from the livers of neonatal chicks. J. Biol. Chem. 248: 1924–1931.

Goodridge, A.G., A. Garay, and P. Silpananta. 1974. Regulation of lipogenesis and the total activities of lipogenic enzymes in a primary culture of hepatocytes from prenatal and early postnatal chicks. J. Biol. Chem. 249: 1469–1475.

Griffin, M.E., M.J. Marcucci, G.W. Cline, K. Bell, N. Barucci, D. Lee, L.J. Goodyear, E.W. Kraegen, M.F. White, and G.I. Shulman. 1999. Free fatty acid-induced insulin resistance is associated with activation of protein kinase C theta and alterations in the insulin signaling cascade. Diabetes 48: 1270–1274.

Grund, V.R., N.D. Goldberg, and D.B. Hunninghake. 1975. Histamine receptors in adipose tissue: Involvement of cyclic adenosine monophosphate and the H2-receptor in the lipolytic response to histamine in isolated canine fat cells. J. Pharmacol. Exp. Ther. 195: 176–184.

Hall, T.R., A. Cheung, S. Harvey, S.K. Lam, and G.S. Spencer. 1986. Somatostatin immunoneutralization affects plasma metabolite concentrations in the domestic fowl. Comp. Biochem. Physiol. V 85: 489–494.

Hamann, M., C.J. Limpus, and J.M. Whittier. 2003. Seasonal variation in plasma catecholamines and adipose tissue lipolysis in adult female green sea turtles (*Chelonia mydas*). Gen. Comp. Endocrinol. 130: 308–316.

Hamano, Y. 2006. Effects of dietary lipoic acid on plasma lipid, *in vivo* insulin sensitivity, metabolic response to corticosterone and *in vitro* lipolysis in broiler chickens. Br. J. Nutr. 95: 1094–1101.

Han, J., D.P. Hajjar, X. Zhou, A.M. Gotto, Jr., and A.C. Nicholson. 2002. Regulation of peroxisome proliferator-activated receptor-g mediated gene expression. J. Biol. Chem. 277: 23582–23586.

Hansen, L.H., B. Madsen, B. Teisner, J.H. Nielsen, and N. Billestrup. 1998. Characterization of the inhibitory effect of growth hormone on primary preadipocyte differentiation. Mol. Endocrinol. 12: 1140–1149.

Harden, R.L., and T.P. Oscar. 1993. Thyroid hormone and growth hormone regulation of broiler adipocyte lipolysis. Poult. Sci. 72: 669–676.

Harmon, J.S., and M.A. Sheridan. 1992. Effects of nutritional state, insulin, and glucagon on lipid mobilization in rainbow trout, *Oncorhynchus mykiss*. Gen. Comp. Endocrinol. 87: 214–221.

Harrison, A.P., A.M. Rowlerson, and M.J. Dauncey. 1996. Selective regulation of myofiber differentiation by energy status during postnatal development. Am. J. Physiol. 270: R667–R674.

Harvey, S., C.G. Scanes, and T. Howe. 1977. Growth hormone effects on *in vitro* metabolism of avian adipose and liver tissue. Gen. Comp. Endocrinol. 33: 322–328.

Hauner, H., P. Schmid, and E.F. Pfeiffer. 1987. Glucocorticoids and insulin promote the differentiation of human adipocyte precursor cells into fat cells. J. Clin. Endocrinol. Metab. 64: 832–835.

Hausman, G.J., and R.L. Richardson. 1998. Newly recruited and pre-existing preadipocytes in cultures of porcine stromal-vascular cells: Morphology, expression of extracellular matrix components, and lipid accretion. J. Anim. Sci. 76: 48–60.

Heald, P.J., P.M. McLachlan, and K.A. Rookledge. 1965. The effects of insulin, glucagon and ACTH on the plasma glucose and free fatty acids of the domestic fowl. J. Endocrinol. 33: 83–95.

Helmrath, T.A., and L.L. Bieber. 1975. Glucagon stimulation of hepatic gluconeogenesis in neonatal pigs. Proc. Soc. Exp. Biol. Med. 150: 561–563.

Hendrick, G.K., R.T. Frizzell, and A.D. Cherrington. 1987. Effect of somatostatin on nonesterified fatty acid levels modifies glucose homeostasis during fasting. Am. J. Physiol. 253: E443–452.

Herd, R.M., and P.F. Arthur. 2009. Physiological basis for residual feed intake. J. Anim Sci. 87: E64–71.

Hill, R.A., A.L. Strat, N.J. Hughes, T.A. Kokta, M.V. Dodson, and A. Gertler. 2004. Early insulin signaling cascade in a model of oxidative skeletal muscle: mouse Sol8 cell line. Biochimica et Biophysica Acta 1693: 205–211.

Hocquette, J.F., I. Ortigues-Marty, D. Pethick, P. Herpin, and X. Fernandez. 1998. Nutritional and hormonal regulation of energy metabolism in skeletal muscles of meat-producing animals. Livestock Production Science 56: 115–143.

Hodnett, D.W., D.A. Fantozzi, D.C. Thurmond, S.A. Klautky, K.G. MacPhee, S.T. Estrem, G. Xu, and A.G. Goodridge. 1996. The chicken malic enzyme gene: Structural organization and identification of triiodothyronine response elements in the 5′-flanking DNA. Arch Biochem. Biophys. 334: 309–324.

Hood, D.A. 2001. Invited Review: Contractile activity-induced mitochondrial biogenesis in skeletal muscle. J. Appl. Physiol. 90: 1137–1157.

Hood, D.A., I. Irrcher, V. Ljubicic, and A.-M. Joseph. 2006. Coordination of metabolic plasticity in skeletal muscle. J. Exp. Biol. 209: 2265–2275.

Hoppeler, H. 1986. Exercise-induced ultrastructural changes in skeletal muscle. Int. J. Sports Med. 7: 187–204.

Houseknecht, K.L., C.P. Portocarrero, S. Ji, R. Lemenager, and M.E. Spurlock. 2000. Growth hormone regulates leptin gene expression in bovine adipose tissue: correlation with adipose IGF-1 expression. J. Endocrinol. 164: 51–57.

Hu, C.Y., Novakofski, J., and H.J. Mersmann. 1987. Hormonal control of porcine adipose tissue fatty acid release and cyclic AMP concentration. J. Anim. Sci. 64: 1031–1037.

Hu, X., M. She, H. Hou, Q. Li, Q. Shen, Y. Luo, and W. Yin. 2007. Adiponectin decreases plasma glucose and improves insulin sensitivity in diabetic swine. Acta Biochim. Biophys. Sin. (Shanghai). 39: 131–136.

Hutley, L.J., A.C. Herington, W. Shurety, C. Cheung, D.A. Vesey, D.P. Cameron, and J.B. Prins. 2001. Human adipose tissue endothelial cells promote preadipocyte proliferation. Am.J. Physiol. 281: E1037–E1044.

Ji, B., B. Ernest, J.R. Gooding, S. Das, A.M. Saxton, J. Simon, J. Dupont, S. Métayer-Coustard, S.R. Campagna, and B.H. Voy. 2012. Transcriptomic and metabolomic profiling of chicken adipose tissue in response to insulin neutralization and fasting. BMC Genomics 13: 441.

Ji, S., P. Cheguru, S. Acharya, and R.A. Hill. 2013. Regulation of adipogenesis and key adipogenic gene expression by nutritional molecules in 3T3 –L1 preadipocytes. Experimental Biology, Boston, MA.

Jia, D., and J.N. Heersche. 2000. Insulin-like growth factor-1 and -2 stimulate osteoprogenitor proliferation and differentiation and adipocyte formation in cell populations derived from adult rat bone. Bone 27: 785–794.

Jia, Y., R. Cong, R. Li, X. Yang, Q. Sun, N. Parvizi, and R. Zhao. 2012. Maternal low-protein diet induces gender-dependent changes in epigenetic regulation of the glucose-6-phosphatase gene in newborn piglet liver. J. Nutr. 142: 1659–1665.

Jiang, K.J., H.C. Jiao, Z.G. Song, L. Yuan, J.P. Zhao, and H. Lin. 2008. Corticosterone administration and dietary glucose supplementation enhance fat accumulation in broiler chickens. Br. Poult. Sci. 49: 625–631.

Johnson, E.M., and R.L. Hazelwood. 1982. Avian pancreatic polypeptide (APP) levels in fasted-refed chickens: Locus of postprandial trigger? Proc. Soc. Exp. Biol. Med. 169: 175–182.

Joseph, J., and A.V. Ramachandran. 1992. Alterations in carbohydrate metabolism by exogenous dexamethasone and corticosterone in post-hatched White Leghorn chicks. Br. Poult. Sci. 33: 1085–1093.

Kersten, S., S. Mandard, N.S. Tan, P. Escher, D. Metzger, P. Chambon, F.J. Gonzalez, B. Desvergne, and W. Wahli. 2000. Characterization of the fasting-induced adipose factor FIAF, a novel peroxisome proliferator-activated receptor target gene. J. Biol. Chem. 275: 28488–28493.

Kim, D.W., M.M. Mushtaq, R.H.K. Kang, J.H. Kim, J.C. Na, J. Hwangbo, J.D. Kim, C.B. Yang, B.J. Park, and H.C. Choi. 2015. Various levels and forms of dietary α-lipoic acid in broiler chickens: Impact on blood biochemistry, stress response, liver enzymes, and antibody titers. Poult. Sci. Epub ahead of print.

Kim, J.B., P. Sarraf, M. Wright, K.M. Yao, E. Mueller, G. Solanes, B.B. Lowell, and B.M. Spiegelman. 1998. Nutritional and insulin regulation of fatty acid synthetase and leptin gene expression through ADD1/SREBP1. J. Clin. Invest. 101: 1–9.

Kim, Y.B., S. Uotani, D.D. Pierroz, J.S. Flier, and B.B. Kahn. 2000. *In vivo* administration of leptin activates signal transduction directly in insulin-sensitive tissues: Overlapping but distinct pathways from insulin. Endocrinology 141: 2328–2339.

Kimmel, J.R., and H.G. Pollock. 1981. Target organs for avian pancreatic polypeptide. Endocrinology 109: 1693–1699.

Kitabgi, P., G. Rosselin, and D. Bataille. 1976. Interactions of glucagon and related peptides with chicken adipose tissue. Horm. Metab. Res. 8: 266–270.

Kobayashi, T., I. Iwai, R. Uchimoto, M. Ohta, M. Shiota, and T. Sugano. 1989. Gluconeogenesis in perfused livers from dexamethasone-treated chickens. Am. J. Physiol. 256: R907–914.

Kokta, T.A., M.V. Dodson, A. Gertler, and R.A. Hill. 2004. Intercellular signaling between adipose tissue and muscle tissue. Domest. Anim. Endocrinol. 27: 303–331.

Kristensen, N.B., and D.L. Harmon. 2004. Splanchnic metabolism of volatile fatty acids absorbed from the washed reticulorumen of steers. J. Anim. Sci. 82: 2033–2042.

Kurima, K., W.L. Bacon, and R. Vasilatos-Younken. 1994a. Effects of somatostatin on plasma growth hormone and metabolite concentrations in fed and feed-deprived young female turkeys. Poult. Sci. 73: 714–723.

Kurima, K., W.L. Bacon, and R. Vasilatos-Younken. 1994b. Effects of glucagon infusion, alone or in combination with somatostatin, on plasma growth hormone and metabolite levels in young female turkeys under different feeding regimens. Poult. Sci. 73: 704–713.

Kuwahara, H., S. Uotani, T. Abe, M. Degawa-Yamauchi, R. Takahashi, A. Kita, N. Fujita, K. Ohshima, H. Sakamaki, H. Yamasaki, Y. Yamaguchi, and K. Eguchi. 2003. Insulin attenuates leptin-induced STAT3 tyrosine-phosphorylation in a hepatoma cell line. Mol. Cell. Endocrinol. 205: 115–120.

Lacasa, P., E.G.D. Santos, and Y. Guidicelli. 2001. Site-specific control of rat preadipocyte adipose conversion by ovarian status. Endocrine 15: 103–110.

Langslow, D.R., and C.N. Hales. 1969. Lipolysis in chicken adipose tissue *in vitro*. J. Endocrinol. 43: 285–294.

Langslow, D.R., E.J. Butler, C.N. Hales, and A.W. Pearson. 1970. The response of plasma insulin, glucose and non-esterified fatty acids to various hormones, nutrients and drugs in the domestic fowl. J. Endocrinol. 46: 243–260.

Langslow, D.R. 1971. The anti-lipolytic action of prostaglandin E on isolated chicken fat cells. Biochim. Biophys. Acta. 239: 33–37.

Langslow, D.R. 1972. The development of lipolytic sensitivity in the isolated fat cells of Gallus domesticus during the foetal and neonatal period. Comp. Biochem. Physiol. B 43: 689–701.

Langslow, D.R. 1973. The action of gut glucagon-like immunoreactivity and other intestinal hormones on lipolysis in chicken adipocytes. Horm. Metab. Res. 5: 428–432.

Lanna, D.P., and D.E. Bauman. 1999. Effect of somatotropin, insulin, and glucocorticoid on lipolysis in chronic cultures of adipose tissue from lactating cows. J. Dairy Sci. 82: 60–68.

Lee, K., D.B. Hausman, and R.G. Dean. 1999. Expression of CCAAT/enhancer binding protein C/EBPa, b and d in rat adipose stromal-vascular cells *in vitro*. Biochimica et Biophysica Acta 1450: 397–405.

Lee, K., J.A. Villena, Y.S. Moon, K.H. Kim, S. Lee, C. Kang, and H.S. Sul. 2003. Inhibition of adipogenesis and development of glucose intolerance by soluble preadipocyte factor-1 (Pref-1). J. Clin. Invest. 111: 453–461.

Lefaucheur, L., and D. Gerrard. 2000. Muscle fiber plasticity in farm mammals. J. Anim. Sci. 77: 1–19.

Lefevre, P., C. Diot, P. Legrand, and M. Douaire. 1999. Hormonal regulation of stearoyl coenzyme-A desaturase 1 activity and gene expression in primary cultures of chicken hepatocytes. Arch. Biochem. Biophys. 368: 329–337.

Lepine, A.J., M. Watford, R.D. Boyd, D.A. Ross, and D.M. Whitehead. 1993. Relationship between hepatic fatty acid oxidation and gluconeogenesis in the fasting neonatal pig. Br. J. Nutr. 70: 81–91.

Lewis, K.J., P.C. Molan, J.J. Bass, and P.D. Gluckman. 1988. The lipolytic activity of low concentrations of insulin-like growth factors in ovine adipose tissue. Endocrinology 122: 2554–2557.

Li, Y.C., X.L. Zheng, B.T. Liu and G.S. Yang. 2010. Regulation of ATGL expression mediated by leptin *in vitro* in porcine adipocyte lipolysis. Mol. Cell. Biochem. 333: 121–128.

Lindsay, D.B., and B.P. Setchell. 1976. The oxidation of glucose, ketone bodies and acetate by the brain of normal and ketonaemic sheep. J. Physiol. 259: 801–823.

Magnoni, L., E. Vaillancourt, and J.M. Weber. 2008. *In vivo* regulation of rainbow trout lipolysis by catecholamines. J. Exp. Biol. 211: 2460–2466.

Margetic, S., C. Gazzola, G.G. Pegg, and R.A. Hill. 2002. Leptin: A review of its peripheral actions and interactions. Int. J. Obes. Relat. Metab. 26: 1407–1433.

Martin, S.E., and E.L. Bockman. 1986. Adenosine regulates blood flow and glucose uptake in adipose tissue of dogs. Am. J. Physiol. 250: H1127–1135.

Mattin, M., D. O'Neill, D. Church, P.D. McGreevy, P.C. Thomson, and D. Brodbelt. 2014. An epidemiological study of diabetes mellitus in dogs attending first opinion practice in the UK. Vet. Rec. 174: 349.

May, S.G., J.W. Savell, D.K. Lunt, J.J. Wilson, J.C. Laurenz, and S.B. Smith. 1994. Evidence for preadipocyte proliferation during culture of subcutaneous and intramuscular adipose tissues from Angus and Wagyu crossbred steers. J. Anim. Sci. 72: 3110–3117.

McClelland, G.B., C.S. Kraft, D. Michaud, J.C. Russell, C.R. Mueller, and C.D. Moyes. 2004. Leptin and the control of respiratory gene expression in muscle. Biochim. Biophys. Acta 1688: 86–93.

McCumbee, W.D., and R.L. Hazelwood. 1977. Biological evaluation of the third pancreatic hormone (APP): Hepatocyte and adipocyte effects. Gen. Comp. Endocrinol. 33: 518–525.

McCumbee, W.D., and R.L. Hazelwood. 1978. Sensitivity of chicken and rat adipocytes and hepatocytes to isologous and heterologous pancreatic hormones. Gen. Comp. Endocrinol. 34: 421–427.

McMurtry, J.P., W. Tsark, L. Cogburn, R. Rosebrough, and D. Brocht. 1996. Metabolic responses of the turkey hen (Meleagris gallopavo) to an intravenous injection of chicken or porcine glucagon. Comp. Biochem. Physiol. C 114: 159–163.

Mei, B., L. Zhao, L. Chen, and H.S. Sul. 2002. Only the large soluble form of preadipocyte factor-1 (Pref-1), but not the small soluble and membrane forms, inhibits adipocyte differentiation: Role of alternative splicing. Biochem. J. 364: 137–144.

Mersmann, H.J., L.J. Brown, B.R. Deuving, and M.C. Arakelian. 1976. Lipolytic activity of swine adipocytes. Am. J. Physiol. 230: 1439–1443.

Mersmann, H.J. 1986. Acute effects of metabolic hormones in swine. Comp. Biochem. Physical. A 83: 653–660.

Migliorini, R.H., J.S. Lima-Verde, C.R. Machado, G.M. Cardona, M.A. Garofalo, and I.C. Kettelhut. 1992. Control of adipose tissue lipolysis in ectotherm vertebrates. Am. J. Physiol. 263: R857–862.

Miller, M.F., H.R. Cross, D.K. Lunt, and S.B. Smith. 1991. Lipogenesis in acute and 48-hour cultures of bovine intramuscular and subcutaneous adipose tissue explants. J. Anim. Sci. 69: 162–170.

Mills, S.E. 1999. Regulation of porcine adipocyte metabolism by insulin and adenosine. J. Anim. Sci. 77: 3201–3207.

Minokoshi, Y., Y.B. Kim, O.D. Peroni, L.G. Fryer, C. Müller, D. Carling, and B.B. Kahn. 2002. Leptin stimulates fatty-acid oxidation by activating AMP-activated protein kinase. Nature 415: 339–343.

Mittelman, S.D., and R.N. Bergman. 2000. Inhibition of lipolysis causes suppression of endogenous glucose production independent of changes in insulin. Am. J. Physiol. 279: E630–637.

Montague, C.T., J.B. Prins, L. Sanders, J.E. Digby, and S. O' Rahilly. 1997. Depot- and sex-specific differences in human leptin leptin mRNA expression: Implications for the control of regional fat distribution. Diabetes 46: 342–347.

Morand, C., C. Yacoub, C. Remesy, and C. Demigne. 1988. Characterization of glucagon and catecholamine effects on isolated sheep hepatocytes. Am. J. Physiol. 255: R539–546.

Morand, C., C. Redon, C. Remesy, and C. Demigne. 1990. Non-hormonal and hormonal control of glycogen metabolism in isolated sheep liver cells. Int. J. Biochem. 22: 873–881.

Muller, W.A., L. Girardier, J. Seydoux, M. Berger, A.E. Renold, and M. Vranic. 1978. Extrapancreatic glucagon and glucagonlike immunoreactivity in depancreatized dogs. A quantitative assessment of secretion rates and anatomical delineation of sources. J. Clin. Invest. 62: 124–132.

Müller, M.J., P.E. Mitchinson, U. Paschen, and H.J. Seitz. 1988. Glucoregulatory function of glucagon in hypo-, eu- and hyperthyroid miniature pigs. Diabetologia 31: 368–374.

Muoio, D.M., G.L. Dohm, F.T. Fiedorek, E.B. Tapscott, and R.A. Coleman. 1997. Leptin directly alters lipid partitioning in skeletal muscle. Diabetes 46: 1360–1363.

Muoio, D.M., G.L. Dohm, E.B. Tapscott, and R.A. Coleman. 1999. Leptin opposes insulin's effects on fatty acid partitioning in muscle isolated from obese ob/ob mice. Am. J. Physiol. 276: E913–E921.

Ng, T.B. 1990. Studies on hormonal regulation of lipolysis and lipogenesis in fat cells of various mammalian species. Comp. Biochem. Physiol. B. 97: 441–446.

Nie, Q., M. Fang, L. Xie, X. Peng, H. Xu, C. Luo, D. Zhang, and X. Zhang. 2009. Molecular characterization of the ghrelin and ghrelin receptor genes and effects on fat deposition in chicken and duck. J. Biomed. Biotechnol. 2009: 567120.

Niesler, C.U., J.B. Prins, S. O' Rahilly, K. Siddle, and C.T. Montague. 2001. Adipose depot-specific expression of cIAP2 in human preadipocytes and modulation of expression by serum factors and TNFa. Int. J. Obes. Relat. Metab. Disord. 25: 1027–1033.

Nishiki, Y., T. Kono, K. Fukao, K. Sato, K. Takahashi, M. Toyomizu, and Y. Akiba. 2008. Nitric oxide (NO) is involved in modulation of non-insulin mediated glucose transport in chicken skeletal muscles. Comp. Biochem. Physiol. B 149: 101–107.

Nishikimi, T., C. Iemura-Inaba, K. Akimoto, K. Ishikawa, S. Koshikawa, and H. Matsuoka. 2009. Stimulatory and inhibitory regulation of lipolysis by the NPR-A/cGMP/PKG and NPR-C/G(i) pathways in rat cultured adipocytes. Regul. Pept. 153: 56–63.

Niswender, K.D., G.J. Morton, W.H. Stearns, C.J. Rhodes, M.G. Myers, Jr., and M.W. Schwartz. 2001. Intracellular signalling. Key enzyme in leptin-induced anorexia. Nature 413: 794–795.

Nkrumah, J.D., J.A. Basarab, Z. Wang, C. Li, M.A. Price, E.K. Okine, D.H. Crews Jr., and S.S. Moore. 2007. Genetic and phenotypic relationships of feed intake and measures of efficiency with growth and carcass merit of beef cattle. J. Anim. Sci. 85: 2711–2720.

O'Hea, K.E., and G.A. Leveille. 1969. Lipid biosynthesis and transport in the domestic chick (*Gallus domesticus*). Comp. Biochem. Physiol. 30: 149–159.

Öhlund, M., T. Fall, B. Ström Holst, H. Hansson-Hamlin, B. Bonnett, and A. Egenvall. 2015. Incidence of diabetes mellitus in insured Swedish cats in relation to age, breed and sex. J. Vet. Intern. Med. 29: 1342–1347.

O'Neill, I.E., and D.R. Langslow. 1978. The action of hydrocortisone, insulin, and glucagon on chicken liver hexokinase and glucose-6-phosphatase and on the plasma glucose and free fatty acid concentrations. Gen. Comp. Endocrinol. 34: 428–437.

Ono, Y., M.B. Solomon, C.M. Evock-Clover, N.C. Steele, and K. Maruyama. 1995. Effects of porcine somatotropin administration on porcine muscles located within different regions of the body. J. Anim. Sci. 73: 2282–2288.

Onoagbe, I.O. 1993. Hormonal control of glycogenolysis in isolated chick embryo hepatocytes. Exp. Cell. Res. 209: 1–5.

Oscar, T.P. 1991. Glucagon-stimulated lipolysis of primary cultured broiler adipocytes. Poult. Sci. 70: 326–332.

Oscar, T.P. 1993. Enhanced lipolysis from broiler adipocytes pretreated with pancreatic polypeptide. J. Anim. Sci. 71: 2639–2644.

Oscar, T.P. 1996a. Down-regulation of glucagon receptors on the surface of broiler adipocytes. Poult. Sci. 75: 1027–1034.

Oscar, T.P. 1996b. Prolonged *in vitro* exposure of broiler adipocytes to somatostatin enhances lipolysis and induces desensitization of antilipolysis. Poult. Sci. 75: 393–401.

Ostaszewski, P., and S. Nissen. 1988. Effect of hyperglucagonemia on whole-body leucine metabolism in immature pigs before and during a meal. Am. J. Physiol. 254: E372–377.

Pagliassotti, M.J., M.C. Moore, D.W. Neal, and A.D. Cherrington. 1992. Insulin is required for the liver to respond to intraportal glucose delivery in the conscious dog. Diabetes 41: 1247–1256.

Palokangas, R., V. Vihko, and I. Nuuja. 1973. The effects of cold and glucagon on lipolysis, glycogenolysis and oxygen consumption in young chicks. Comp. Biochem. Physiol. A 45: 489–495.

Pethick, D.W., N. Harman, and J.K. Chong. 1987. Non-esterified long-chain fatty acid metabolism in fed sheep at rest and during exercise. Aust. J. Biol. Sci. 40: 221–234.

Philipps, A.F., J.W. Dubin, P.J. Matty, and J.R. Raye. 1983. Influence of exogenous glucagon on fetal glucose metabolism and ketone production. Pediatr. Res. 17: 51–66.

Picard, B., L. Lefaucheur, C. Berri, and M.J. Duclos. 2002. Muscle fibre ontogenesis in farm animal species. Reprod. Nutr. Dev. 42: 415–431.

Prigge, W.F., and F. Grande. 1971. Effects of glucagon, epinephrine and insulin on *in vitro* lipolysis of adipose tissue from mammals and birds. Comp. Biochem. Physiol. B 39: 69–82.

Ramachandran, R., O.M. Ocón-Grove, and S.L. Metzger. 2007. Molecular cloning and tissue expression of chicken AdipoR1 and AdipoR2 complementary deoxyribonucleic acids. Domest. Anim. Endocrinol. 33: 19–31.

Ramsay, T.G. 2001. Porcine leptin alters insulin inhibition of lipolysis in porcine adipocytes *in vitro*. J. Anim. Sci. 79: 653–657.

Ramsay, T.G. 2003. Porcine leptin inhibits lipogenesis in porcine adipocytes. J. Anim. Sci. 81: 3008–3017.

Ramsay, T.G., and R.W. Rosebrough. 2003. Hormonal regulation of postnatal chicken preadipocyte differentiation *in vitro*. Comp. Biochem. Physiol. B 136: 245–253.

Ramsay, T.G. 2004. Porcine leptin alters isolated adipocyte glucose and fatty acid metabolism. Domest. Anim. Endocrinol. 26: 11–21.

Ramsay, T.G. 2005. Porcine preadipocyte proliferation and differentiation: A role for leptin? J. Anim. Sci. 83: 2066–2074.

Richardson, E.C., and R.M. Herd. 2004. Biological basis for variation in residual feed intake in beef cattle. 2. Synthesis of results following divergent selection. Aust. J. Exp. Agric. 44: 431–440.

Richardson, R.L., G.J. Hausman, and H.R. Gaskins. 1992. Effect of transforming growth factor-beta on insulin-like growth factor 1- and dexamethasone-induced proliferation and differentiation in primary cultures of pig preadipocytes. Acta Anat. (Basel) 145: 321–326.

Richter, W.O., H. Robl, and P. Schwandt. 1989. Human glucagon and vasoactive intestinal polypeptide (VIP) stimulate free fatty acid release from human adipose tissue *in vitro*. Peptides 10: 333–335.

Roden, M., T.B. Price, G. Perseghin, K.F. Petersen, D.L. Rothman, G.W. Cline, and G.I. Shulman. 1996. Mechanism of free fatty acid-induced insulin resistance in humans. J. Clin. Invest. 97: 2859–2865.

Rudas, P., and C.G. Scanes. 1983. Influences of growth hormone on glucose uptake by avian adipose tissue. Poult. Sci. 62: 1838–1845.

Sato, M., and A. Hiragun. 1988. Demonstration of 1 alpha, 25-dihydroxyvitamin D3 receptor-like molecule in ST 13 and 3T3 L1 preadipocytes and its inhibitory effects on preadipocyte differentiation. J. Cell. Physiol. 135: 545–550.

Saulnier-Blache, J.S., M. Dauzats, D. Daviaud, D. Gaillard, G. Ailhaud, R. Négrel, and M. Lafontan. 1991. Late expression of alpha 2-adrenergic-mediated antilipolysis during differentiation of hamster preadipocytes. J. Lipid Res. 32: 1489–1499.

Schenkel, F.S., S.P. Miller, and J.W. Wilton. 2004. Genetic parameters and breed differences for feed efficiency, growth, and body composition traits of young beef bulls. Can. J. Anim. Sci. 84: 177–185.

Seki, Y., K. Sato, T. Kono, H. Abe, and Y. Akiba. 2003 Broiler chickens (Ross strain) lack insulin-responsive glucose transporter GLUT4 and have GLUT8 cDNA. Gen. Comp. Endocrinol. 133: 80–87.

Sells, R.A., R.Y. Calne, V. Hadjiyanakis, and V.C. Marshall. 1972. Glucose and insulin metabolism after pancreatic transplantation. Br. Med. J. 3: 678–681.

Scanes, C.G. 2015. Carbohydrate metabolism. pp. 421–441. *In*: C.G. Scanes (ed.). Sturkie's Avian Physiology. 6th Edition Elsevier, Amsterdam.

Sengenès, C., A. Zakaroff-Girard, A. Moulin, M. Berlan, A. Bouloumié, M. Lafontan, and J. Galitzky. 2002. Natriuretic peptide-dependent lipolysis in fat cells is a primate specificity. Am. J. Physiol. 283: R257–265.

Serr, J., Y. Suh, S.A. Oh, S. Shin, M. Kim, J.D. Latshaw, and K. Lee. 2011. Acute up-regulation of adipose triglyceride lipase and release of non-esterified fatty acids by dexamethasone in chicken adipose tissue. Lipids 46: 813–820.

Sharp, P.J., I.C. Dunn, D. Waddington, and T. Boswell. 2008. Chicken leptin. Gen. Comp. Endocrinol. 158: 2–4.

Sheridan, M.A., and H.A. Bern. 1986. Both somatostatin and the caudal neuropeptide, urotensin II, stimulate lipid mobilization from coho salmon liver incubated *in vitro*. Regul. Pept. 14: 333–344.

Simmons, M.A., M.D. Jones, F.C. Battaglia, and G. Meschia. 1978. Insulin effect on fetal glucose utilization. Pediatr. Res. 12: 90–92.

Simon, J., and M.P. Dubois. 1983. Subtotal pancreatectomy in the chicken: Evidence of a diabetic state and effect on the somatostatin cells of the digestive tract. Diabete Metab. 9: 75–82.

Simon, J., M. Derouet, and C. Gespach. 2000. An anti-insulin serum, but not a glucagon antagonist, alters glycemia in fed chickens. Horm. Metab. Res. 32: 139–141.

Smith, C.J., V. Vasta, E. Degerman, P. Belfrage, and V.C. Manganiello. 1991. Hormone-sensitive cyclic GMP-inhibited cyclic AMP phosphodiesterase in rat adipocytes. Regulation of insulin- and cAMP-dependent activation by phosphorylation. J. Biol. Chem. 266: 13385–13390.

Smith, D.J., and J.P. McNamara. 1989. Lipolytic response of bovine adipose tissue to alpha and beta adrenergic agents 30 days pre- and 120 days postpartum. Gen. Pharmacol. 20: 369–374.

Smith, P.J., L. Wise, R. Berkowitz, C. Wan, and C.S. Rubin. 1988. Insulin-like growth factor-I is an essential regulator of the differentiation of 3t3-L1 adipocytes. J. Biol. Chem. 263: 9402–9408.

Solinas, G., S. Summermatter, D. Mainieri, M. Gubler, L. Pirola, M.P. Wymann, S. Rusconi, J.P. Montani, J. Seydoux, and A.G. Dulloo. 2004. The direct effect of leptin on skeletal muscle thermogenesis is mediated by substrate cycling between *de novo* lipogenesis and lipid oxidation. FEBS Lett. 577: 539–544.

Sollevi, A., and B.B. Fredholm. 1981. The antilipolytic effect of endogenous and exogenous adenosine in canine adipose tissue *in situ*. Acta Physiol. Scand. 113: 53–60.

Sorisky, A. 1999. From preadipocyte to adipocyte: Differentiation-directed signals of insulin from the cell surface to the nucleus. Crit. Rev. Clin. Lab. Sci. 36: 1–34.

Steiner, K.E., R.W. Stevenson, B.A. Adkins-Marshall, and A.D. Cherrington. 1991. The effects of epinephrine on ketogenesis in the dog after a prolonged fast. Metabolism 40: 1057–1062.

Strat, A.L., T.A. Kokta, M.V. Dodson, A. Gertler, Z. Wu, and R.A. Hill. 2005. Early signaling interactions between the insulin and leptin pathways in bovine myogenic cells. Biochim. Biophys. Acta 1744: 164–175.

Strosser, M.T., D. Di Scala-Guenot, B. Koch, and P. Mialhe. 1983. Inhibitory effect and mode of action of somatostatin on lipolysis in chicken adipocytes. Biochim. Biophys. Acta 763: 191–196.

Sun, J.M., M.P. Richards, R.W. Rosebrough, C.M. Ashwell, J.P. McMurtry, and C.N. Coon. 2006. The relationship of body composition, feed intake, and metabolic hormones for broiler breeder females. Poult. Sci. 85: 1173–1184.

Suniga, R.G., and T.P. Oscar. 1994. Triiodothyronine attenuates somatostatin inhibition of broiler adipocyte lipolysis. Poult. Sci. 73: 564–570.

Sweazea, K.L., and E.J. Braun. 2006. Glucose transporter expression in English sparrows (Passer domesticus). Comp. Biochem. Physiol. B 44: 263–270.

Szanto, I., and C.R. Kahn. 2000. Selective interaction between leptin and insulin signaling pathways in a hepatic cell line. Proc. Natl. Acad. Sci. U.S.A. 97: 2355–2360.

Takeuchi, S., and S. Takahashi. 1998. Melanocortin receptor genes in the chicken—tissue distributions. Gen. Comp. Endocrinol. 112: 220–231.

Takeuchi, S., K. Teshigawara, and S. Takahashi. 1999. Molecular cloning and characterization of the chicken pro-opiomelanocortin (POMC) gene. Biochim. Biophys. Acta 1450: 452–459.

Takeuchi, S., K. Teshigawara, and S. Takahashi. 2000. Widespread expression of Agouti-related protein (AGRP) in the chicken: a possible involvement of AGRP in regulating peripheral melanocortin systems in the chicken. Biochim. Biophys. Acta 1496: 261–269.

Tang, Q.Q., T.C. Otto, and M.D. Lane. 2002. CAAT/enhancer-binding protein b is required for mitotic clonal expansion during adipogenesis. PNAS 100: 850–855.

Tokushima, Y., K. Takahashi, K. Sato, and Y. Akiba. 2005. Glucose uptake *in vivo* in skeletal muscles of insulin-injected chicks. Comp. Biochem. Physiol. B 141: 43–48.

Trayhurn, P., and J.H. Beattie. 2001. Physiological role of adipose tissue: White adipose tissue as an endocrine and secretory organ. Proc. Nutr. Soc. 60: 329–339.

Valet, P., M. Berlan, M. Beauville, F. Crampes, J.L. Montastruc, and M. Lafontan. 1990. Neuropeptide Y and peptide YY inhibit lipolysis in human and dog fat cells through a pertussis toxin-sensitive G protein. J. Clin. Invest. 85: 291–295.

Van den Bergh, R., W. Oelofsen, R.J. Naudé, and S.E. Terblanche. 1992. The effect of exercise and *in vivo* treatment with ACTH and norepinephrine on the lipolytic responsiveness of guinea pig (*Cavia porcellus*) adipose tissue. Comp. Biochem. Physiol. B. 101: 553–557.

van Harmelen, V., A. Dicker, M. Rydén, H. Hauner, F. Lönnqvist, E. Näslund, and P. Arner. 2002. Increased lipolysis and decreased leptin production by human omental as compared with subcutaneous preadipocytes. Diabetes 51: 2029–2036.

Vannucci, S.J., F. Maher, and I.A. Simpson. 1997. Glucose transporter proteins in brain: Delivery of glucose to neurons and glia. Glia 21: 2–21.

Vernon, R.G., A. Faulkner, W.W. Hay, D.T. Calvert, and D.J. Flint. 1990. Insulin resistance of hind-limb tissues *in vivo* in lactating sheep. Biochem. J. 270: 783–786.

Vianen, J., P. Obels, G. van den Thillart, and J. Zaagsma. 2002. β-Adrenoceptors mediate inhibition of lipolysis in adipocytes of tilapia (*Oreochromis mossambicus*). Am. J. Physiol. 282: E318–E325.

Voet, D., J.G. Voet, and C.W. Pratt. 1999. Fundamentals of Biochemistry. John Wiley & Sons, Inc., New York.

Walton, P.E., and T.D. Etherton. 1986. Stimulation of lipogenesis by insulin in swine adipose tissue: Antagonism by porcine growth hormone. J. Anim. Sci. 62: 1584–1595.

Walton, P.E., T.D. Etherton, and C.M. Evock. 1986. Antagonism of insulin action in cultured pig adipose tissue by pituitary and recombinant porcine growth hormone: Potentiation by hydrocortisone. Endocrinology 118: 2577–2581.

Wang, X., H. Lin, Z. Song, and H. Jiao. 2010. Dexamethasone facilitates lipid accumulation and mild feed restriction improves fatty acids oxidation in skeletal muscle of broiler chicks (*Gallus gallus domesticus*). Comp. Biochem. Physiol. C 151: 447–454.

Watt, P.W., E. Finley, S. Cork, R.A. Clegg, and R.G. Vernon. 1991. Chronic control of the beta- and alpha 2-adrenergic systems of sheep adipose tissue by growth hormone and insulin. Biochem. J. 273: 39–42.

Welch, C.M., J.K. Ahola, J.B. Hall, J.I. Szasz, L. Keenan, and R.A. Hill. 2009. Physiological drivers of variation in feed efficiency in Red Angus-sired calves. J. Anim. Sci. 87(E Suppl. 2): 322.

Welch, C.M., M. McGee, T.A. Kokta, and R.A. Hill. 2012. Muscle and adipose tissue: Potential roles in driving variation in feed efficiency. pp. 175–198. *In*: R.A. Hill (ed.). Feed Efficiency in the Beef Industry. Wiley-Blackwell. Ames, Iowa.

Welch, K.C., Jr., A. Allalou, P. Sehgal, J. Cheng, and A. Ashok. 2013. Glucose transporter expression in an avian nectarivore: The ruby-throated hummingbird (*Archilochus colubris*). PLoS One 8: e77003.

Wray-Cahen, D., A.W. Bell, R.D. Boyd, D.A. Ross, D.E. Bauman, B.J. Krick, and R.J. Harrell. 1995. Nutrient uptake by the hindlimb of growing pigs treated with porcine somatotropin and insulin. J. Nutr. 125: 125–135.

Yamamoto, H., S. Kurebayashi, T. Hirose, H. Kouhara, and S. Kasayama. 2002. Reduced IRS-2 and GLUT4 expression in PPARg2-induced adipocytes derived from C/EBPB and C/EBPd-deficient mouse embryonic fibroblasts. J. Cell Sci. 115: 3601–3607.

Yorek, M.A., G.A. Rufo, Jr., and P.D. Ray. 1980. Gluconeogenesis in rabbit liver. III. The influences of glucagon, epinephrine, alpha- and beta-adrenergic agents on gluconeogenesis in isolated hepatocytes. Biochim. Biophys. Acta 632: 517–526.

Yu, H.Y., T. Inoguchi, M. Kakimoto, N. Nakashima, M. Imamura, T. Hashimoto, F. Umeda, and H. Nawata. 2001. Saturated non-esterified fatty acids stimulate *de novo* diacylglycerol synthesis and protein kinase c activity in cultured aortic smooth muscle cells. Diabetologia 44: 614–620.

Yu, Z.K., and G.J. Hausman. 1998. Expression of CCAAT/enhancer binding proteins during porcine preadipocyte differentiation. Exp. Cell. Res. 245: 343–349.

Yuan, L., H. Lin, K.J. Jiang, H.C. Jiao, and Z.G. Song. 2008. Corticosterone administration and high-energy feed results in enhanced fat accumulation and insulin resistance in broiler chickens. Brit. Poult. Sci. 49: 487–495.

Zhang, Y., K.Y. Guo, P.A. Diaz, M. Heo, and R.L. Leibel. 2001. Determinants of leptin gene expression in fat depots of lean mice. Am. J. Physiol. Regulatory Integrative Compl. Physiol. 282: 226–234.

Zhao, J.P., H. Lin, H.C. Jiao, and Z.G. Song. 2009. Corticosterone suppresses insulin- and NO-stimulated muscle glucose uptake in broiler chickens (*Gallus gallus domesticus*). Comp. Biochem. Physiol. C 149C: 448–454.

Zhao, J.P., J. Bao, X.J. Wang, H.C. Jiao, Z.G. Song, and H. Lin. 2012. Altered gene and protein expression of glucose transporter1 underlies dexamethasone inhibition of insulin-stimulated glucose uptake in chicken muscles. J. Anim. Sci. 90: 4337–4345.

Zimmerman, A.W., and J.H. Veerkamp. 2002. New insights into the structure and function of fatty acid-binding proteins. Cell Mol. Life Sci. 59: 1096–1116.

Section C
Reproduction

CHAPTER-8

Reproduction in Poultry
An Overview

Murray R. Bakst

llllllllllllllllllllllllllllllllllllllllllllllllllllllllllllllllllllllllllllllllllllllllllllllllllllllllllllllllllllllllllllllllllllllllll

INTRODUCTION

This brief overview will provide to the reader a fundamental understanding of the anatomy and physiology of reproduction in birds, with a strong emphasis on poultry. An attempt was made to cite the most recent publications describing male and female reproductive functions while including more comprehensive reviews for those who wish to delve more deeply into a particular subject area.

Structure and Function of the Male Reproductive System

As with all vertebrates, the reproductive system in male birds consist of paired testes that produce sperm and a duct system, in birds referred to as the excurrent ducts, that transports the sperm to a copulatory apparatus found on the ventral floor of the cloaca for transfer to the female at copulation (Figure 1). From a reproductive perspective, fundamental differences exist between male livestock and male poultry (in this chapter, poultry will only include domesticated chickens and turkeys). Roosters and toms possess internal testes characterized by an accelerated rate of spermatogenesis, have no accessory sex glands associated with their excurrent ducts, engage in numerous copulations with hens in the absence of an estrus (no synchronization of ovulation and copulation), and semen transfer at copulation is accomplished with a non-intromittent (non-penetrating) phallus formed on the ventral floor of their cloacae.

Retired from the Agricultural Research Service, U.S. Department of Agriculture, Beltsville, Maryland 20705.
Email: murray.bakst@outlook.com

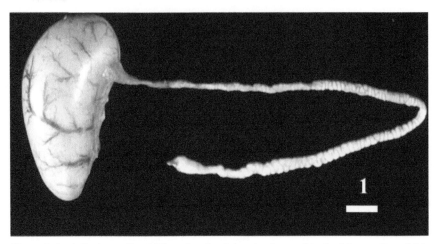

Figure 1: A relatively small epididymal region is barely observed in the hilar portion of this turkey testis. The densely coiled ductus (d.) deferens is the primary sperm storage site in the male tract. The wart-like, reddish papilla at the distal end of the d. deferens projects into the cloaca's urodeum (bar = 10 mm).

Spermatogenesis

Located at the anterior ends of the kidneys, the paired testicles are suspended from the dorsal wall of the abdominal cavity. Sperm are produced by the seminiferous epithelium lining the lumen of the seminiferous tubules (Figure 2). The highly convoluted and anastomosing seminiferous tubules fills most of the volume of the testes. Located in the space between the seminiferous tubules, referred to as the interstitial space, are connective tissue cells and fibers, blood vessels, nerves, and the Leydig cells responsible for androgen production. Two cell types comprise the seminiferous epithelium, the Sertoli cell, which is a somatic cell, and the germ cells (Figure 2). The difference between somatic cells throughout the body and germ cells localized to the seminiferous epithelium is that only germ cells undergo meiosis. Collectively, the germ cell population represents the sequential transformation of the round-ovoid spermatogonium to the morphologically mature sperm. For more detailed and comprehensive reviews of avian spermatogenesis, see Jones and Lin (1992) and Aire (2007).

Germ cells are intimately associated with the Sertoli cells that line the lumen of the seminiferous tubules. Sertoli cells are elongated and extend from the basement membrane of the seminiferous epithelium to the lumen of the seminiferous tubule. Also referred to as sustentacular cells, Sertoli cells not only serve as scaffolding for the morphing germ cells, but provide nutrients and the cell signals regulating spermatogenesis. In addition, in its basal region, tight Sertoli cell-to-cell junctions form what is referred to as the blood testes barrier. The blood-testes barrier limits blood and lymph

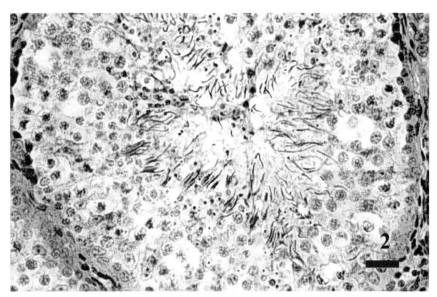

Figure 2: A cross section of a seminiferous tubule with germ cells in the different stages of differentiation is observed. The smaller denser cells lining the periphery of the seminiferous tubule are the spermatogonia. The larger diameter cells internal to the spermatogonia are the spermatocytes while the round spermatids possess the smallest nuclei, the narrow condensing nuclei of the elongating are highly distinguishable. The rust colored reaction product (wheat germ agglutinin) highlights Sertoli cell cytoplasm (bar = 10 μm).

products from entering the adluminal compartment of the seminiferous tubule thus creating a unique environment for the spermatocytes and spermatids to morphologically mature.

Except for the kinetics (rate of germ cell divisions) of the germ cells comprising the seminiferous epithelium, spermatogenesis in male livestock and poultry is basically quite similar. Spermatogonia are aligned on the inside wall of the seminiferous tubules (Figure 2). Based on their structural characteristics, spermatogonia are divided into four subpopulations of cells: Dark type A (A_d); Pale type B1 (A_{p1}); Pale type B2 (A_{p2}); and, Type B spermatogonia (Lin and Jones, 1992). The A_d are the spermatogonial stem cells (SSC) that not only is self-renewing, forming another A_d with each mitotic division, but also gives rise to a single A_{p1} spermatogonia that undergoes two additional divisions to form two A_{p2} spermatogonia and then four Type B spermatogonia. It is worth noting that partly because of the technical difficulties of accessing the germinal disc on the surface of an ovulated ovum immediately after ovulation for the production of transgenic poultry, there has been considerable interest in the isolation, culture, *in vitro* propagation, and transfection of the avian SSC. Such transfected SSC can be transferred into sterilized recipient testes, repopulate the seminiferous epithelium and eventually produce sperm carrying the transgene. It also should be noted that little is known regarding the molecular signaling and control mechanisms of germ cell differentiation in birds. In contrast, control mechanism leading to the differentiation of mammalian spermatogonia has received far more attention (Busada and Geyer, 2016).

Type B spermatogonia give rise to primary spermatocytes, which marks the meiotic phase of spermatogenesis. Primary spermatocytes are characterized by their large nuclei with chromosomes in various states of condensation and arrangements during first meiotic prophase [see reviews by Jones and Lin (1992) and Aire (2007) for more details]. As a result the original 2n number (diploid) of chromosomes in primary spermatocytes is reduced to 1n number (haploid) of chromosomes that characterize the secondary spermatocytes. The secondary spermatocytes rapidly give rise to the round spermatids marking the beginning of spermiogenesis.

Spermiogenesis is that phase of spermatogenesis when the round spermatids are structurally reorganized into the characteristic filiform shape of poultry sperm. Aire (2007) summarized the steps of spermiogenesis comparing what had been observed both in passerine and non-passerine species. The transformation of the round spermatid to an elongated spermatid to a morphologically mature sperm is characterized by the following significant events: a loss of cytoplasm; condensation and reshaping of the nucleus with concurrent replacement of the nuclear histones with protamines (thought to better stabilize sperm DNA); formation of the sperm tail (similar in structure to a cilium); the migration of mitochondria to form the sperm midpiece; and, the formation of the acrosome and perforatorium located at the anterior end of the sperm head. Unlike testicular sperm from livestock, a small percentage of morphologically mature testicular sperm from chickens are capable of progressive motility and capable of fertilizing an ovum *in vitro*.

Morphologically mature sperm released from the seminiferous epithelium (spermiation), are filiform shaped, enveloped by a plasmalemma, and are subdivided into the head, which includes the acrosome, perforatorium, and nucleus, and the tail consisting of the neck, midpiece, and principal piece (Figures 3 and 4). Situated at the anterior tip of the sperm head, the cone-shaped acrosome contains a hydrolytic enzyme (acrosin) that digests a path through the inner perivitelline layer (IPL), an acellular investment around the ovum at ovulation. Capped by the acrosome, the lanceolate-shaped perforatorium is tightly situated between the acrosome and a concavity in the cranial end of the nucleus. Given its persistent association with the nucleus in damaged frozen-thawed poultry sperm, it is assumed that the perforatorium affords some protection to the sperm nucleus during its transit through the hydrolyzed IPL. The elongated haploid sperm nucleus consists of tightly condensed chromation and at its distal end articulates with the beginning of the tail at the sperm neck. In

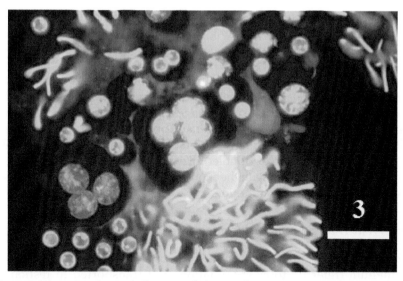

Figure 3: Dispersing the seminiferous epithelium and staining the nuclei with a nuclear fluorescence dye (bisbenzimide) highlights the shape and chromatin content of the nuclei. Small, round nuclei and the condensed elongated nuclei are the round and elongated spermatids, respectively. The larger nuclei across the middle are the spermatocytes. A single spermatogonium nucleus (arrow) and a single Sertoli cell nucleus (arrowhead) are also observed (bar = 10 μm).

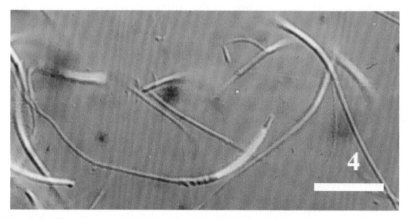

Figure 4: Poultry sperm are filiformed shaped with the head region being about 0.5 μm wide. The nuclear fluorescent dye highlights the sperm nuclei and contrasts with the acrosome at the anterior tip of the sperm and the midpiece, just distal to nucleus. The mitochondrial comprising the midpiece of one sperm are abnormally swollen and have a cobblestone like appearance. The tail is unremarkable (bar = 10 μm).

chicken and turkey sperm, the neck consists of a short proximal centriole oriented perpendicular to the long axis of the sperm and an elongated distal centriole aligned with the sperm's long axis. The sperm midpiece is characterized by a mitochondrial sheath that surrounds the neck's distal centriole and extends distally to the annulus. This narrow band-like constriction separates the midpiece from the principal piece band, the sperm tail. The principal piece is the longest part of the sperm and contains the axoneme, a collection of longitudinally oriented microtubules arrange with a central pair of doublet microtubules surrounded circumferentially by nine additional pairs of doublet microtubules (designated as a 9 + 2 configuration). Associated with the axoneme doublets are protein complexes responsible for the tail beat motion.

Excurrent Duct System and Cloaca

Morphologically mature sperm are liberated from the seminiferous epithelium (referred to as spermiation) and transported in testicular fluid through the seminiferous tubules to the rete testes. The rete testes are characterized by intra- and extra-testicular ducts and lacunae that collectively serve as a conduit between the seminiferous tubules and the excurrent duct system. The excurrent duct of male poultry consists of two segments: the epididymal region and the ductus (d.) deferens (Figure 1). Except for a short stem leading to the ductus deferens, the epididymal region is attached to the concave, medial surface (hilus) of the testis. The d. deferens extends parallel and adjacent to the ureters and terminates in the central compartment (urodeum) of the cloacal (Figure 5) as wart-like projections, the papillae.

Sperm are transported from the rete testis to the epididymal efferent ductules. Based on histological differences, the efferent ductules are subdivided into the more voluminous proximal efferent ducts and the less voluminous distal efferent ducts. From the efferent ducts, sperm are transported through the collecting ductules which collectively feed into the d. epididymides. Functionally, epididymal region, particularly the proximal efferent ducts, is involved in fluid re-absorption, thus increasing the concentration of sperm entering the d. deferens. Functional maturation of sperm in transit through the epididymal region is expressed as increases in both the mean sperm swimming velocity and the percentage of motile sperm (Nixon et al., 2014). These authors attributed the augmentation of sperm motility characteristics in the epididymal region to a secreted protein identified as hemogloblin. The d. deferens is the primary sperm storage region of the excurrent ducts: its role in post-testicular sperm maturation remains unclear.

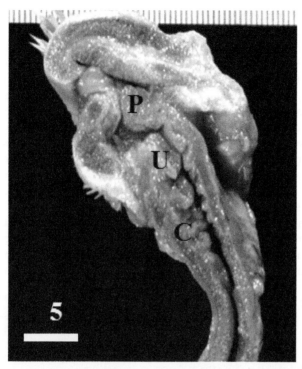

Figure 5: A longitudinal view of the cloaca from the cloacal lips through the coprodeum (C), the most cranial compartment of the cloaca and the most distal portion of gut. The slit-like proctodeum is separated from the underlying urodeum by the uroproctodeal fold (P). A wart-like papilla can be seen inside the urodeal cavity. The coprodeum is separated from the urodeum by the urocoprodeal fold (U) (bar = 10 mm).

The cloaca serves as the common opening for the deposition of excretory and digestive wastes and the reproductive tracts in both the male and female birds (Figure 5). It is divided into three compartments: the cranial compartment, the coprodeum, which is an extension of the large intestine; the central compartment, the urodeum, where the excretory ducts (ureters), vagina, and the d. deferens terminate and release their contents; and the caudal compartment, the proctodeum. The dorsal roof of the proctodeum and the ventral floor of the proctodeum, which in the male forms the phallus, are bound externally by the dorsal and ventral lips of the cloaca (Figure 6). When the dorsal and ventral lips are parted, the opening of the cloaca (the vent) is apparent.

Ejaculation is in response to visual and behavioral cues during courtship behavior or manual stimulation of the abdominal/cloacal region in order to collect semen for artificial insemination (Burrows and Quinn, 1939). While the phallus non-protrudens is not an intromittent organ in poultry, ejaculation is immediately preceded by phallic tumescence. Tumescence is result of the engorgement of lymph vessels and channels with a blood derived lymph-like fluid that originate from the vascular bodies, paired structures located in the connective tissue where the ureters and d. deferens enter interior wall of the cloaca (Knight et al., 1984). Upon sexual stimulation, blood hydrostatic pressure increases within the dense capillary network characterizing the vascular bodies. These capillaries are lined with a fenestrated endothelium, facilitating the rapid transfer of blood-derived fluid and proteins from the capillaries into the interstitial spaces. The fluid collects in the lymph vessels and ducts in the vascular body that continue into the phallus producing a transient tumescence. After ejaculation the fluid flows back toward the vascular body to eventually drain into the circulatory system.

With the possible exception of the foam gland found in the roof of the quail proctodeum, there are no accessory sex glands associated with the reproductive tract of male birds. However, at the time of ejaculation poultry semen is diluted with a variable volume of transparent-fluid.

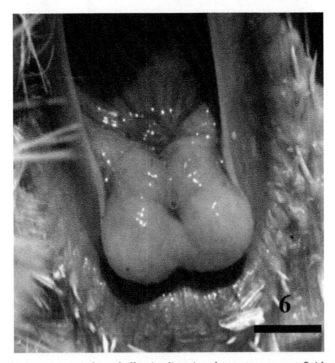

Figure 6: This tumescent turkey phallus is glistening due to transparent fluid, a vascular transudate responsible the tumescence. The phallus is part of the floor of the protodeum with the two lateral lymph folds merging medially and giving rise to the more bulbous pair of lateral phallic bodies that overhang, the ventral lip of the cloaca. By parting the cloacal (venting) the uroproctodeal fold is observed (bar = 10 mm).

Transparent-fluid is likely the same fluid responsible for phallic tumescence and originates as a transudate from both the phallus and floor of the proctodeum during tumescence (Figure 6). While phallic tumescence is a neural reflex initiated by visual and behavioral cues or manual stimulation, actual ejaculation during manual semen collection is achieved by squeezing the abdominal region surrounding the cloaca during phallic tumescence (referred to as a cloacal stroke). The semen collector should perform no more than two cloacal strokes during each collection. The reasons for this is that nearly 80% of the sperm reserve in the d. deferens has been collected and additional cloacal strokes may lead to the contamination of the collected semen with excretory debris or excessive transparent fluid.

Structure and Function of the Female Reproductive System

When the chicken and turkey hen reach about 20 and 26 wk of age, respectively, maintained on an appropriate nutritional plane, and are exposed to more than 12 hr of light (photostimulation), the left ovary and oviduct will begin to mature and reach functional maturation in about 2 wk (Figure 7). Maturation of the ovary and oviduct result from the interplay

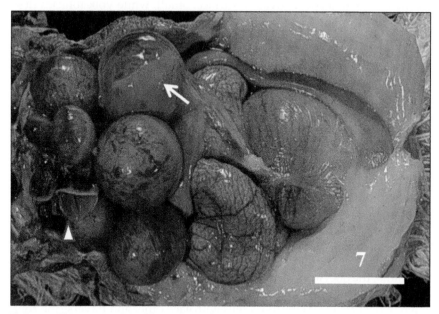

Figure 7: The ovary and oviduct of the turkey hen in egg production occupy a significant portion of the abdominal cavity. The ovarian follicular hierarchy (larger to smaller follicular oocytes) is observed with two post-ovulatory follicular shealths (white triangle). The fimbriated region was laid over the F1 follicle. A shell-membrane bound egg mass is in the uterus and the abdominal fat pad is masking the coiled vagina bound in connective tissue (bar = 3.5 cm).

between both the gonadotrophic and endocrine hormones [see Johnson (2000, 2014) for comprehensive reviews].

The hen's ovary is attached to dorsal wall of the abdominal cavity near the cranial end of the left kidney. Anatomically and functionally the ovary is divided into the cortical region, where all the oocytes are localized, and the medullary region, consisting of connective tissues, vascular systems, and nerves that support rapid follicular maturation during egg production. In the hen in egg production, there is a hierarchy of maturing yellow yolk follicular oocytes with the largest, and next to ovulate, designated as F1, the second largest being F2, continuing to where the follicular oocytes are about 1 cm in diameter. Collectively, these follicles are in the rapid growth phase of oocyte maturation accumulating yolk proteins (vitellogenin) and lipoproteins that, under the regulation of the estradiol, are synthesized in the liver and transported to the ovary via the vascular system.

During the rapid growth phase, follicular oocytes are suspended from the ovary by the stalk-like pedicule. Histologically, the pedicule appears to be an extension of the medullary region of the ovary containing blood and lymph vessels, nerves, and bundles of smooth muscles cells. Follicular oocytes are composed of concentric cell layers referred to as the follicular sheath. The external layer of the pedicule is a continuation of the germinal epithelium which covers ovary's surface. Small, white yolk follicles (less than 2 mm in diameter) visible on the ovary's surface also populate the pedicle. Subjacent to the germinal epithelium is a superficial layer (tunica) characterized by connective tissue and large venous blood vessels. Internal to and seemingly continuous with the tunica is the theca external formed by concentric layers of connective tissue and smooth muscle cells. Also observed are clusters of steroid hormone producing cells referred to as the theca gland cells. The thecal gland cells synthesize most of the circulating estradiol. The theca interna is more vascularized and its connective tissue and smooth muscle cell layers less densely arranged than that of the theca externa.

Internal to the thecal layers is the granulosa cell layer. During the rapid growth phase, the granulosa consists of a single layer of cuboidal to columnar cells enveloping the oocyte's surface. The granulosa cell layer is characterized by relatively large intercellular spaces. Such spaces facilitate the movement of vitellogenin and lipoproteins derived from thecal layer's vasculature to zona radiata, the gap between the granulosa cells and the oolemma, the oocyte's plasma membrane. Originating from the lateral and apical surfaces of the granulose cells are long cytoplasmic processes that interconnect adjacent granulosa cells with each other and traverse the zona radiata to interact with the oolemma. Small vesicles, transosomes, observed in granulosa cells and characterized by a dense rim around its outer face, are transported across the zona radiata to the oocyte. Whether transosomes are vehicles transporting RNA into maturing oocyte has not be definitively determined. However, RNA does accumulate in the oocytes during the rapid growth phase (Olszanska and Stepinska, 2008). The presence of junctional complexes where cytoplasmic processes terminate on another granulose cell plasma membrane or the oolemma suggests cell-to-cell communication, possibly affording a coordinated response to hormonal signaling.

Early in the rapid growth phase of follicular development, granulosa cells begin to synthesize the proteinaceous rod-like fibers that will eventually intermesh to form the inner perivitelline layer (IPL), an investment that anatomically corresponds to the zona pellucida of the mammalian oocyte. The IPL is an acellular investment about 2 μm thick surrounding the oocyte, except overlying the germinal disc where it is about 1.5 μm thick. The interstices between the IPL fibers are occupied by an amorphous ground substance. The granulosa cell cytoplasmic processes radiate through the developing IPL before reaching the zona radiata and oolemma. Unfortunately, the nomenclature describing the investments of the oocyte prior to ovulation and the ovum after its transport through the infundibulum is not consistent and has led to confusion. It is suggested that the oocyte's plasma membrane, the primary investment of the oocyte, be referred to as the oolemma, as the term 'vitelline membrane' originally intended by Wyburn et al. (1965) to denote the oolemma has often been used to describe the oolemma and IPL complex. The term 'perivitelline membrane' is a misnomer because, as described above, the IPL is a fibrous reticulum formed by the granulosa during the oocytes rapid growth phase, and not a plasma membrane. Since the IPL is elaborated by the granulosa cells, this is considered a secondary investment. Following ovulation, the tertiary layer, the outer perivitelline layer (OPL), forms around the ovum as a result of the secretion of albumen proteins secreted by the subepithelial tubular glands of the mid to distal infundibulum and proximal magnum as the ovum traverses these segments.

Another conspicuous feature of the follicular oocyte during the rapid growth phase is the stigma, a narrow, almost clear band due to the absence of medium to large blood vessels and the presence of a more attenuated thecal layer (Figure 7). The stigma generally curves around the pole opposite the pedicle and is the site where the follicular sheath spits at ovulation releasing the ovum. At ovulation, the ovum is surrounded only by the IPL. The granulosa cells are retained by the post-ovulatory follicular sheath that gradually atrophies over the next few days. Unlike mammals, there is no corpus luteum formation.

In the preovulatory oocyte, the germinal disc is a 3 mm diameter disc subjacent to the oolemma. It whitish color is due to the aggregation of white yolk spheres and cell organelles including the germinal vesicle, the haploid nucleus. The germinal vesicle is a fluid filled sphere located in the center of the germinal disc and is barely visible to the eye as a black dot. Dispersed chromosomes are suspended in the fluid component of the germinal vesicle that throughout follicular maturation was arrested in the first prophase stage of meiosis. About 4–6 hr prior to ovulation, the F1 follicle responds to peaking levels of luteining hormone (LH) by coordinated retraction of the granulosa cell cytoplasmic processes from adjacent cells and the oolemma and by the initiation of germinal vesicle breakdown and resumption of meiosis. By the time the ovulated ovum reaches the site of fertilization in the oviduct, the infundibulum, the haploid nucleus is in the metaphase stage of the 2nd meiotic division.

Fertilization

Fertilization in poultry is complex, multi-step process limited to the germinal disc region (see Stepinska and Bakst, 2007 for review). While the molecular interactions between sperm and the IPL have been summarized (Nishio et al., 2014; Ichikawa et al., 2016) similar information regarding the events leading to the initial cleavage division is not available. Briefly, fertilization begins when one or more sperm bind to sperm-specific ZP receptors associated with the IPL overlying the germinal disc (Figures 8, 9). This initiates a breakdown of the acrosome and the release of acrosin, an enzyme that hydrolyzes the fibers of the thus forming a path or hole for the sperm to reach the oolemma (Figure 10). A staining technique [for procedure see Bramwell and Donoghue (2010)] to visualize sperm-holes in the IPL overlying the germinal disc in the laid egg (Figure 11) is used by both poultry scientists and commercial breeding farms personnel to

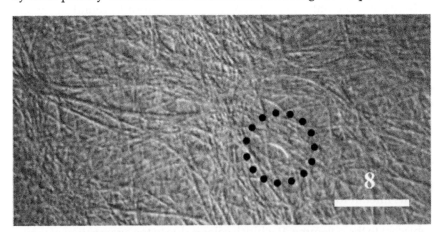

Figure 8: At ovulation the ovum is enveloped by the inner perivitelline layer. This acellular investment is composed of fibers and a ground substance filling between the fibers. Sperm must hydrolyze the IPL overlying the germinal disc to fertilize the ovum. A sperm stained with a nuclear fluorescence dye is observed on the IPL surface (bar = 20 μm).

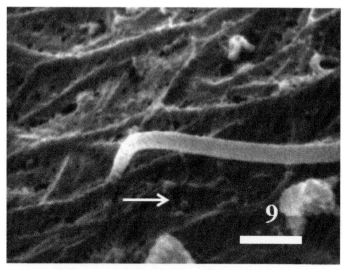

Figure 9: A scanning electron micrograph showing what is observed in Figure 8 but at a higher resolution. Remnant cytoplasmic processes (arrow) lay across the surface of the IPL (bar = 1 μm).

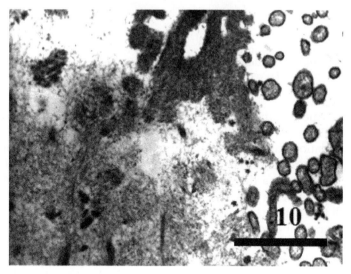

Figure 10: Sperm interacting with receptors on the surface of the IPL undergo an acrosome reaction, acrosin is released, and the fibers of the IPL are hydrolyzed. The IPL fibers breakdown to microfibrils creating a path for the sperm. Cross sections of microvilli are observed in the right side of the image indicating that the IPL was removed from the germinal disc region (bar = 0.5 μm).

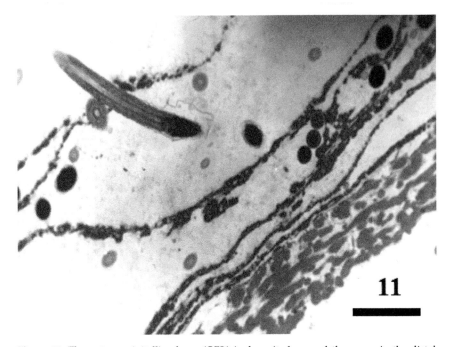

Figure 11: The outer perivitelline layer (OPL) is deposited around the ovum in the distal infundibulum and proximal magnum. The OPL servers as block to pathological polyspermy by trapping sperm between the layers of the fibrous albumen proteins. A tangential section of the sperm midpiece and several cross sections of sperm are observed in the OPL. The fibers of the IPL remain intact (bar = 1.5 μm).

determine true fertility and to estimate how many sperm interact with the germinal disc region at the time of fertilization.

At ovulation, the oolemma enveloping the ovum is discontinuous and relatively featureless except at the germinal vesicle. Here the oolemma is characterized by a dense array of microvilli and other cytoplasmic elaborations. Sperm passing through the IPL contact the oolemma microvilli that appear to embrace and subsequently internalize the sperm into the germinal disc. Perry (1987) observed the following sequence of nuclear events following sperm incorporation into the ovum: less than 1 hr after ovulation sperm nuclei decondensed and formed pronuclei; by 3 hr, pronuclei throughout the germinal disc had enlarged; at the same time, in the center of the germinal disc were a pair of apposed pronuclei, presumably the haploid male and female pronuclei; 4 hr after ovulation, mitotic figures were observed in the center of the germinal disc and assumed to be the zygote. One assumes that the diploid number of chromosomes was reconstituted, a process known as syngamy, between 3 and 4 hr after oviposition. It has been suggested that the supernumerary sperm in germinal disc are degraded by DNases (Olszanska and Stepinska, 2008).

Polyspermy, more than one sperm penetrating the IPL and entering the ovum is normal in birds. Given the mass of the germinal disc, multiple sperm entry into the germinal disc may be necessary to facilitate activation of maternal RNAs initiating development. However, while uncommon, a single sperm penetration site in isolated IPL's overlying the germinal disc have been observed in fertilized ova (unpublished observation) supporting the suggestion that polyspermic fertilization may not be obligatory in birds (Stepinska and Bakst, 2007).

While one sperm or hundreds of sperm penetrating the IPL may result in fertilization, there is some maximum number of sperm interacting with the ovum that will lead to pathological polyspermy. Pathological polyspermy can be due to the destruction of the germinal disc by the excessive hydrolytic activity of the sperm or early embryonic mortality. Unless there is an aberration in the oviductal sperm transport or sperm storage mechanisms, pathological polyspermy is likely rare following natural mating. In contrast, when a hen is artificially or surgically inseminated anterior to the vagina, large numbers of sperm are transported directly to the infundibulum. This by-passes both the vagina, where there is a selective reduction in sperm numbers and the uterovaginal junction (UVJ) where 'selected' sperm are sequestered in sperm storage tubules (SSTs) to be release over the daily ovulatory cycle. In the routine artificial insemination of commercial turkey hens, pathological polyspermy may be observed if the hen is inseminated shortly after oviposition. At this time the vagina is more flaccid due to the recent passage of the egg mass. If not carefully vagina is not carefully everted for proper placement of the insemination straw, the distal end of the vagina may project into the cloaca (insemination crews refer to this type of vaginal eversion as a "rose" due to the appearance of the exteriorized vaginal mucosa). With such an eversion, semen in the insemination straw semen may be deposited in the anterior end of the vagina (UVJ) or uterus thus by-passing the vagina. Abnormally large numbers of sperms engorge the infundibular sperm storage sites and subsequently interact with the ovulated ovum. Pathological polyspermy may also be observed when co-incubating a freshly ovulated ovum with sperm (Howarth and Digby, 1973).

Unlike mammalian ova where cortical granules are released from the ovum with the initial sperm penetration blocking further sperm penetration, the block to pathological polyspermy in hens is the OPL (see previous section). That sperm have been observed embedded at varying levels of this tertiary investment suggests that the fibrous proteins act as a physical barrier to sperm reaching the IPL (Figure 12). Therefore, to reach the surface of the IPL prior to OPL accretion, sperm must be in the luminal mucosa of the fimbriated region of the infundibulum, which lacks the sub-epithelial tubular glands in the distal half of the infundibulum that synthesize the albumen proteins forming the OPL. It is assumed that sperm are on the

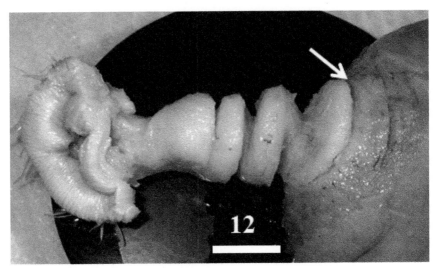

Figure 12: If the excised vagina and uterus are fixed as one segment and the connective tissue dissected away, it is apparent that the vagina is a highly coiled segment with the UVJ (arrow) actually coiled and attached on the uterine wall (bar = 10 mm).

surface mucosa of the fimbria and may be transferred to the surface of the ovum either during the retrieval of the ovulated ovum or as the ovum enters the slit-like opening (ostium) of the infundibulum.

A question that remains unanswered is how sperm preferentially localize on the IPL overlying the germinal disc. Possibly a chemotaxic mechanism (Howarth and Digby, 1973) attracts sperm to the germinal disc where associated sperm receptors (Bakst and Howarth, 1977) bind the sperm to the IPL overlying the germinal disc. Whether chemotaxis is responsible for attracting sperm to the germinal disc is still not known. However, Nishio et al. (2014) suggested that an abundance of ZP2 at the germinal disc region may be the basis for preferential sperm attachment to the IPL and the initiation of the acrosome reaction.

Oviduct

The hen's oviduct is a remarkable organ not only for its assembly-line like formation of the laid egg but its more subtle functions in the fertilization process and sustained fertility. (Fertilization, as described in the previous section, should not be confused with fertility, which is the ratio of the number of eggs fertilized to the number of eggs laid.) The interest in the hen's oviduct goes beyond that of egg production: the oviduct has served for decades as a model system for studies addressing the cellular and molecular mechanisms of steroid hormones on cell differentiation, particularly in the magnum, the albumen producing segment of the oviduct. The primary focus of this section will be on the roles of the oviduct in sperm selection, sperm storage, and sustained fertility.

Anatomically, the oviduct is divided into five functionally segments (Jacobs and Bakst, 2007). The most cranial segment, the infundibulum, is the adjacent to the ovary and has two regions. The fimbriated region, often referred to as the funnel, is thin walled and lacks sub-epithelial tubular glands. Its function in the process of fertilization was addressed in the previous section. The funnel portion gradually tapers to form the tubular neck region of infundibulum. Here sub-epithelial tubular glands begin to differentiate as invaginations of the surface epithelium and progressively become more densely distributed in the distal infundibulum. Functionally, the OPL restricts further sperm interaction with the IPL and also serves as a barrier preventing bacteria from reaching the ovum.

While the infundibulum is considered a secondary sperm storage site (Bakst et al., 1994), its role in oviductal sperm storage is not comparable to that of the primary site, the UVJ-SSTs (Bakst, 2003). Reference to the infundibulum as a secondary sperm storage site is based on the observations that sperm released from the UVJ-SSTs during the daily ovulatory cycle ascend the oviduct and are then stored in the distal infundibulum. Sperm must also reach the fimbriated region of the infundibulum to interact with the IPL enveloping the ovulated ovum. Other sperm are temporarily stored in the lumina of sub-epithelial tubular glands and between apposed folds comprising infundibular mucosa. When the ovulated ovum reaches the distal infundibulum it distends the luminal mucosal as well as induces the release of the secretory material accrued by the epithelial cells forming the sub-epithelial glands. It is during this secretory activity induced by the passage of the ovum that sperm residing within the sub-epithelial glands are transported out along with the albumen-like proteins and most likely constitute the population of sperm embedded in the OPL.

As the density of the sub-epithelial tubular glands in the distal infundibulum continues to increase, the muscoal folds visibly increase in height and width and gradually become indistinguishable from the mucosal folds of the magnum. The magnal sub-epithelial tubular glands synthesize and secrete the albumen that accumulates around the yolk during its transport toward the uterus. The sub-epithelial tubular glands in the isthmus mucosa synthesize and secrete the proteins that form the shell membrane enveloping the albumen and ovum, collectively referred to as the oviductal egg mass (not to be confused with the laid egg). How active secretion of both the albumen and the shell membrane proteins and other components impact abovarian transport has been the subject of limited speculation. However, with the increasing mass of the egg mass and the consistency and composition of the albumen in particular, it has been assumed that such physical barriers may impede the transport of sperm to the infundibulum (Bakst et al., 1994). Birkhead and Brillard (2007) provide a comprehensive review of the barriers sperm confront in the oviduct leading up to fertilization.

With completion of the shell membrane, the egg mass progresses from the isthmus to the tubular red region of the uterus, then proceeds to the uterine pouch where it resides until completion of eggshell formation. In the past decade, there has been a resurgence of interest not only in the cellular but the molecular mechanisms orchestrating eggshell formation. This is summarized by Brionne et al. (2014) and Marie et al. (2015). Brionne et al. (2014) stated that all the precursors required for shell formation are secreted by the uterine mucosa over the 20 hr period the egg mass is in the uterus including large amounts of calcium and bicarbonate. During the period of shell formation two other events relative to fertile egg production are also in progress. First, the fertilized germinal disc, which began cleavage in the isthmus, continues to develop in the uterus and at oviposition consists of a 50,000 to 80,000 cell blastoderm. Second, the egg mass in the uterine pouch distends the uterine as well as the UVJ mucosa, causing the later to become becoming contiguous with the uterine mucosa (Bakst and Akuffo, 2009). Thus, sperm released from the SSTs are immediately exposed the calcium and bicarbonate rich fluids in the uterine lumen presumably potentiating sperm motility at the time of transport to the infundibulum.

At the completion of shell formation the egg mass is pushed into the vagina and quickly expelled, a process known as oviposition. Oviposition is a coordinated response to the contractions of the smooth muscle layers of the uterus and vagina that are primarily regulated by arginine vasotocin and the neurotransmitter galanin (Li et al., 1996). Although speculated upon since the discovery of the UVJ SSTs in the 1960s, there remains no consensus regarding the influence of oviposition on sperm release from the SSTs (Bakst et al., 1994; Bakst, 2011). While there is no muscular sphincter restraining the uterine egg mass from entering the vagina during shell formation, the cranial portion of the vagina is tightly coiled and/or folded and bound to the distal third of the uterus by the connective tissue envelop (see next section). At oviposition, hormonal induced uterine smooth muscle contractions and, one assumes, a relaxation of the connective tissues enveloping the UVJ and vagina, pushes the egg mass into the cranial vaginal orifice for expulsion.

Role of Vagina in Sustained Fertility

The vagina is more than just a conduit for the hard shelled egg at oviposition. Its roles in sperm selection and transport from the site of semen transfer to the UVJ and sperm storage in the SSTs are integral to the maintenance of sustained fertility in domestic and nondomestic birds laying one or more clutches of eggs. When the abdominal cavity of a hen in egg production is

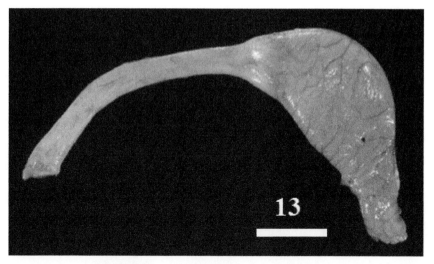

Figure 13: If the excised vagina and uterus are dissected free of connective tissue without fixation, the vagina is straight and about 9 cm long. The UVJ appears to be slightly dilated compared to the remainder of the vagina. (bar = 20 mm).

observed the caudal end of the uterus appears to be continuous with a short mass of connective tissue that merges with the cloacal region (Figure 7). If the uterus-connective tissue mass-cloaca is isolated as a single segment, placed in fixative, and then dissected free of the cloacal tissue and the connective tissue surrounding the caudal end of the uterus and the vagina, the coiled configuration of the vagina is apparent (Bakst and Akuffo, 2009) (Figure 13). However, if the same segment is unfixed and then dissected out, the turkey vagina will be straight and about 8 cm long (Figure 14). When cut longitudinally from the cloacal orifice to the caudal end of the uterus, the straight parallel primary mucosal folds of the vaginal mucosa are observed. At the anterior 2 cm of the vagina, which is the UVJ, the parallel folds are slightly less voluminous than the remainder of the vagina. The mucosal folds of the UVJ gradually transition to more wavy folds prior to morphing into the irregular folds of the uterine mucosa.

In the light of more recent observations regarding oviductal sperm motility and sperm storage in- and release from the SSTs (Sasanami et al., 2013), a more comprehensive description of the fate of sperm in the hen's oviduct than previously reported (Bakst, 2011) is warranted. Following semen transfer to the distal vagina either by natural mating or artificial insemination (AI), a small number of sperm (no more than 5 million under optimal conditions in the turkey following AI) are transported to UVJ to enter the SSTs by a combination of their intrinsic mobility (measured as the capacity to move through a viscous medium at 40C; Froman et al., 1999) and

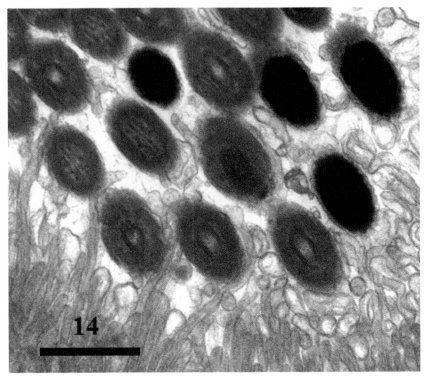

Figure 14: Sperm (predominantly cross sections of sperm nuclei and midpieces) are closely associated with the microvilli from the apical surface of the SST epithelium. The tips of some of the microvilli have formed microvillous blebs that pinch off and become closely associated with the sperm (bar = 1 μm).

cilia beat activity (Bakst, 2011). Based on histology of the vaginal mucosal folds following AI, adovarian sperm transport is most likely between apposed folds as sperm in the central lumen have been observed trapped in a fibrous secretory material, particularly in the distal vagina.

Holt and Fazeli (2015) in a comparative review of sperm storage in vertebrate females recognized that ejaculated semen consists of subpopulations of sperm that may exhibit variation in sperm morphology and responsiveness to signaling molecules. This concept certainly applies to poultry with respect to sperm motility characteristics. Individual ejaculates from toms and roosters possess sperm exhibiting different mobility rates and those males with the highest sperm mobility scores produce the most offspring (Froman et al., 2006). The subpopulation of sperm reaching the SSTs are characterized by having a high rate of mobility (Froman et al., 2006) and possessing specific glycoconjugates associates with the plasmalemmal surface (Steele and Wishart, 1996; Wishart and Horrocks, 2000; Pelaez and Long, 2008). The failure of sperm from other species to be transported to the UVJ following insemination is most likely due the non-recognition of the surface glycoconjugates (Steele and Wishart, 1992) by the vagina. However, when co-incubated with UVJ explants containing SSTs, Steele and Wishart (1992) observed such hetereologous sperm within the SSTs. Other factors also influence successful sperm transport to the UVJ including the stage of the ovulatory cycle at the time of semen transfer (important when performing AI) and the timing of the AI relative to the onset of egg production. Optimal fertility in commercial turkey hens is realized when the initial inseminations are performed prior to the onset of egg production. This is common to most birds studied following copulation as well (Hemmings et al., 2015). Interestingly, it has been suggested that following mating the red junglefowl hen may select sperm from males with a dissimilar major histocompatibility complex (Løvlie et al., 2013) that would, in effect, minimize inbreeding. This process of post-mating sperm selection by the female is referred to as cryptic female choice and has been reported both in invertebrate and vertebrate females (Eberhard, 1996).

How sperm survive within the SSTs has been the object of much speculation since their discovery in the 1960s (Bakst et al., 1994; Sasanami et al., 2013). Although a foreign body in the hen, there is no evidence of an immunological response initiated by the vaginal mucosa to the sperm in the SSTs in healthy hens. Das et al. (2008) summarized evidence that the SSTs are immuno-privileged sites for sperm and that this immune-suppression is mediated by transforming growth factor-β (TGFβ). Specifically, expression of TGFβ increased only with the presence of sperm in the SSTs and that TGFβ suppresses local T and B cell immune-responses to sperm (Das et al., 2006). In addition to immune-suppression, it has long been suggested that certain sperm functions need to be reversible suppressed (sperm metabolism and motility) or stabilized (sperm plasmalemma and acrosome) (Bakst et al., 1994) for successful sperm storage. Using Japanese quail, Matsuzaki et al. (2015) provided evidence that low oxygen levels and high lactic acid concentrations suppressed sperm motility both in the SSTs and under *in vitro* conditions. These authors also suggested that the suppressed sperm motility resulted in lower ATP consumption and subsequently reduced formation of reaction oxygen species. This is intriguing since the major impediment to successful semen storage *in vitro*, where sperm are actively motile, is lipid peroxidation. Referencing other work (Ito et al., 2011), Matsuzaki et al. (2015) indicated that the quiescent sperm in the SSTs of Japanese quail are expelled from the SSTs by a progesterone triggered mechanism. This mechanism may be due to a progesterone-induced, coordinated contraction of the prominent actin band observed in the apical tight junction zone subjacent to the apical microvilli in SST epithelial cells (Freedman et al., 2001). While progesterone may initiate the expulsion of sperm from the SSTs, progesterone has little impact on sperm motility (author's observations).

In the SSTs, lactic acid appears to render sperm quiescent. However, when sperm are released into the UVJ, it appears that heat shock protein 70 (HSP70), which is expressed by the UVJ in Japanese quail, may augment sperm motility in the UVJ lumen for transport to the infundibulum, where HSP70 is also expressed (Hiyama et al., 2014).

In contrast to the observations observed in the Japanese quail, Froman (2003) proposed that sperm in the SST lumen swim against a current generated by the SST toward the UVJ. As long as the sperm velocity is at least as great as the current passing through the SST, the sperm are retained in the SST lumen. Only when the sperm swimming velocity drops below the velocity of the SST fluid do sperm escape into the UVJ lumen. Furthermore, Froman et al. (2011) indicated that chicken sperm rely on endogenous fatty acids to enter the SSTs and on the exogenous lipids released from the SST epithelium to survive within the SST. The source of the exogenous lipids along with proteins and signaling molecules appear to be microvillous blebs (MvB) released from the microvilli forming the apical border of the SST epithelial cells (Bakst and Bauchan, 2015) (Figure 14). These authors suggested that the MvBs were a variant of the shedding vesicles released by the luminal epithelium lining the mammalian epididymis and prostate gland (known as epididymosomes and prostasomes, respectively) that have been shown contribute to post-testicular sperm maturation, sperm membrane modifications, protection sperm against oxidative stress, sperm capacitation and decapacitation, and other sperm functions (see review by Sullivan and Saez, 2013).

In the turkey with an egg mass in the uterus, the UVJ folds containing SSTs are contiguous with the uterine folds (Bakst, 2011) (Figure 15). It is likely that during their storage in the SSTs, intracellular sperm calcium concentrations are reduced thus contributing to the sperm's quiescent in the SSTs (see above). In contrast, sperm released from the SSTs during that extended period of the daily ovulatory cycle when the uterus contains an egg mass would be exposed to the calcium-rich uterine fluids. It has been shown repeatedly that calcium augments chicken and turkey sperm motility *in vitro* and that calcium acts as an activator of sperm function prior to fertilization in mammals. If HSP70 augments chicken and turkey motility as it does in Japanese quail, what is the role of the calcium rich environment in the uterine, more vigorous motility? Regardless, sperm motility is activated upon release from the SSTs for transport to the infundibulum.

It is likely that the sperm exiting the SSTs have been modified at both the cellular and molecular level during their residence in the SSTs. A select population of actively motile sperm enter the SSTs and are reversibly rendered quiescent during storage in the SSTs. Upon their egress from the SST into the UVJ and uterine lumina sperm regain their active motility. This sequence is reminiscent of mammalian sperm following semen transfer

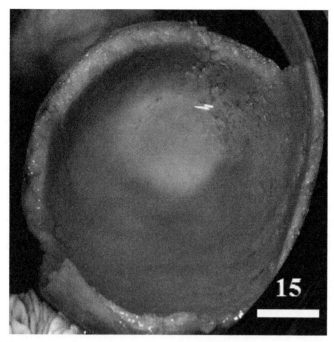

Figure 15: The vagina and uterus containing a hard-shelled egg mass was excised and fixed. When cut transversely, the egg mass removed, the UVJ mucosa (white rounding area) is contiguous with the uterine mucosa lining the uterine pouch. This indicates that sperm exiting the SSTs are immediately exposed to the calcium rich uterine fluid, thus augmenting sperm motility (bar = 10 mm).

where a select population of motile sperm reach isthmus, associate with its epithelial cells and become quiescent, and then are released after some time period in an activated state. However, unlike avian sperm that ascend to the infundibular secondary sperm storage sites, activated mammalian sperm ascend to the site of fertilization in near synchrony with impending ovulation. It is a matter of time before we better understand the cellular and molecular events that transpire and influence sperm storage in the SSTs. It is likely that the plasmalemma of sperm within the SST lumen is modified to accommodate storage for days and weeks. If at the time of release from the SSTs sperm motility is active, do the sperm becomes quiescent again in the infundibular sperm storage sites, and then activates them once again at the time of ovulation. This is feasible as calcium has been detected in the fluid removed from the ovarian pocket around the time of ovulation (Ashizawa and Wishart, 1992).

Keywords: spermatogenesis, sperm maturation, ovary, oviduct, oviductal sperm storage, fertilization

References Cited

Aire, T.A. 2007. Spermatogenesis and testicular cycles. pp. 279–348. *In*: B.G.M. Jamieson (ed.). Reproductive Biology and Phylogeny of Birds—Part A. Science Publishers, Enfield, NH.

Ashizawa, K., and G.J. Wishart. 1992. Factors from fluid of the ovarian pocket that stimulate sperm motility in domestic hens. J. Reprod. Fert. 95: 855–860.

Bakst, M.R., and B. Howarth, Jr. 1977. Hydrolysis of the hen's perivitelline layer by cock sperm *in vitro*. Biol. Reprod. 17: 370–379.

Bakst, M.R., G.J. Wishart, and J.P. Brillard. 1994. Oviducal sperm selection, transport, and storage in poultry. Poult. Sci. Rev. 5: 117–143.

Bakst, M.R. 2003. Oviducal Sperm storage in turkeys: The infundibulum as a secondary sperm storage site, or is it? pp. 447–450. *In*: A. Legakis, S. Sfenthourakis, R. Polymeni, and M. Thessalou-Legaki (eds.). The New Panorama of Animal Evolution: Proceedings XVIII International Congress of Zoology. Athens, Greece.

Bakst, M.R., and V. Akuffo. 2009. Morphology of the turkey vagina with and without an egg mass in the uterus. Poult. Sci. 88: 631–635.

Bakst, M.R. 2011. Physiology and endocrinology symposium: Role of the oviduct in maintaining sustained fertility in hens. J. Anim. Sci. 89: 1323–1329.

Bakst, M.R., and G. Bauchan. 2015. Apical blebs on sperm storage tubule epithelial cell microvilli: Their release and interaction with resident sperm in the turkey hen oviduct. Theriogenology 83: 1438–1444.

Birkhead, T.R., and J.P. Brillard. 2007. Reproductive isolation in birds: Postcopulatory prezygotic barriers. Trends Eco. Evo. 22: 266–272.

Bramwell, R.K., and A.M. Donoghue. 2010. Predicting fertility: Section 4. Determination of holes made by sperm in the perivitelline layer of laid eggs: The sperm penetration assay. pp. 90–94. *In*: M.R. Bakst, and J.A. Long (eds.). Techniques for Semen Evaluation, Semen Storage, and Fertility Determination. 2nd Edition. The Midwest Poultry Federation, Buffalo, MN.

Brionne, A., Y. Nys, C. Hennequet-Antier, and J. Gautron. 2014. Hen uterine gene expression profiling during eggshell formation reveals putative proteins involved in the supply of minerals or in the shell mineralization process. BMC Genomics 15: 220.

Burrows, W.H., and J.P. Quinn. 1939. Artificial insemination of chickens and turkeys. U.S. Department of Agriculture, Washington, D.C. Circular No. 525: 1–13.

Busada, J.T., and C.B. Geyer. 2016. The role of retinoic acid (RA) in spermatogonial differentiation. Biol. Repro. 94: 1–10.

Das, S.C., N. Isobe, M. Nishibori, and Y. Yoshimura. 2006. Expression of transforming growth factor-β isoforms and their receptors in utero-vaginal junction of hen oviduct in presence or absence of resident sperm with reference to sperm storage. Reproduction 132: 781–790.

Das, S.C., N. Isobe, and Y. Yoshimura. 2008. Mechanism of prolonged sperm storage and sperm survivability in the hen's oviduct: A review. Am. J. Reprod. Immunol. 60: 477–481.

Deviche, P., L. Hurley, and H.B. Fokidis. 2011. Avian testicular structure, function, and regulation. Horm. Repro. Vert. 4: 27–69.

Eberhard, W.G. 1996. Selection on cryptic female choice. pp. 44–79. Female Control: Sexual Selection by Cryptic Female Choice. Princeton University Press, NJ.

Freedman, S., V. Akuffo, and M.R. Bakst. 2001. Evidence for the innervation of sperm-storage tubules in the oviduct of the turkey (*Meleagris gallopavo*). Reproduction 121: 809–814.

Froman, D.P., A.J. Feltmann, M.L. Rhoads, and J.D. Kirby. 1999. Sperm mobility: A primary determinant of fertility in the domestic fowl (*Gallus domesticus*). Bio. Repro. 61: 400–405.

Froman, D.P. 2003. Deduction of a model for sperm storage in the oviduct of the domestic owl (*Gallus domesticus*). Biol. Reprod. 69: 248–253.

Froman, D.P., J.C. Wardell, and A.J. Feltmann. 2006. Sperm mobility: Deduction of a model explaining phenotypic variation in roosters (*Gallus domesticus*). Biol. Reprod. 74: 487–491.

Froman, D.P., A.J. Feltmann, K. Pendarvis, A.M. Cooksey, S.C. Burgess, and D.D. Rhoads. 2011. A proteome-based model for sperm mobility phenotype. J. Anim. Sci. 89: 1330–1337.

Hemmings, N., T.R. Birkhead, J.P. Brillard, P. Froment, and S. Briere. 2015. Timing associated with oviductal sperm storage and release after artificial insemination in domestic hens. Theriogenology 83: 1174–1178.

Hiyama, G., M. Matsuzaki, S. Mizushima, H. Dohra, K. Ikegami, T. Yoshimura, K. Shiba, K. Inaba, and T. Sasanami. 2014. Sperm activation by heat shock protein 70 supports the migration of sperm released from sperm storage tubules in Japanese quail (*Coturnix japonica*). Reproduction 147: 167–178.

Holt, W.V., and A. Fazeli. 2015. Sperm storage in the female reproductive tract. Annual Rev. Anim. Biosci. 4: 291–310.

Howarth, B., and S.T. Digby. 1973. Evidence for the penetration of the vitelline membrane of the hen's ovum by a trypsin-like acrosomal enzyme. J. Repro. Fert. 33: 123–125.

Ichikawa, Y., M. Matsuzaki, G. Hiyama, S. Mizushima, and T. Sasanami. 2016. Sperm-egg interaction during fertilization in birds. J. Poult. Sci. doi: 10.2141/jpsa.0150183.

Ito, T., N. Yoshizaki, T. Tokumoto, H. Ono, T. Yoshimura, A. Tsukada, N. Kansaku, and T. Sasanami. 2011. Progesterone is a sperm-releasing factor from the sperm-storage tubules in birds. Endocrinology 152: 3952–3962.

Jacobs, M., and M.R. Bakst. 2007. Anatomy of the female reproductive tract. pp. 149–179. *In*: B.G.M. Jamieson (ed.). Reproductive Biology and Phylogeny of Birds—Part A. Science Publishers, Enfield, NH.

Johnson, A.L. 2000. Reproduction in the female. pp. 569–596. *In*: P.D. Sturkie (ed.). Avian Physiology. Academic Press, San Diego, CA.

Johnson, A.L. 2014. The avian ovary and follicle development: Some comparative and practical insights. Turkish J. Vet. Anim. Sci. 38: 660–669.

Jones, R.C., and M. Lin. 1992. Spermatogenesis in birds. Oxford Rev. Reprod. Biol. 15: 233–264.

Knight, C.E., M.R. Bakst, and H.C. Cecil. 1984. Anatomy of the Corpus vasculare paracloacale of the male turkey. Poult. Sci. 63: 1883–1891.

Li, D., K. Tsutsui, Y. Muneoka, H. Minakata, and K. Nomoto. 1996. An oviposition-inducing peptide: Isolation, localization, and function of avian galanin in the quail oviduct. Endocrinology 137: 1618–1626.

Lin, M., and R.C. Jones. 1992. Renewal and proliferation of spermatogonia during spermatogenesis in the Japanese quail, *Coturnix coturnix japonica*. Cell and Tissue Res. 267: 591–601.

Løvlie, H., M.A. Gillingham, K. Worley, T. Pizzari, and D.S. Richardson. 2013. Cryptic female choice favours sperm from major histocompatibility complex-dissimilar males. Proc. Royal Soc. London B: Biol. Sci. 280: 1296–2013.

Marie, P., V. Labas, A. Brionne, G. Harichaux, C. Hennequet-Antier, A.B. Rodriguez-Navarro, Y. Nys, and J. Gautron. 2015. Quantitative proteomics provides new insights into chicken eggshell matrix protein functions during the primary events of mineralisation and the active calcification phase. J. Proteomics 126: 140–154.

Matsuzaki, M., S. Mizushima, G. Hiyama, N. Hirohashi, K. Shiba, K. Inaba, T. Suzuki, H. Dohra, T. Ohnishi, Y. Sato, and T. Kohsaka. 2015. Lactic acid is a sperm motility inactivation factor in the sperm storage tubules. Sci. Rep. 5: 17643.

Nishio, S., Y. Kohno, Y. Iwata, M. Arai, H. Okumura, K. Oshima, D. Nadano, and T. Matsuda. 2014. Glycosylated chicken ZP2 accumulates in the egg coat of immature oocytes and remains localized to the germinal disc region of mature eggs. Biol. Repro. 91: 1–10.

Nita, H., Y. Osawa, and J.M. Bahr. 1991. Multiple steroidogenic cell populations in the thecal layer of preovulatory follicles of the chicken ovary. Endocrinology 129: 2033–2040.

Nixon, B., K.A. Ewen, K.M. Krivanek, J. Clulow, G. Kidd, H. Ecroyd, and R.C. Jones. 2014. Post-testicular sperm maturation and identification of an epididymal protein in the Japanese quail (*Coturnix coturnix japonica*). Reproduction 147: 265–277.

Olszanska, B., and U. Stepinska. 2008. Molecular aspects of avian oogenesis and fertilisation. Internat. J. Dev. Biol. 52: 187–194.

Pelaez, J., and J.A. Long. 2008. Characterizing the glycocalyx of poultry spermatozoa: II. *In vitro* storage of turkey semen and mobility phenotype affects the carbohydrate component of sperm membrane glycoconjugates. J. Androl. 29: 431–439.

Perry, M.M. 1987. Nuclear events from fertilisation to the early cleavage stages in the domestic fowl (*Gallus domesticus*). J. Anat. 150: 99–109.

Sasanami, T., M. Matsuzaki, S. Mizushima, and G. Hiyama. 2013. Sperm storage in the female reproductive tract in birds. J. Repro. Dev. 59: 334–338.

Steele, M.G., and G.J. Wishart. 1992. Evidence for a species-specific barrier to sperm transport within the vagina of the chicken hen. Theriogenology 38: 1107–1114.

Steele, M.G., and G.J. Wishart. 1996. Demonstration that the removal of sialic acid from the surface of chicken spermatozoa impedes their transvaginal migration. Theriogenology 46: 1037–1044.

Stepinska, U., and M.R. Bakst. 2007. Fertilization. pp. 553–587. *In*: B.G.M. Jamieson (ed.). Reproductive Biology and Phylogeny of Birds—Part A. Science Publishers, Enfield, NH.

Sullivan, R.F., and F. Saez. 2013. Epididysomes, prostasomes, and liposomes: their roles in mammalian reproductive physiology. Reproduction 146: R21–R35.

Wishart, G.J., and A.J. Horrocks. 2000. Fertilization in birds. pp. 193–222. *In*: J.J. Tarin, and A. Caro (eds.). Fertilization in Protozoa and Metazoan Animals. Springer, NY.

Wyburn, G.M., R.N. Aitken, and H.S. Johnston. 1965. The ultrastructure of the zona radiata of the ovarian follicle of the domestic fowl. J. Anat. 99: 469–484.

Section D
Animal Stress and Welfare

CHAPTER-9

Effects of Stress on Growth and Development
From Domestication to Factory Farming

W.M. Rauw* and *L. Gomez Raya*

INTRODUCTION

This chapter presents an overview of the pathways through which stress may affect growth and development. We begin by going back in time to the moment that men started "one of the greatest biological experiments" (Belyaev, 1979): the domestication of animals. Domestication meant artificial selection for tame behavior with pronounced consequences for the growth and development of the stress response and the phenotype. A reduction in aggressive behavior, fear, and the removal of resource demanding factors that were no longer needed in the captive environment, such as searching for food, must have resulted in food resources that became now available for use by other processes. Subsequent selection for growth and other production traits resulted in an unprecedented increase in production levels of livestock species. However, we can expect domesticated species to be again limited by their environment such that increased energy demand through the activation of the stress response is expected to trade-off with growth; several examples are given. In the last few decades it has become documented that stress may not just have pronounced effects on the development and stress response of the individual, but also on its descendants through the effects of epigenetic modifications and telomere length. Prolonged or severe activation of the stress response negatively affects animal welfare and livestock production, and should therefore be avoided.

Departamento de Mejora Genética Animal, Instituto Nacional de Investigación y Tecnología Agraria y Alimentaria (INIA), Ctra. de La Coruña km 7, 28040 Madrid, Spain.
* Corresponding author: rauw.wendy@inia.es

Domestication: Selection for a Reduced Stress Response

Most of the modern livestock species have their origin in the domestication of their respective ancestors several thousands of years ago. Domestication has been described as a process of "the capture and taming by man of animals of a species with particular behavioral characteristics, their removal from their natural living area and breeding community, and their maintenance under controlled breeding conditions for mutual benefits" (Bökönyi, 1986). During the initial phase of domestication, phenotypic traits were selected that facilitated the process of domestication itself, whereas once populations became established, a relaxation of natural selective pressure allowed for the appearance of mutations that could subsequently be selected for, both intentionally and unintentionally (Larson and Fuller, 2014). As Darwin (1883) writes: "Although man does not cause variability and cannot prevent it, he can select, preserve, and accumulate the variations given to him by the hand of nature almost in any way which he chooses; and thus he can certainly produce a great result. (…) Man may select and preserve each successive variation, with the distinct intention of improving and altering a breed, in accordance with a preconceived idea; and by thus adding up variations, often so slight as to be imperceptible by an uneducated eye, he has effected wonderful changes and improvements. (…) [D]omesticated breeds show adaptation to his wants and pleasures (…) [they are] modified not for their own benefit, but for that of man". A consequence of preserving and separating animals displaying pronounced deviations of (natural) structures was not only a much greater phenotypic variability but also a higher frequency of monstrosities observed under domestication than in nature (Darwin, 1883). In other words, man can preserve and further select the variants that otherwise have no chance of survival.

In an attempt to explain this phenomenon of increased phenotypic variation in domesticated animals, Belyaev (1979) discussed what he termed "destabilizing selection" as opposed to "stabilizing selection" that was proposed earlier by Schmalhausen in 1949 explaining the process that would result in genotypes that are less sensitive or "canalyzed" to environmental fluctuations (Rauw and Gomez Raya, 2015). Under *de*stabilizing selection, new stressful factors or usual stresses that increase in strength result in changes in the regulation of genes (timing and amount of gene expression), and break up previously integrated ontogenetic systems leading to multiple phenotypic effects (Belyaev, 1979; Trut, 1998). These disruptions open the box of the accumulated selectively neutral underlying genetic variation and mutation accumulation that was hitherto absorbed in the canalyzed phenotypes. Subsequently, artificial selection, by increasing the mean values of selected traits results in further loss of coadaptation of the system (Trut, 1988).

However, although phenotypic variation under domestication is greater than that observed in nature, this variation is homologous with respect to many phenotypic features in different domesticated species. This "domestication syndrome" was well acknowledged by Vavilov in an extensive work published in 1922: "One can observe some regularities in their varietal diversity, in spite of their enormous polymorphism". Phenotypic regularities include neoteny, loss of strict seasonal patterns of reproduction, increased fertility, variations in coat color and texture, docility, alterations in skull shape, and floppy ears; domesticated birds and even fish share some components of this spectrum of traits (Belyaev, 1979; Larson and Fuller, 2014; Wilkins et al., 2014). Based on this observation, Dmitry Belyaev formulated the hypothesis that the homologous variation is a consequence of selection for the domesticated type of *behavior* at the very beginning of selection (Belyaev, 1979). According to his hypothesis, the main criterion for domestication was the ability of animals not to be afraid of man. To test this hypothesis he started a selection experiment in silver foxes selecting for a low expression of the defense behavior towards man, or tamability. This resulted in a population of foxes that were not only not afraid of people, they tried actively to attract attention by wagging their tails and reacted positively to human contact by licking their hands and faces. And indeed, as hypothesized, they showed homologous variation in several characteristics as a correlated effect to changes in behavior: extra-seasonal activation of reproductive function, prolongation of the reproductive season, increased fertility in females, an increase in the time of moulting, and an increased frequency of newly appearing morphological traits such as floppy ears, piebaldness, and short and curly tails (Belyaev, 1979).

Further experiments showed that selection for tame behavior had brought about important changes in both the central and the peripheral mechanisms of the neuro-endocrine control of ontogeny. An important observation in the neuro-endocrine response to domestication is a reduced size and function of the adrenal glands that are responsible for fear, stress and adaptation. In the tame silver foxes, in generation 45, basal and stress-induced blood cortisol levels were three- to fivefold lower than in farm-bred foxes (Trut et al., 2009). This decrease in the functional state of the hypothalamic-pituitary-adrenal (HPA) axis results in a relatively immature emotional response to social threat. In addition, the delayed adrenal gland maturation extends the socialization window allowing human caretakers to be recognized as low-threat stimuli before the HPA-axis is mature (Wilkins et al., 2014). Wilkins et al. (2014) posit that adrenal hypofunction results from a developmental reduction in neural crest input, which in turn may result from multiple genetic changes of moderate, quantitative effect. Indeed most (but not all) of the modified traits included in the domestication syndrome can be explained as direct consequences of such deficiencies. For example, a diminished function of the hypothalamic-pituitary-gonadal axis may be responsible for accelerated reproductive maturation and reduced interbirth intervals (Wilkins et al., 2014). In addition, selection for reduced wariness and low reactivity to external stimuli has had a significant impact on the size, organization and function of the brains of domesticated animals, and this may be an indirect result of modified cranial neural crest cells (Zeder, 2012; Wilkins et al., 2014). In particular, a profound reduction in the size of structures within the limbic system can be tied to raising the behavioral thresholds for aggression, fight, and flight (Zeder, 2012).

Since activity levels, aggression, and boldness all appear to be systematically reduced during the process of domestication, it can be hypothesized that domestication resulted in the selection of an entire suit of behaviors, referred to as "personality". Animal personality expresses itself as a coping strategy that is consistent across contexts by suites of correlated behaviors or "behavioral types" (Koolhaas et al., 1999; Sih et al., 2004). Coping styles are closely related to individual fitness, since they form general adaptive response patterns in reaction to everyday challenges and stress. As opposed to reactive (passive, shy) animals, proactive (active, bold) animals attack or flee from an opponent, are aggressive, fast exploring, impulsive, actively manipulate events, score high in frustration tests, and are risk takers and novelty seekers (Benus, 1988; Coppens et al., 2010). In non-social situations, proactive copers are less affected by changes in the environment; their behavior is more intrinsically organized ('brain-engrained') and they tend to develop routine behavior. In contrast, reactive copers are more guided by external stimuli and hence their behavior is more flexible and adaptable (Benus, 1988; Coppens et al., 2010). In humans, Folkman and Lazarus (1980) called this 'problem' focused coping, i.e., to actively attempt to remove the source of stress or remove oneself from the source of stress, versus 'emotion focused' coping, i.e., to reduce the emotional impact of stress. Since nature has not favored any coping strategy in particular, both strategies may have benefits and should be considered as alternative strategies to cope with environmental demands: relying on previous experience is fast, but may be inaccurate in new situations, therefore, proactive animals may thrive better in stable environmental conditions, whereas reactive individuals may do better under variable and unpredictable environmental conditions (Benus et al., 1991; Coppens et al., 2010). Coping strategies are also shown to differ in the animal's physiological response to stress: the response of proactive animals is dominated by an enhanced sympathetic and (nor-) adrenergic response, whereas that of reactive animals is dominated by enhanced parasympathetic activation and a high HPA response (Carere et al., 2010). Although these differences are quite consistently observed in different species, the animals with different coping strategies do not always show this typical stress response (Herborn et al., 2011).

Proactive mice and rats have furthermore been shown to have lower levels of release of serotonin than reactive animals, which has been associated with an increase in aggression and impulsive behavior at the level of the prefrontal cortex (Coppens et al., 2010). A brain circuit in which serotonin neurons moderate coping behavior was recently presented by Puglisi-Allegra and Andolina (2015). Indeed, selection for tameness in silver foxes (Popova et al., 1991a) and low aggressiveness to man in Norway rats (Popova et al., 1991b) has resulted in animals that show a higher serotonin neurotransmitter level in the midbrain and hypothalamus, and selection for low fear of humans in Junglefowl (Agnvall et al., 2015) resulted in males that had higher plasma levels of serotonin. This suggests that the brain serotonergic system is involved in the mechanism of domestication, and that selection may have been for docile, reactive behavioral suits of personalities. Whereas reactive personalities are often found to have an *increased* HPA-reactivity to stress, domestication meant selection for tameness, i.e., a combination of low aggression and low fearfulness. This must indicate that the behavioral and the physiological profile of personalities are indeed not necessarily correlated. The behavioral and physiological profile of domesticated animals must have influenced their ability to adapt to the new, captive environment.

Life History Strategies and Resource Allocation Theory

Wolf et al. (2007) showed that personalities are related to risk-taking behavior and expected future fitness. Indeed, a meta-analysis performed by Smith and Blumstein (2008) showed that bolder individuals have increased reproductive success, but this incurs a survival cost such that shy individuals that have reduced short-term reproductive success but live longer may have the same overall fitness as bold individuals. Careau et al. (2010) found that docile dogs live longer than bold ones; in addition, aggressive breeds have higher energy needs than unaggressive ones. Because activity, exploration, boldness and aggression are energetically costly, they are part of the life history strategy and can be expected to trade off against other life history traits such as growth, reproduction, and immune function (Careau et al., 2008; Rauw, 2012).

As described by Glazier (2009b) trade-offs between life functions may result from the allocation of limited resources, ecological constraints, opposite independent effects of an environmental factor, morphological constraints, physiological limits, functional conflicts, multiple trait interactions, physiological regulation, genetic regulation and conflicts over gene expression. Trade-offs resulting from the allocation of limited resources amongst growth, reproduction, maintenance and storage is a fundamental assumption of life history (Stearns, 1989). The life-history trajectory of an individual or population follows from its resource investment choices over its lifetime: fitness is a function of the life history. Its optimization must always be considered from the perspective of the entire lifespan; different actions cannot be considered in isolation (McNamara et al., 2001; Kozlowski, 2006). Maintenance is essential for survival, growth is the construction of the body's machinery and enhances both survival and reproductive ability, reproduction directly relates to fitness though the production of offspring, and storage is a buffer against predictable and unpredictable challenges, or resources for future reproduction. Because these processes are costly, the investment follows a cost-profit function that changes during the lifetime and may differ for each organism, species, sex and individual (Gadgil and Bossert, 1970; Glazier, 2009a). Allocation models around life history theory aim to find the optimal proportions of the resources allocated into different processes (Kozlowski, 2006). It can be theorized that as the life-history traits that underlie the fitness function change during the lifetime, so will the priorities for allocation of resources to those variables change, and consequently also the optimal allocation pattern. Dynamic modelling of allocation patterns of assimilated resources optimally allocated to several body subsystems is extensively described by Perrin and Sibly (1993) and McNamara et al. (2001).

Whereas in nature it is generally assumed that trade offs occur between life history (i.e., fitness) traits that compete for limited resources, in livestock production the production goal is not necessarily to improve overall fitness. Instead, natural selection is now redefined to the traits that are included in the breeding goal which is particularly aimed at increasing the levels of output traits (growth, milk, eggs, wool). Resource allocation models that are designed to predict nutrient partitioning to different life functions and its consequences on health, reproduction and longevity in livestock production are developed and reviewed by Friggens et al. (2013). Although the captive environment during the first phases of domestication must have introduced new stressors, it can be assumed that several important resource demanding processes were removed, such as the need to be wary of predators and to search for food, which meant that those resources now became available for use by other purposes (Beilharz et al., 1993). In addition, the decrease in aggressive behavior by selection for tameness may have further released available resources (Careau et al., 2008). As a consequence, after animals became domesticated, they could now be further selected for objectives specifically involving performance, such as growth, milk yield, egg production, wool production, superior feed efficiency and low fatness. This resulted in an unprecedented increase in production levels in every livestock species (Rauw et al., 1998). However, we can assume that production has increased to such levels that resources in farm environments have again become limited, and that any further use of resources to support an increase in production must necessarily result in trade-offs with other life functions (Beilharz, 1998). This is supported by the observation that improved production in livestock species in many cases results in behavioral, physiological and immunological problems (Rauw et al., 1998).

From the resource allocation models it can be predicted that resources needed to respond to stress will trade-off with those needed for growth and development. In order to describe these effects from a resource allocation perspective, it is of interest to first consider the resource demand of these processes individually. This is the aim of the following sections. We will then present a brief overview of how stress is shown to affect growth and development in domesticated livestock species.

Resource Allocation and Trade-Offs

Growth and Development and its Resource Demands

In 1908, Robertson proposed that growth followed from some sort of master chemical autocatalytic reaction. He developed the logistic growth function, based on an earlier model developed for chemical acid-cathalized reactions (in Parks, 1982). A few years later, Thompson (1917) opposed that the analogy with a chemical reaction was a physico-chemical one, since a chemical reaction results in chemical equilibrium, whereas organic growth results in a very different kind of equilibrium due to the gradual differentiation of the organism into parts. Instead he proposed the analogy was a mathematical one, where growth follows from the action of physical forces. He wrote, "it is in obedience to the laws of physics that [the cells and tissues during growth] have been moved, moulded and conformed. (...) Their problems of form are in the first instance mathematical problems, their problems of growth are essentially physical problems". He defined growth as the complex and slow process "resulting from chemical, osmotic and other forces, by which material is introduced into the organism and transferred from one part to another". The changes of form during development results from the observation that every growing organism, and every part thereof, has its own specific rate of growth. These specific growth rates change with time such that the "curve of growth" has a characteristic curvature. This occurs in such an orderly way, he thought, that growth tends to follow a smooth, beautifully regular S-shaped line, with a point of inflection, and giving a bell-shaped derivative of time increments from minimal to maximal growth rate and back again. Growth could be described as a 'time-energy' diagram resulting from the absorption of energy beyond maintenance until the individual accumulates no longer. He proposed that it may be "fallacious to measure all [living organisms] alike by the common timepiece of the sun", instead, he suggested, life has a varying time-scale of its own. Indeed, when comparing growth curves of different species, many of the S-shaped curves look very similar but at a different time scale. In 1964, Taylor developed the concept of one uniform time-scaling unit for maturation intervals and generalized this in 1980 with the genetic size-scaling theory, i.e., a set of scaling rules which relate all traits associated with growth and metabolism to a single genetic size factor (mature body weight, A) that can be used as a quantitative description of the growth and development of a typical mammal. His methods are extensively described in Taylor (2009).

Several different mathematical descriptions of growth as the change of size, live weight or biomass with time have been proposed. These models generally are sigmoid-shaped and include the mathematical property of an asymptote to mature weight (A) such that the degree of maturity is a proportion thereof (BW/A). The sigmoid shape changes from curving upwards to curving downwards at a certain age t' and body weight BW', and growth rate is maximum at this inflection point (Parks, 1982). Notable growth functions following these properties are those developed by Gompertz in 1825 (but later used to describe growth), Robertson (the logistic function) in 1908, Brody in 1937, Von Bertalanffy in 1938, and Richards in 1959 (in Parks, 1982; Höök et al., 2011). In addition, the multiphasic growth model proposed by Koops (1986) implements the notion that parts of the body grow in different phases at different rates and mature at different time points. Models describing changes in the general composition of the body with the progression of live weight and time are furthermore particularly well described for pigs by Kyriazakis and Whittemore (2006).

However, what all these widely used 'output-only' growth functions lack, Parks (1982) complained, was any consideration for the input and allocation of food resources. He writes: "There is a strong reliance on the assumption that the energy and the raw materials of growth are never limiting, therefore negating the need for explicit expression of the source of the nutrients". A very early attempt of describing the dynamic process of growth as a function of energy input is given by Hatai (1911). He described growth as an increase in volume and weight until attainment of an inherited, intrinsic, mature size that is accomplished by means of interaction with external agencies. This inherited character of mature size, which differs between species, the author hypothesized, must result from a certain amount of 'potential growth energy' that is stored within the fertilized ovum before any growth takes place. As growth proceeds and the individual thus gains weight, growth energy diminishes gradually at a fixed rate, until its value is either zero or negligibly small. Hatai (1911) showed that this occurs following Maupertius' law of least action, i.e., at a maximum rate and most economically with the least loss of energy. He thus proposed the following provisional definition of growth: "An organism during growth tends to form the greatest amount of mass with least loss of growing capacity". Although Hatai's notion of 'growth energy' was a naïve conception of the relationship between energy and growth, at least it did not forget or neglect the energetic aspect of growth. Parks (1982) developed a model that described growth of animals as input-output systems where growth is described as a function of feed intake, such as that applied in the work of Whitehead and Parks (1988). Models predicting growth from parameters that relate resource intake and partitioning to maintenance and potential deposition of body protein and lipid were developed by Knap et al. (2003) for pigs. As Whittemore (2009) proposes, it may be argued that the individual passively adapts to the availability of nutrient resources, however, it can also be hypothesized that growth as a function of food resource acquisition and allocation may be viewed as a result of *active* decisions aimed towards the retention in the body of protein mass, lipid mass, water mass, and mineral mass, and the satisfaction of the processes of maintenance of body functions and productive activity. This hypothesis is supported by Emmans (1987): an animal actively eats to attain its potential growth rate and desired fatness. Pigs know what and how much to eat to make correct dietary choices when given a choice that will meet their protein requirements for growth (Kyriazakis et al., 1991).

In a life history resource allocation model it is usually assumed that maintenance costs are paid first, and that growth results from the allocation of surplus energy to increase in size (Kozlowski, 2006; Glazier, 2009a). When food is limited, maintenance (survival or longevity) is usually given precedence over growth, as indicated by the stress-related reduction in growth and subsequent compensatory growth when resources again become available and are again allocated to this process. The following section will give an overview of the very well described theories on resource allocation to the stress response and the origin of its trade-off with other life functions.

The Stress Response and its Resource Demands: Allostatic (over)load

Moberg (2000) developed a resource allocation model describing trade-offs between stress and other biological functions. The model suggests the existence of a reserve budget that can be used to deal with stress.

The biological cost of stress depends on the duration (acute vs. chronic) and on the severity of the stressor, and on the number of stressors (or repeated exposure to the same stressor). When the cost can be met by the reserve, stress has no impact on other biological functions, however, when the costs are larger than those available in the reserve, resources must be reallocated away from other biological functions which now become impaired. The animal enters a prepathological-pathological state and experiences *distress*, until the animal is able to replenish its biological resources (Moberg, 2000; McNamara and Buchanan, 2005).

Stress occurs when a discrepancy between a homeostatic setpoint and the actual level of a variable elicits compensatory responses to decrease the discrepancy. Homeostasis, as coined by Cannon in 1932 and referring to the steady-state of the 'milieu intérieur', in reality depends on a tremendous array of homeostatic systems or 'homeostats' that monitor variables and detect deviations from a set of steady-states (Goldstein, 2003). Whereas the principle of homeostasis concentrates around the constancy of the internal environment around a more or less invariable set point, organisms must also be able to (temporarily) change defended homeostatic levels when this is needed to adjust to a new or variable environment. Such newly (temporarily) defended levels are influenced by learning, experience, and anticipatory responding, i.e., they are feed-forward regulation mechanisms that function in addition to homeostatic negative feedback mechanisms (Korte, 2001; Ramsay and Woods, 2014). The regulation around this steady-state is the essence of 'allostasis', which means "maintaining stability through change" (Sterling and Eyer, 1988; Goldstein, 2003). The triggering of mediators of allostasis form a non-linear complex network and vary with age, gender, genotype, social status, health, body condition, the environment, etc. (McEwen and Wingfield, 2010).

The concept of allostasis is deeply integrated into the concept of life history theory and resource allocation. Central to the model is the concept of energy; energy demanded by the mediators determine 'allostatic load' which is a function of the metabolic demand of daily and seasonal routines and unpredictable perturbations of the environment. As 'reserve' in Moberg's (2000) model, Romero et al. (2009) in their Reactive Scope Model use the metaphor 'financial savings' for the amount of energy that is available to deal with everyday challenges and allostatic load, a financial buffer that may change with time. Different homeostats and their mediators may be activated intermittently, frequently, or continuously, but it is when the system is inefficient and results in wear-and-tear that 'allostatic load' turns into 'allostatic overload'. Or using Romero et al.'s (2009) metaphor: the individual will require more money than is currently in its financial savings. Prolonged overload resulting from chronic or repeated stress will leave no margin for responding to additional challenges, no opportunity for relaxation, and no capacity for more responsiveness, and will eventually lead to dysregulation amongst mediators, homeostatic failure, breakdown, and pathophysiology (Koob and Le Moal, 2001; Romero et al., 2009; McEwen and Wingfield, 2010). Eventually, the final manifestations of failure of allostasis is aging, senescence and death (Rattan, 2006). Romero et al. (2009) suggest that it may not always be possible to return to the original level. Prolonged allostatic load, away from adaptive levels, may result in the establishment of new, maladaptive set points (Koob and Le Moal, 2001). Allostatic overload has also been thought to result in the accumulated exposure to oxidative damage to proteins, lipids, and nucleic acids caused by reactive oxygen species, resulting in the acceleration of the aging process (Selman et al., 2012). Allocation of resources to the functions of the stress response is regulated by the demand of the respective functions which is achieved through neural or hormonal communication with the organs involved (Råberg et al., 1998). These allostatic mediators can be assessed in animals to investigate allostatic load.

The Components of the Stress Response

"Stressors, like beauty, lie in the eye of the beholder", indeed, it is only when a potential threat of disruption of homeostasis is perceived that a multiaxial stress response is initiated (Everly and Lating, 2013). What follows is activation of the autonomic nervous system response and the neuroendocrine response. Nervous system activation is the most direct and quickest of all pathways and includes the sympathetic and parasympathetic branches that are activated by the hypothalamus, in addition, a multitude of hormones are involved in the stress response, involving virtually every endocrine system.

One of the best known responses involves the hypothalamic-pituitary-adrenal (HPA) axis (Matteri et al., 2000). In response to stress, hypophysiotropic neurons located in the hypothalamus synthesize and secrete corticotropin-releasing factor, which is released into hypophysial portal vessels, binds to its receptor in the anterior pituitary gland, induces the release of adrenocorticotropic hormone (ACTH) into the systemic circulation, which stimulates the synthesis and secretion of glucocorticoids in the adrenal cortex. They regulate physiological changes through ubiquitously distributed intracellular receptors found in every nucleated cell (Miller and O'Callaghan, 2002; Smith and Vale, 2006). Glucocorticoids (either cortisol or corticosterone) may bind to two dramatically different receptors that differ in their neuroanatomical distribution, but also in their affinity and binding capacity: the high-affinity mineralocorticoid receptor (MR), which is almost saturated under basal conditions, and the low-affinity glucocorticoid receptor (GR), which becomes occupied only during stress and at the circadian peak (Korte, 2001). Homeostatic equilibrium is maintained via these two receptor types. Chronical dysregulation of the MR/GR balance through stress may result in maladaptive corticosteroid responses with elevated baseline plasma corticosteroid levels and levels that remain high longer after peak stress (Korte, 2001). Glucocorticoids eventually inhibit the production of corticotropin-releasing factor through a feed-back mechanism (Miller and O'Callaghan, 2002). They are involved in the development of the brain, lungs, and other organ systems, in the potentiation of fear, in learning and memory consolidation, in osmoregulation, and they are metabolic hormones that respond to energetic needs resulting from resource limitation or increased resource demands (Bonier et al., 2009; Denver, 2009). Glucocorticoid-induced physiological changes include gluconeogenesis by stimulating the liver to convert fat and protein to glucose for energy providing the fuel necessary for the increased metabolic demands of an emergency situation, and the stimulation of feeding behavior following a stress response to replenish depleted energy stores (Matteri et al., 2000; Miller and O'Callaghan, 2002; Denver, 2009). Glucocorticoids are largely immunosuppressive which could be an adaptive mechanism to reallocate resources from the maintenance cost, to avoid the cost of mounting an immune response, or both (Råberg et al., 1998). Kadarmideen and Janss (2007) estimated that cortisol is highly genetically determined with heritabilities of 0.40 to 0.70.

When investigated, glucocorticoids are generally measured in two contexts: as (chronic) baseline activity, and as peak activity in (acute) response to stress-induced situations which occurs some minutes after the stressful event. Baseline concentrations are related to an organism's energetic state that corresponds to variation in energetic demand varying during the animal's life cycle (Crespi et al., 2013). In addition, there is evidence that up-regulation of HPA-axis activity is implicated in mediating the physiology and behavior underlying transitions between life-history stages, such as growth and reproduction, acting as a modulator of developmental plasticity in vertebrates. Crespi et al. (2013) thus proposes that the function of the HPA-axis can be described in three tiers: (1) as a primary mediator of energy balance homeostasis, (2) as a master organizer of life-history transitions, and (3) as an integral responder to stressors. Although it is generally assumed that animals when measured under the same circumstance that show higher baseline levels of glucocorticoids are in worse condition and have reduced relative fitness, baseline circulating glucocorticoids and HPA-activity appear to have a non-linear relationship with reproduction, fitness and survival and may depend on the specific life-history context in which they are measured (Bonier et al., 2009; Crespi et al., 2013).

Stress also activates distinct peripheral catecholamine systems, each with different effectors, regulation, and roles: the sympathetic nervous system

(SNS), the adrenomedullary hormonal system, and the DOPA-dopamine autocrine/paracrine system (Goldstein, 2009). The main neurotransmitter in sympathetic nervous system regulation is norepinephrine (noradrenaline), which derives from sympathetic nerve endings, organs, and glands, and affects the function of virtually all body organs (Goldstein, 2010). It is responsible for tonic and reflexive changes in cardiovascular tone, maintaining appropriate blood flow to the brain, body temperature, and delivery of metabolic fuel to body organs. Although it was thought that the SNS system was active only in emergencies, norepinephrine plasma levels are detectable even under resting conditions, with a continuous drainage to most organs (Goldstein, 2003).

Goldstein (2008) describes the adrenal medulla as the filling of the "adrenal bonbon" to which the aforementioned glucocorticoid-producing cortex is the outer layer. The main hormone of the adrenomedullary hormonal system is epinephrine (adrenaline), which is a key determinant of responses to metabolic or global challenges to homeostasis and emotional distress (Goldstein, 2008). It can markedly increase blood glucose levels through stimulation of gluconeogenesis, lipolysis, hepatic glycogenolysis and decreasing peripheral glucose uptake, and raise body temperature and increase heart rate and cardiac output for a vigorous 'fight or flight' response (Matteri et al., 2000; Wurtman, 2002). The cortex and the medulla, which have very different embryological origins, are arranged such that the medulla is bathed in very high levels of steroids, such as glucocorticoids (Goldstein, 2008). Indeed, glucocorticoid secreted in the cortex contributes importantly to the control of epinephrine synthesis, thereby affecting its secretion (Wurtman, 2002; Goldstein, 2008).

Just as the sympathetic branch of the nervous system is associated with energy expenditure, the parasympathetic branch generally promotes actions of opposite nature, i.e., energy conservation, relaxation, and restorative functioning through the release of acetylcholine, which decreases heart rate, ventilation, muscle tension, and several other functions (Everly, 2013). Both systems control skeletal and smooth muscles of the vasculature, heart, gut, and other organs in coordination with the endocrine and behavioral aspects of the stress response (Denver, 2009). Although both systems are partially active at all times, since they act as the accelerator and break to the processes they affect they are mutually exclusive such that they cannot dominate their activity at the same time (Everly and Lating, 2013; Everly, 2013). Parasympathetic tempering of the excitation of the sympathetic branch helps control the duration of the autonomic response (Ulrich-Lai and Herman, 2009).

The multi-axial stress response further includes activation of the somatotrophic axis, the lactotrophic axis, the gonadotrophic axis, and the thyrothrophic axis, as reviewed by Matteri et al. (2000).

Trade-offs between Stress and Growth in Livestock Production

The influence of a diverse array of stressors on livestock growth, in particularly as measured by the release of glucocorticoids, has been described in the literature. For example, in pigs, in the study of Hemsworth et al. (1981, 1987), an unpleasant handling treatment resulted in higher corticosteroid concentrations, a slower growth rate, and a reduced feed efficiency in juvenile females resulting from both acute and chronic stress. Four-hour shipping of pigs resulted in elevated plasma cortisol compared with resident control pigs and a 5.1% reduction in body weight; pigs that had higher plasma cortisol levels lost more body weight ($r = -0.34$; McGlone et al., 1993). Pigs housed in a commercial environment had a slower growth rate during the rearing and finishing period and had higher cortisol concentration of saliva after weaning than those raised in a specific-stress-free environment where pigs were not mixed or transported (Ekkel et al., 1995). Subjecting pigs to high cycling temperature, restricted space allowance, or regrouping reduced growth rates by 10, 16, and 11%, respectively, whereas a combination of the three stressors resulted in a 31% reduction (Hyun et al., 1998). Average daily gain and average daily food intake were reduced by an additive interaction between high stocking density and regrouping in growing pigs in the study of Leek et al. (2004). Pigs housed in a dirty environment with significantly elevated ammonia, carbon dioxide and dust levels had reduced feed intake and lower growth rates than those housed in a clean environment (Lee et al., 2005). Sutherland et al. (2006) showed that pigs subjected to a chronic stressor had lower growth rates than control pigs.

In bull calves, castration increased plasma cortisol concentration and decreased growth rate and feed intake in the study of Fisher et al. (1996, 1997). Brahman calves who had higher serum concentrations of cortisol upon exiting a working chute had lower body weight gains in the study of Burdick et al. (2009). Fell et al. (1999) showed that nervous cattle had significantly higher cortisol levels at weaning and at the feedlot and had significantly lower daily gains. Turner et al. (2014) showed that a calm response in a crush score test was associated with a greater average daily gain during fattening, whereas a calm response during an isolation test was associated with a greater average daily gain and cold carcass weight in Limousin and Aberdeen crossbred beef cattle.

In chickens, McFarlane et al. (1989) observed that animals subjected to multiple stressors, including aerial ammonia, beak trimming, coccidosis infection, intermittent electric shock, heat stress, and continuous noise, had lower weight gain, increased coefficient of interindividual variation in gain, and decreased feed intake and feed conversion efficiency. Chickens responded to each stressor in the same way regardless of whether a stressor occurred singly or concurrently with up to five others. Layer chickens housed in groups of five per cage had significantly lower growth rates and higher plasma corticosterone levels than single housed layers in the study of Onbaşilar and Aksoy (2005). A 12 h light/12 h dark light regime in broiler chickens resulted in increased plasma corticosterone concentration and suppressed body weight by 10% compared to the control in the study of Abbas et al. (2008). Artificial administration of corticosteroids or ACTH in chickens resulted in a sharp reduction in body weight gain in several studies (Virden and Kidd, 2009; Scanes and Braun, 2013). Conversely, white leghorn chickens genetically selected for high group productivity and survivability showed lower levels of plasma dopamine and corticosterone in response to social stress than animals selected for low productivity and survivability, suggesting that selection for high productivity resulted in animals with a better coping capability to social stress (Cheng et al., 2002).

Probably the best described stressor and its effect on growth and other production traits is heat stress, which results in an activation of the hypothalamic-pituitary-adrenal axis and an increase in plasma glucocorticoid concentrations. A reduction in feed intake explains a major part of the performance decreases in growing animals, however in addition, heat stress markedly alters the hierarchy of tissue synthesis (Baumgard and Rhoads, 2013). Extreme heat results in allostatic overload and failure of homeostasis, resulting in reduced productivity or even death. The effects of heat stress on livestock growth have been reviewed by several authors (e.g., Fuquay, 1981; Baumgard and Rhoads, 2013), and specifically in cattle (Blackshaw and Blackshaw, 1994), buffalos (Marai and Haeeb, 2010), pigs (Ross et al., 2015), sheep (Marai et al., 2007), and poultry (Lara and Rostagno, 2013).

Transgenerational Effects of Stress: Epigenetics and Telomere Length

In the last few decades, in particular in human research and in animal models, it has become documented that early life severe allostatic overload, through maladaptive overstimulation of the stress response, can have pronounced consequences for later-life stress response activity, growth and development. In addition, this may severely affect the individual as well as the offspring in subsequent generations. The consequences of early life stress on later life is particularly pronounced when stress occurs during perinatal life, which is a period of increased brain sensitivity and plasticity, and permanent organization or imprinting of physiological systems which serve for the adaptation of the individual after birth (Darnaudéry and Maccari, 2007; Lupien et al., 2009). As a consequence, early life experiences may have effects that persist beyond the period of stress exposure. In addition to prenatal or early postnatal life stages, also the adolescent (pubertal) period is a critical phase in which the body, including the brain, is reshaped by hormones. Stressors experienced during this period also may have life-long effects, including neural remodeling, impaired learning and memory, and altered emotional behaviors (Crews and Gore, 2014).

These effects are found to be mediated by epigenetic mechanisms that respond dynamically to environmental cues. 'Epigenetic' means "over or above" genetic, because changes are not mediated though changes in the DNA sequence, but through modification of the DNA structure and histones such that DNA sequence transcription and expression are modulated (Champagne, 2010). A critical epigenetic mechanism involves the methylation of DNA at the 5'-position of cytosine. When this occurs in the proximity of a gene, this often reduces or silences its expression. Other epigenetic mechanisms involve alteration of chromatin packaging via histone modifications or variants, which influences how tightly chromatin is packaged, and consequently gene accessibility and expression. In addition, as proposed by McGowan and Szyf (2010), chromatin configuration by chromatin-modifying enzymes gates the accessibility of genes to DNA methylation and demethylation reactions. A third form of epigenetic regulation affects gene expression after transcription through small non-coding RNA molecules (Sibille et al., 2012; Rozek et al., 2014).

In humans, later life risk of physical and psychiatric disease resulting from *in utero* exposure to drugs, nutrient restriction, and maternal psychosocial stress has been well documented (Champagne, 2010). Maccari et al. (2003) indicate that prenatal (maternal) restraint stress in rats induces higher levels of anxiety, drug addiction, emotionality, depressive-like behaviors, memory impairment, HPA-axis hyperactivity resulting in an increased stress response, and circadian and sleep functions in adult offspring. It was shown that hyperactivity of the HPA axis in adult offspring was directly related to the high levels of maternal corticosterone secretion during restraint. Meaney and Szyf (2005) showed that parental care during early postnatal development in rats affects the epigenetic status of the glucocorticoid receptor gene promotor and consequently the stress response of offspring throughout their lives. Kinnally et al. (2011) showed that female bonnet macaques that had higher average DNA methylation resulting from stress early in life exhibited the greatest behavioral reactivity to stress as adults. Their results showed that not only DNA methylation of candidate genes but also of whole genomes may be a risk factor to individuals that experience early life stress. In humans, a number of epigenetic modifications have been recognized that influence brain development and growth patterns, leading to altered development and disease states (Hussain, 2012; Bale, 2015).

Epigenetic modifications have also been shown to affect telomere length. Telomeres are composed of tandem repeats of DNA at the ends of chromosomes that protect them from degradation and uncontrolled cellular division. Proper functioning requires a minimum length of repeats since critically short telomeres become senescent. However, cell division and oxidative stress results in telomere shortening. This shortening is compensated by the ribonucleoprotein telomerase, which adds new repeats onto the chromosome ends after each cell division of germ line and stem cells. However, this protein complex is generally inactive in somatic cells, resulting in progressive telomere erosion with age in most tissues (Blasco, 2007; Asok et al., 2013; Sanders and Newman, 2013). In vertebrates, telomeres do not contain genes, whereas subtelomeric regions located adjacent to telomeres contain a low density of genes. Recent studies indicate that epigenetic histone modifications of telomeric chromatin and DNA methylation of subtelomeric DNA are related to telomere-length integrity and deregulation (Blasco, 2007).

Telomere shortening, like epigenetic modifications, can result from physical or psychological stress, and an increased number of adverse experiences or a greater duration of stress appears to be associated with accelerated telomere shortening (Price et al., 2013; Asok et al., 2013). Dysregulation of the HPA axis, resulting in increased circulating levels of glucocorticoids and an increased exposure to oxidative stress, may represent one mechanism through which stress leads to telomere shortening (Epel et al., 2004; Asok et al., 2013). Since telomeres do not appear to have evolved to 'healthier' longer lengths, Eisenberg (2011) proposed that there must be a trade-off with maintenance requirements. His "thrifty telomere model" proposes that telomere length in itself serves to regulate maintenance efforts through costs and trade-offs of cellular proliferation and maintenance of the soma. Shorter telomeres are likely to cause a decrease in maintenance requirements, therefore, telomere function is a life history marker that trades-off with resources that are demanded for growth and reproduction. In humans telomere erosion has been linked to cardiovascular diseases, hypertension, cancer, diabetes, stroke, and autoimmune disease (Price et al., 2013).

Epigenetic modification in response to stress may be expressed spanning three generations when it affects the gestating female, her fetus, and the fetal (maternal or paternal) germline. However, in addition to genomic imprinting, epigenetic modifications may result in transgenerational inheritance by the next generation in the absence of direct environmental influences (Tollefsbol, 2014). In human studies, also telomere length was found to have a high and very consistent heritability (Broer et al., 2013), in addition, Haussmann and Heidinger (2015) laid out evidence, similar to that available for epigenetic modifications, that exposure to stressors in the parental generation influences telomere dynamics in offspring and potentially subsequent generations.

As Rutherford et al. (2012) indicate, also in livestock a number of experimental studies have shown that prenatal or early life experiences can have a substantial impact on health, productivity, and the way the individual copes with their social and physical environment later in life.

Conclusions

Stress is part of life, there is no way to prevent stressful events from occurring ever, however, every life form is equipped with mechanisms that detect stress and react upon it in order to minimize its impact. Although the captive environment during domestication introduced new stressors, to a more or lesser extent it reduced the influence of others that were no longer needed in captivity, such as the need to search for food, deal with food insecurity, fight off predators, finding shelter, protect offspring, etc. At the same time, successful adaptation depended on the similarities between the natural habitat and the captive environment (space allowance, presence or absence of key stimuli, food regime, and social environment) allowing for the development and expression of species-typical behavioral patterns (Price, 1999). The captive environment meant usually a comparatively small-scaled habitat that provided a high level of environmental stability by shielding animals from risk conditions and providing a rather constant group composition and nutritional supply, but which also meant grouping individuals in unnaturally high numbers or unnatural social composition (Brust and Guenther, 2015). Importantly, there is little to no evidence that domestication has resulted in the loss of behaviors or the addition of new behaviors, instead that differences are quantitative in character, affecting the relative expression in frequency or magnitude. This is important since animals in captivity may still feel the need to perform behavior even though it is no longer necessary for survival, such that the inability to fulfill this need may result in frustration stress (Duncan, 1998; Bracke and Hopster, 2006). Differences between the natural 'needs' and the commercial livestock production environment have become particularly large with the introduction of 'factory farming' as so effectively well described by Harrison (1964) in her book "Animal Machines: the New Factory Farming Industry". This initiated such a social debate that it resulted a year later in the Brambell Report, which specified the five freedoms: farm animals should have the freedom to stand up, lie down, turn around, groom themselves and stretch their limbs. This changing attitude towards livestock animals in society, in combination with modern livestock production systems that presented livestock with a large number of stressors that are oftentimes chronic in nature, resulted in a discussion about animal welfare issues that is very lively today (Rauw, 2015).

Stress is not inherently bad. Indeed, not only too much stress, but also too little is undesirable, because a certain degree of 'optimal' stress is essential for maintaining normal biological functioning (Zulkifli and Siegel, 1995). Indeed, apart from the negative effects of prolonged overexpression of the HPA axis, corticosteroids have positive effects on traits related to robustness and adaptation, and it has even been suggested that one strategy to improve robustness in livestock is to select animals with *higher* HPA activity (Mormède et al., 2011). In addition, there are ethical considerations involved when animals would be selected that are less responsive with fewer desires or reduced sentience (D'Eath et al., 2010). Because modern, very high producing livestock are found to be more at risk for behavioral, physiological and immunological problems, we may expect that their buffer to respond to additional resource demands is limited. Selection for increased growth and other production traits may reduce this ability to adequately respond to unexpected stresses and challenges even further, in addition, any demand of resources necessary to deal with stress must result in reduced growth and development. Therefore, livestock production needs to focus on breeding robust, adapted, and healthy animals. The demands from society regarding animal production will be satisfied when high levels of production are combined with proper animal welfare, thus selecting for optimum rather than maximum levels of production.

Keywords: Growth, development, domestication, artificial selection, behavior, phenotype, aggressive behavior, fear, energy demand, epigenetic modification, telomere length, stress response, animal welfare

References

Abbas, A.O., A.K.A. El-Dein, A.A. Desoky, and M.A.A. Magda. 2008. The effects of photoperiod programs on broiler chicken performance and immune response. Int. J. Poultry Sci. 7: 665–671.

Agnvall, B., R. Katajamaa, J. Altimiras, and P. Jensen. 2015. Is domestication driven by reduced fear of humans? Boldness, metabolism and serotonin levels in divergently selected red junglefowl (Gallus gallus). Biol. Lett. 11: 20150509.

Asok, A., K. Bernard, T.L. Roth, J.B. Rosen, and M. Dozier. 2013. Parental responsiveness moderates the association between early-life stress and reduced telomere length. Dev. Psychopathology 25: 577–585.

Bale, T.L. 2015. Epigenetic and transgenerational reprogramming of brain development. Nature Rev. Neurosci. 16: 332–344.

Baumgard, L.H., and R.P. Rhoads. 2013. Effects of heat stress on postabsorptive metabolism and energetics. Annu. Rev. Anim. Biosci. 1: 7.1–7.27.

Beilharz, R.G., B.G. Luxford, and J.L. Wilkinson. 1993. Quantitative genetics and evolution: Is our understanding of genetics sufficient to explain evolution? J. Anim. Breed. Genet. 110: 161–170.

Beilharz, R.G. 1998. Environmental limit to genetic change. An alternative theorem of natural selection. J. Anim. Breed. Genet. 115: 433–437.

Belyaev, D.K. 1979. Destabilizing selection as a factor in domestication. J. Hered. 70: 301–308.

Benus, R.F. 1988. Aggression and Coping. Differences in Behavioural Strategies between Aggressive and Non-Aggressive Male Mice. Ph.D. Thesis, Rijksuniversiteit Groningen, The Netherlands.

Benus, R.F., B. Bohus, J.M. Koolhaas, and G.A. Van Oortmerssen. 1991. Heritable variation for aggression as a reflection of individual coping strategies. Experientia 47: 1008–1019.

Blackshaw, J.K., and A.W. Blackshaw. 1994. Heat stress in cattle and the effects of shade on production and behavior: A review. Austr. J. Exp. Agric. 34: 285–295.

Blasco, M.A. 2007. The epigenetic regulation of mammalian telomeres. Nature Rev. Genet. 8: 299–309.

Bökönyi, S. 1986. Definitions of animal domestication. pp. 22–27. *In*: J. Clutton-Brock (ed.). The Walking Larder: Patterns of Domestication, Pastoralism, and Predation. Unwin Hyman, London, UK.

Bonier, F., P.R. Martin, I.T. Moore, and J.C. Wingfield. 2009. Do baseline glucocorticoids predict fitness? Trends Ecol. Evol. 24: 634–342.

Bracke, M.B.M., and H. Hopster. 2006. Assessing the importance of natural behavior for animal welfare. J. Agric. Environm. Ethics 19: 77–89.

Broer, L., V. Codd, D.R. Nyholt, J. Deelen, M. Mangino, G. Willemsen, E. Albrecht, N. Amin, M. Beekman, E.J.C. De Geus, A. Henders, C.P. Nelson, C.J. Steves, M.J. Wright, A.J.M. De Craen, A. Isaacs, M. Matthews, A. Moayyeri, G.W. Montgomery, B.A. Oostra, J.M. Vink, T.D. Spector, P.E. Slagboom, N.G. Martin, N.J. Samani, C.M. Van Duijn, and D.I. Boomsma. 2013. Meta-analysis of telomere length in 19 713 subjects reveals high heritability, stronger maternal inheritance and a paternal age effect. Eur. J. Hum. Genet. 21: 1163–1168.

Brust, V., and A. Guenther. 2015. Domestication effects on behavioural traits and learning performance: Comparing wild cavies to guinea pigs. Anim. Cogn. 18: 99–109.

Burdick, N.C., J.P. Banta, D.A. Neuendorff, J.C. White, R.C. Vann, J.C. Laurenz, T.H. Welsh, and R.D. Randel. 2009. Interrelationships among growth, endocrine, immune, and temperament variables in neonatal Brahman calves. J. Anim. Sci. 87: 3202–3210.

Careau, V., D. Thomas, M.M. Humphries, and D. Réale. 2008. Energy metabolism and animal personality. Oikos 117: 641–653.

Careau, V., D. Réale, M.M. Humphries, and D.W. Thomas. 2010. The pace of life under artificial selection: Personality, energy expenditure, and longevity are correlated in domestic dogs. Am. Nat. 175: 753–758.

Carere, C., D. Caramaschi, and T.W. Fawcette. 2010. Covariation between personalities and individual differences in coping with stress: Converging evidence and hypotheses. Curr. Zool. 56: 728–740.

Champagne, F.A. 2010. Early adversity and developmental outcomes: Interaction between genetics, epigenetics, and social experience across the life span. Perspect. Psychol. Sci. 18: 564–574.

Cheng, H.W., P. Singleton, and W.M. Muir. 2002. Social stress in laying hens: Differential dopamine and corticosterone responses after intermingling different genetic strains of chickens. Poultry Sci. 81: 1265–1272.

Coppens, C.M., S.F. De Boer, and J.M. Koolhaas. 2010. Coping styles and behavioural flexibility: Towards underlying mechanisms. Phil. Trans. R. Soc. B 365: 4021–4028.

Crespi, E.J., T.D. Williams, T.S. Jessop, and B. Delehanty. 2013. Life history and the ecology of stress: How do glucocorticoid hormones influence life-history variation in animals? Funct. Ecol. 27: 93–106.

Crews, D., and A.C. Gore. 2014. Transgenerational epigenetics: Current controversies and debate. pp. 371–390. *In*: T.O. Tollefsbol (ed.). Transgenerational Epigenetics. Evidence and Debate. Academic Press, London, UK.

Darnaudéry, M., and S. Maccari. 2007. Epigenetic programming of the stress response in male and female rats by prenatal restraint stress. Brain Res. Rev. 57: 571–585.

Darwin, C. 1883. The Variation of Animals and Plants under Domestication. Second Edition. D. Appleton & Co, New York.

D'Eath, R.B., J. Conington, A.B. Lawrence, I.A.S. Olsson, and P. Sandøe. 2010. Breeding for behavioural change in farm animals: Practical, economic and ethical considerations. Anim. Welfare 18: 17–27.

Denver, R.J. 2009. Structural and functional evolution of vertebrate neuroendocrine stress systems. Trends Comp. Endocr. Neurobiol. 1163: 1–16.

Duncan, I.J.H. 1998. Behavior and behavioral needs. Poultry Sci. 77: 1766–1772.

Eisenberg, D.T.A. 2011. An evolutionary review of human telomere biology: The thrifty telomere. Hypothesis and notes on potential adaptive paternal effects. Am. J. Hum. Biol. 23: 149–167.

Ekkel, E.D., C.E. Van Doorn, M.J. Hessing, and M.J. Tielen. 1995. The specific-stress-free housing system has positive effects on productivity, health, and welfare of pigs. J. Anim. Sci. 73: 1544–1551.

Emmans, G.C. 1987. Growth, body composition and feed intake. World's Poultry Sci. J. 43: 208–227.

Epel, E.S., E.H. Blackburn, J. Lin, F.S. Dhabhar, N.E. Adler, J.D. Morrow, and R.M. Cawthon. 2004. Accelerated telomere shortening in response to life stress. PNAS 101: 17312–17315.

Everly, G. 2013. Physiology of stress. pp. 34–48. In: B.L. Seaward (ed.). Managing Stress. Principles and Strategies for Health and Well-Being. Jones and Bartlett Learning, Burlington, MA.

Everly, G.S., and J.M. Lating. 2013. The anatomy and physiology of the human stress response. pp. 17–51. In: G.S. Everly, and J.M. Lating (eds.). A Clinical Guide to the Treatment of the Human Stress Response. Springer Science & Business Media, New York.

Fell, L.R., I.G. Colditz, K.H. Walker, and D.L. Watson. 1999. Associations between temperament, performance and immune function in cattle entering a commercial feedlot. Austr. J. Exp. Agric. 39: 795–802.

Fisher, A.D., M.A. Crowe, M.E. Alonso de la Varga, and W.J. Enright. 1996. Effect of castration method and the provision of local anesthesia on plasma cortisol, scrotal circumference, growth, and feed intake of bull calves. J. Anim. Sci. 74: 2336–2343.

Fisher, A.D., M.A. Crow, E.M. Ó'Nualláin, M.L. Monaghan, J.A. Larkin, P. O'Kiely, and W.J. Enright. 1997. Effects of cortisol on *in vitro* interferon-γ production, acute-phase proteins, growth, and feed intake in a calf castration model. J. Anim. Sci. 75: 1041–1047.

Folkman, S., and R.S. Lazarus. 1980. An analysis of coping in a middle-aged community sample. J. Health Soc. Behav. 21: 219–239.

Friggens, N.C., L. Brun-Lafleur, P. Faverdin, D. Sauvant, and O. Martin. 2013. Advances in predicting nutrient partitioning in the dairy cow: Recognizing the central role of genotype and its expression through time. Animal 7: 89–101.

Fuquay, J.W. 1981. Heat stress as it affects animal production. J. Anim. Sci. 52: 164–174.

Gadgil, M., and W.H. Bossert. 1970. Life historical consequences of natural selection. Am. Nat. 104: 1–24.

Glazier, D.S. 2009a. Resource allocation patterns. pp. 22–43. In: W.M. Rauw (ed.). Resource Allocation Theory Applied to Farm Animal Production. CABI Publishing, Wallingford, UK.

Glazier, D.S. 2009b. Trade-offs. pp. 44–60. In: W.M. Rauw (ed.). Resource Allocation Theory Applied to Farm Animal Production. CABI Publishing, Wallingford, UK.

Goldstein, D.S. 2003. Catecholamines and stress. Endocr. Reg. 37: 69–80.

Goldstein, D.S. 2008. Adrenaline and the Inner World: An Introduction to Scientific Integrative Medicine. JHU Press, Baltimore, MD.

Goldstein, D.S. 2009. Sympathetic noradrenergic and adrenomedullary hormonal systems in stress and disease. pp. 399–405. In: G. Fink (ed.). Stress Science: Neuroendocrinology. Academic Press, San Diego, CA.

Goldstein, D.S. 2010. Adrenaline and noradrenaline. pp. 1–9. In: Encyclopedia of Life Sciences (ELS). John Wiley & Sons, Ltd., Chichester, UK.

Harrison, R. 1964. Animal Machines. The New Factory Farming Industry. Vincent Stuart Ltd., London, UK.

Hatai, S. 1911. An interpretation of growth curves from a dynamical standpoint. Anat. Rec. 5: 373–382.

Haussmann, M.F., and B.J. Heidinger. 2015. Telomere dynamics may link stress exposure and ageing across generations. Biol. Lett. 11: 20150396.

Hemsworth, P.H., J.L. Barnett, and C. Hansen. 1981. The influence of handling by humans on the behavior, growth, and corticosteroids in the juvenile female pig. Horm. Behav. 15: 396–403.

Hemsworth, P.H., J.L. Barnett, and C. Hansen. 1987. The influence of inconsistent handling by humans on the behavior, growth and corticosteroids of young pigs. Appl. Anim. Behav. Sci. 17: 245–252.

Herborn, K.A., J. Coffey, S.D. Larcombe, L. Alexander, and K.E. Arnold. 2011. Oxidative profile varies with personality in European greenfinches. J. Exp. Biol. 214: 1732–1739.

Höök, M., J. Li., N. Oba, and S. Snowden. 2011. Descriptive and predictive growth curves in energy system analysis. Nat. Resour. Res. 20: 103–116.

Hussain, N. 2012. Epigenetic influences that modulate infant growth, development, and disease. Antioxidants & Redox Signaling 17: 224–236.

Hyun, Y., M. Ellis, G. Riskowski, and R.W. Johnson. 1998. Growth performance of pigs subjected to multiple concurrent environmental stressors. J. Anim. Sci. 76: 721–727.

Kadarmideen, H.N., and L.G. Janss. 2007. Population and systems genetic analyses of cortisol in pigs divergently selected for stress. Physiol. Genomics 29: 57–65.

Kinnally, E.L., C. Feinberg, D. Kim, K. Ferguson, R. Leibel, J.D. Coplan, and J. John Mann. 2011. DNA methylation as a risk factor in the effects of early life stress. Brain Behav. Immun. 25: 1548–1553.

Knap, P.W., R. Roehe, K. Kolstad, C. Pomar, and P. Luiting. 2003. Characterization of pig genotypes for growth modeling. J. Anim. Sci. 81(E. Suppl. 2): E187–E195.

Koob, G.F., and M. Le Moal. 2001. Drug addiction, dysregulation of reward, and allostasis. Neuropsychopharmacol. 24: 97–129.

Koolhaas, J.M., S.M. Korte, S.F. De Boer, B.J. Van der Vegt, C.G. Van Reenen, H. Hopster, I.C. De Jong, M.A. Ruis, and H.J. Blokhuis. 1999. Coping styles in animals: Current status in behavior and stress-physiology. Neurosci. Biobehav. R. 23: 925–935.

Koops, W.J. 1986. Multiphasic growth curve analysis. Growth 50: 169–177.

Korte, S.M. 2001. Corticosteroids in relation to fear, anxiety and psychopathology. Neurosci. Biobehav. Rev. 25: 117–142.

Kozlowski, J. 2006. Why life histories are diverse. Polish J. Ecol. 54: 585–605.

Kyriazakis, I., G.C. Emmans, and C.T. Whittemore. 1991. The ability of pigs to control their protein intake when fed in three different ways. Physiol. Behav. 50: 1197–1203.

Kyriazakis, I., and C.T. Whittemore. 2006. Growth and body composition changes in pigs. pp. 65–103. *In*: C.T. Whittemore (ed.). Whittemore's Science and Practice of Pig Production. 3rd ed. Blackwell Publishing Ltd., Oxford, UK.

Lara, L.J., and M.H. Rostagno. 2013. Impact of heat stress on poultry production. Animals 3: 356–369.

Larson, G., and D.Q. Fuller. 2014. The evolution of animal domestication. Annu. Rev. Ecol. Evol. Syst. 66: 115–136.

Lee, C., L.R. Giles, W.L. Bryden, J.L. Downing, P.C. Owens, A.C. Kirby, and P.C. Wynn. 2005. Performance and endocrine responses of group housed weaner pigs exposed to the air quality of a commercial environment. Livest. Prod. Sci. 93: 255–262.

Leek, A.B.G., B.T. Sweeney, P. Duffy, V.E. Beattie, and J.V. O'Doherty. 2004. Effect of stocking density and social regrouping stressors on growth performance, carcass characteristics, nutrient digestibility and physiological stress responses in pigs. Anim. Sci. 79: 109–119.

Lupien, S.J., B.S. McEwen, M.R. Gunnar, and C. Heim. 2009. Effects of stress throughout the lifespan on the brain, behavior and cognition. Nature Rev. Neurosci. 10: 434–445.

Maccari, S., M. Darnaudery, S. Morley-Fletcher, A.R. Zuena, C. Cinque, and O. Van Reeth. 2003. Prenatal stress and long-term consequences: implication of glucocorticoid hormones. Neurosci. Biobehav. Rev. 27: 119–127.

Marai, I.F.M., A.A. El-Darawany, A. Fadiel, and M.A.M. Abdel-Hafez. 2007. Physiological traits as affected by heat stress in sheep—A review. Small Rumin. Res. 71: 1–12.

Marai, I.F.M., and A.A.M. Haeeb. 2010. Buffalo's biological functions as affected by heat stress. Livest. Sci. 127: 89–109.

Matteri, R.L., J.A. Carroll, and C.J. Dyer. 2000. Neuroendocrine responses to stress. pp. 43–76. *In*: G.P. Moberg, and J.A. Mench (eds.). The Biology of Animal Stress. Basic Principles and Implications for Animal Welfare. CAB International, Wallingford, UK.

McEwen, B.S., and J.C. Wingfield. 2010. What's in a name? Integrating homeostasis, allostasis and stress. Horm. Behav. 57: Article 105, pp. 1–16.

McFarlane, J.M., S.E. Curtis, R.D. Shanks, and S.G. Carmer. 1989. Multiple concurrent stressors in chicks. Poultry Sci. 68: 501–509.

McGlone, J.J., J.L. Salak, E.A. Lumpkin, R.I. Nicholson, M. Gibson, and R.L. Norman. 1993. Shipping stress and social status effects on pig performance, plasma cortisol, natural killer cell activity, and leukocyte numbers. J. Anim. Sci. 71: 888–896.

McGowan, P.O., and M. Szyf. 2010. The epigenetics of social adversity in early life: Implications for mental health outcomes. Neurobiol. Dis. 39: 66–72.

McNamara, J.M., A.I. Houston, and E.J. Collins. 2001. Optimality models in behavioral biology. SIAM Rev. 43: 413–466.

McNamara, J.M., and K.L. Buchanan. 2005. Stress, resource allocation, and mortality. Behav. Ecol. 16: 1008–1017.

Meaney, M.J., and M. Szyf. 2005. Maternal care as a model for experience-dependent chromatin plasticity? Trends Neurosci. 28: 456–463.

Miller, D.B., and J.P. O'Callaghan. 2002. Neuroendocrine aspects of the response to stress. Metabolism 51: 5–10.

Moberg, G.P. 2000. Biological response to stress: implications for animal welfare. pp. 1–22. *In*: G.P. Moberg, and J.A. Mench (eds.). The Biology of Animal Stress. Basic Principles and Implications for Animal Welfare. CAB International, Wallingford, UK.

Mormède, P., A. Foury, E. Terenina, and P.W. Knap. 2011. Breeding for robustness: The role of cortisol. Animal 5: 651–657.

Onbaşilar, E.E., and F.T. Aksoy. 2005. Stress parameters and immune response of layers under different cage floor and density conditions. Livest. Prod. Sci. 95: 255–263.

Parks, J.R. 1982. A Theory of Feeding and Growth in Animals. Springer-Verlag, New York.

Perrin, N., and R.M. Sibly. 1993. Dynamic models of energy allocation and investment. Annu. Rev. Ecol. Syst. 24: 379–410.

Popova, N.K., N.N. Voitenko, A.V. Kulikov, and D.F. Avgustinovich. 1991a. Evidence for the involvement of central serotonin in mechanism of domestication of silver foxes. Pharmac. Biochem. Behav. 40: 751–756.

Popova, N.K., A.V. Kulikov, E.M. Nikulina, E.Y. Kozlachkova, and G.B. Maslova. 1991b. Serotonin metabolism and serotonergic receptors in Norway rats selected for low aggressiveness to man. Aggr. Behav. 17: 207–213.

Price, E.O. 1999. Behavioral development in animals undergoing domestication. Appl. Anim. Behav. Scie. 65: 245–271.

Price, L.H., H.T. Kao, D.E. Burgers, L.L. Carpenter, and A.R. Tyrka. 2013. Telomeres and early-life stress: An overview. Biol. Psychiatry 73: 15–23.

Puglisi-Allegra, S., and D. Andolina. 2015. Serotonin and stress coping. Behav. Brain Res. 277: 58–67.

Råberg, L., M. Grahn, D. Hasselquist, and E. Svensson. 1998. On the adaptive significance of stress-induced immunosuppression. Proc. R. Soc. Lond. B. 265: 1637–1641.

Ramsay, D.S., and S.C. Woods. 2014. Clarifying the roles of homeostasis and allostasis in physiological regulation. Psychol. Rev. 121: 225–247.

Rattan, S.I.S. 2006. Theories of biological aging: Genes, proteins, and free radicals. Free Radical Res. 40: 1230–1238.

Rauw, W.M., E. Kanis, E.N. Noordhuizen-Stassen, and F.J. Grommers. 1998. Undesirable side effects of selection for high production efficiency in farm animals: a review. Livest. Prod. Sci. 56: 15–33.

Rauw, W.M. 2012. Immune response from a resource allocation perspective. Front. Genet. 3: Article 267, pp. 1–14.

Rauw, W.M. 2015. Philosophy and ethics of animal use and consumption: From Pythagoras to Bentham. CAB Rev. 10(16): 1–25.

Rauw, W.M., and L. Gomez Raya. 2015. Genotype by environment interaction and breeding for robustness in livestock. Front. Genet. 6: Article 310, pp. 1–15.

Romero, L.M., M.J. Dickens, and N.E. Cyr. 2009. The reactive scope model—A new model integrating homeostasis, allostasis, and stress. Hormones Behav. 55: 375–389.

Ross, J.W., B.J. Hale, N.K. Gabler, R.P. Rhoads, A.F. Keating, and L.H. Baumgard. 2015. Physiological consequences of heat stress in pigs. Anim. Prod. Sci. 55: 1381–1390.

Rozek, L.S., D.C. Dolinoy, M.A. Sartor, and G.S. Omenn. 2014. Epigenetics: Relevance and implication for human health. Annu. Rev. Public Health 35: 105–122.

Rutherford, K.M.D., R.D. Donald, G. Arnott, J.A. Rooke, L. Dixon, J.J.M. Mehers, J. Turnbull, and A.B. Lawrence. 2012. Farm animal welfare: Assessing risks attributable to the prenatal environment. Anim. Welf. 12: 419–429.

Sanders, J.L., and A.B. Newman. 2013. Telomere length in epidemiology: A biomarker of aging, age-related disease, both, or neither? Epidemiologic Rev. 35: 112–131.

Scanes, C.G., and E. Braun. 2013. Avian metabolism: Its control and evolution. Front. Biol. 8: 134–159.

Schmalhausen, I.I. 1949. Factors of Evolution: The Theory of Stabilizing Selection. Oxford: Blakiston.

Selman, C., J.D. Blount, D.H. Nussey, and J.R. Speakman. 2012. Oxidative damage, ageing, and life-history evolution: Where now? Trends Ecol. Evol. 27: 570–577.

Sibille, K.T., L. Witek-Janusek, H.L. Mathews, and R.B. Fillingim. 2012. Telomeres and epigenetics: Potential relevance to chronic pain. Pain 153: 1789–1793.

Sih, A., A. Bell, and J.C. Johnson. 2004. Behavioral syndromes: An ecological and evolutionary overview. Trends Ecol. Evol. 19: 372–378.

Smith, B.R., and D.T. Blumstein. 2008. Fitness consequences of personality: A meta-analysis. Behav. Ecol. 19: 448–455.

Smith, S.M., and V.W. Vale. 2006. The role of the hypothalamic-pituitary-adrenal axis in neuroendocrine responses to stress. Dialogues Clin. Neurosci. 8: 383–395.

Stearns, S.C. 1989. Trade-offs in life-history evolution. Funct. Ecol. 3: 259–268.

Sterling, P., and J. Eyer. 1988. Allostasis: A new paradigm to explain arousal pathology. pp. 629–649. In: S. Fisher, and J. Reason (eds.). Handbook of Life Stress, Cognition and Health. John Wiley & Sons, New York.

Sutherland, M.A., S.R. Niekamp, S.L. Rodriguez-Zas, and J.L. Salak-Johnson. 2006. Impacts of chronic stress and social status on various physiological and performance measures in pigs of different breeds. J. Anim. Sci. 84: 588–596.

Taylor, St. C.S. 2009. Genetic size-scaling. pp. 147–168. In: W.M. Rauw (ed.). Resource Allocation Theory Applied to Farm Animal Production. CABI Publishing, Wallingford, UK.

Thompson, D.W. 1917. On Growth and Form. Cambridge University Press, Cambridge, UK.

Tollefsbol, T.O. 2014. Transgenerational epigenetics. pp. 1–8. In: T.O. Tollefsbol (ed.). Transgenerational Epigenetics. Evidence and Debate. Academic Press, London, UK.

Trut, L.N. 1988. The variable rates of evolutionary transformations and their parallelism in term of destabilizing selection. J. Anim. Breed. Genet. 105: 81–90.

Trut, L.N. 1998. The evolutionary concept of destabilizing selection: *status quo*. J. Anim. Breed. Genet. 115: 415–431.

Trut, L.N., I. Oskina, and A. Kharlamova. 2009. Animal evolution during domestication: The domesticated fox as a model. Bioessays 31: 349–360.

Turner, S.P., E.A. Navajas, J.J. Hyslop, D.W. Ross, R.I. Richardson, N. Prieto, M. Bell, M.C. Jack, and R. Roehe. 2014. Associations between response to handling and growth and meat quality in frequently handled Bos Taurus beef cattle. J. Anim. Sci. 89: 4239–4248.

Ulrich-Lai, Y.M., and J.P. Herman. 2009. Neural regulation of endocrine and autonomic stress response. Nature Reviews—Neuroscience 10: 397–409.

Vavilov, N.I. 1922. The law of homologous series in variation. J. Genet. 12: 47–89.

Virden, W.S., and M.T. Kidd. 2009. Physiological stress in broilers: Ramifications on nutrient digestibility and responses. J. Appl. Poultry Sci. 18: 338–347.

Whitehead, C.C., and J.R. Parks. 1988. The growth to maturity of lean and fat lines of broiler chickens fed diets of different protein content: Evaluation of a model to describe growth and feeding characteristics. Animal Science 46: 469–478.

Whittemore, C.T. 2009. Allocation of resources to growth. pp. 130–146. *In*: W.M. Rauw (ed.). Resource Allocation Theory Applied to Farm Animal Production. CABI Publishing, Wallingford, UK.

Wilkins, A.S., R.W. Wrangham, and W. Tecumseh Fitch. 2014. The "domestication syndrome" in mammals: A unified explanation based on neural crest cell behavior and genetics. Genetics 197: 795–808.

Wolf, M., G.S. Van Doorn, O. Leimar, and F.J. Weissing. 2007. Life-history trade-offs favour the evolution of animal personalities. Nature 447: 581–584.

Wurtman, R.J. 2002. Stress and the adrenocortical control of epinephrine synthesis. Metabolism 51: 11–14.

Zeder, M.A. 2012. Pathways to animal domestication. pp. 227–259. *In*: P. Gepts, T.R. Famula, R.L. Bettinger, S.B. Brush, A.B. Damania, P.E. McGuire, and C.O. Qualset (eds.). Biodiversity in Agriculture: Domestication, Evolution, and Sustainability. Cambridge University Press, Cambridge, UK.

Zulkifli, I., and P.B. Siegel. 1995. Is there a positive side to stress? World's Poultry Science Journal 51: 63–76.

CHAPTER-10

Biology of Stress in Livestock and Poultry

Colin G. Scanes,* Yvonne Vizzier-Thaxton and
Karen Christensen

||||||||||||||||||||||||||||||||||||||||||||||||||||||||||||||||||||||||||||||||||||||||||||||||||||||||||||||||||||||||||||||||||||

INTRODUCTION

The word stress (meaning "hardship" or "adversity") is thought to have entered the English language in about the year 1300 as a shortening of the word 'distress' (Online Etymology Dictionary, 2015). In turn, the word 'distress' derives from the French word *'estrice'* (meaning "narrowness" or "oppression") and this in turn from the Latin *"strictus"* (meaning "drawn tight"). By the middle of 16th century, the word stress had come to mean a physical strain on an object and in the 19th century as a concept in mechanics (Online Etymology Dictionary, 2015). Hans Selye (1936) advanced the physiological model of stress—the *"General Adaptation Syndrome"*. This was extended to stress evokes increased secretion of glucocorticoid hormones from the adrenal cortex (Sapolsky et al., 2000).

Production Responses to Stress

In a thoughtful review, Temple Grandin (1997) discussed assessment of stress in cattle related to some production practices. Stress is accompanied by reductions important production indices in the following:

- Growth rate (transient or long-term)
- Production of the following:
  - Milk (cattle, goats and sheep)

Center for Food Animal Wellbeing and Department of Poultry Science, University of Arkansas, Fayetteville, AR 72701.
* Corresponding author: cscanes@uark.edu

- Eggs (chickens with ducks and geese important in China)
- Wool (sheep)

- Feed intake (and hence feed: gain another production index)

Growth rate and other production indices have multiple causalities and might even be considered as being non-specific. For instance, growth rate is influenced by food intake, nutritional status, toxicants, pathogenic diseases and parasites in addition to stress with the concomitant elevated glucocorticoids also depressing growth (discussed below). For instance, following castration, lambs exhibit weight loss and have increased circulating concentrations of cortisol (F) (Paull et al., 2008). There are examples of the effects of stress on milk, egg and wool production. For example, wool production is depressed by nutritional deprivation (e.g., Naqvi and Rai, 1990; Oliver and Oliver, 2010) and heat stress (Thwaites, 1968).

Production stress might be expressed in the following equation.

Equation 1
Decrease in production parameter = Index of nutritional restriction
+ Index of infectious disease
+ Index of STRESS
+ Index of other factors, etc.

It should be noted that stresses frequently depress feed intake. There is reduced feed intake in pigs exposed to elevated environmental temperatures (e.g., Spencer et al., 2003). While, the stress undoubtedly results in multiple physiological effects, *Occam's razor* would indicate feed intake would be one of the major mediators of the effect of the stress. It is therefore recommended that pair-fed controls be included in studies of stress.

Biological Responses to Stress

The biology of the responses of livestock and poultry to stress is addressed below. The first response to stress that will be considered will be the hypothalamo-pituitary-adrenocortical (HPA) axis and the secretion of glucocorticoids—F in livestock and corticosterone (CORT) in poultry (see section 2). Equation 2 encapsulates the model of stress with stress increasing F or CORT secretion and circulating concentrations of F or CORT reflecting the level of stress.

Equation 2
Stress = Constant x Integrated plasma (or urinary) concentrations of glucocorticoids

Unfortunately, plasma concentrations of cortisol are reported usually at one to three time points. It is possible that F in hair or CORT in feathers readily provide a measure of integrated plasma (or urinary) concentrations of glucocorticoids. The glucocorticoids act on metabolism, growth and immune function and thereby mediate the response to the stress by:

- Counter-acting the adverse effects of stress
- Preventing the over-reaction to the stressor (e.g., reviewed: Sapolsky et al., 2000)

It is possible that there are other effects of glucocorticoids.

Other indices of stresses employed in livestock and poultry include the following:

- Behavioral indices including tonic immobility (chickens: Cunningham et al., 1988) and other abnormal behaviors
- Central nervous indices including E, NE, DA and 5HT concentrations in regions of the brain (Cheng and Fahey, 2009)
- Hypothalamo-Pituitary-Adrenocortical (HPA) Axis activation:
  - Circulating concentrations of cortisol (F) [or corticosterone (CORT) in poultry] and adrenocorticotropic hormone (ACTH)
  - Urinary/fecal/hair/feathers concentrations of F or CORT
  - Expression of enzymes that synthesize glucocorticoids and/or the precursors of ACTH, i.e., pro-opiomelanocortin (POMC)
- Adrenal medulla activity together with sympathetic activity release of catecholamines. Indices including:
  - Circulating concentrations of epinephrine (E), norepinephrine (NE), dopamine (DA) or serotonin (5HT) in the blood (e.g., Yan et al., 2014)
  - Urinary concentrations of metabolites of E and NE (Hay and Mormède, 1998)
  - Tyrosine hydroxylase activity (EC 1.14.16.2) in the adrenal medulla (Chobotská et al., 1998)
  - Chromogranin concentrations in the saliva (Ott et al., 2014); chromogranin being a peptide produced by the adrenal medulla
- Sympathetic innervation of the heart indices including heart rate, ventricular ectopic beats, heart weight (e.g., chickens: Cunningham et al., 1988) and other heart related variables together with respiration rate
- Leukocyte parameters
  - Numbers/concentration in the blood
  - Neutrophil: lymphocyte ratio (N:L) in mammals [or the equivalent in poultry - heterophil:lymphocyte ratio (H:L)]
- Other indices including circulating concentrations of hormones [e.g., thyroid hormones: Spencer, 1994 and growth hormone (GH)], of cytokines or of intermediary metabolites (glucose and lactate) or of

potassium or hematocrit (Neubert et al., 1996; Peeters et al., 2005) and concentrations of haptoglobin in plasma (Paull et al., 2008) and saliva (Ott et al., 2014)

Rarely are more than two indices measured.

What is Stress?

On an anthropomorphic basis, it would seem that stress is unpleasant stimulus. This certainly goes along with the origin of the word coming from mechanics. Examples of types of stresses examined in studies with livestock are summarized in Table 1. What is missing is the breaking the circle of stress increases F or CORT (or other indicator of stress) and level of F or CORT indicating stress. Moreover, there is a danger of selectivity of the literature. The effects of adverse stimuli on indicators of stress are reported but the effects of benign stimulator are frequently neglected. For example, ether elevates circulating concentrations of CORT in chickens (see below). Further, it would seem that the corollary of maximal production is the absence of stress. However, there is likely to be a cut-off with some low level of stress not influencing production and/or recovery following stress.

Hypothalamo-Pituitary-Adrenocortical (HPA) Axis

Overview

The predominant control of F or CORT release from the adrenal cortical cells is the brain and, specifically, the hypothalamus with release of corticotropin release hormone (CRH) into the hypophyseal portal vessels. The release of adrenocorticotropic hormone (ACTH) from the corticotropes in the anterior pituitary gland is stimulated by CRH, together with vasopressin, in mammals or arginine vasotocin in poultry. Production of F or CORT is stimulated by ACTH. Secretion of CRH and AVP is regulated by noradrenergic and NPY neural pathways (Turner et al., 2002a). The HPA axis in livestock and poultry is discussed below (also see Figure 1). There are shifts in the hypothalamic-pituitary-adrenal (HPA) axis during late fetal development with increases in POMC expression but decreases in pituitary responsiveness to CRH (Lü et al., 1994).

Corticotropin Release Hormone (CRH)

CRH was first characterized from pig hypothalami (Patthy et al., 1985). The ability of CRH to stimulate ACTH release in livestock and poultry has been as demonstrated in pigs by, for instance, following intracerebroventricular injection (Johnson et al., 1994) and studies with chicken anterior pituitary

Table 1: Stresses in livestock.

A. Production stresses

    i. Branding (freeze or hot iron) (cattle: Lay et al., 1992a,b)

    ii. Castration (sheep: Mellor and Murray, 1989a; horses: Ayala et al., 2012)

    iii. Crowding/space (chickens: Cunningham et al., 1988; pigs: Khafipour et al., 2014)

    iv. Dehorning (cattle: Laden et al., 1985; Wohlt et al., 1994; Sutherland et al., 2002)

    v. Early weaning stress (pigs: Smith et al., 2010)

    vi. Machine milking (Tancin et al., 2000)

    vii. Maternal deprivation (pigs: Schwerin et al., 2005)

    viii. Perch availability (laying hens: Yan et al., 2014)

    ix. Restraint, e.g., snare of pigs (Ciepielewski et al., 2013)

    x. Social or peck order (chickens: Cunningham et al., 1988)

    xi. Tail docking (sheep: Mellor and Murray, 1989a)

    xii. Training (horses: McCarthy et al., 1991)

    xiii. Transportation stress (horses: Oikawa et al., 2004; pigs: Dalin et al., 1993a; McGlone et al., 1993)

    xiv. Weaning (cattle: Lefcourt and Elsasser, 1995)

B. Simulated Production/Stresses

    i. Vibration (Perremans et al., 2001)

    ii. Simulated loading (Spencer, 1994)

    iii. Enterotoxigenic *E. coli* challenge (pigs: Khafipour et al., 2014)

    iv. Lipopolysaccharide (LPS) challenge (Collier et al., 2011)

    v. Immobilization (stress)

C. Diseases

    i. Acute abdomen syndrome (horses: Ayala et al., 2012)

    ii. Laminitis (horses: Ayala et al., 2012)

D. Environmental stresses

    i. Cold (Liu et al., 2014)

    ii. Heat (Liu et al., 2014)

E. Social stresses

    i. Social stress (mixing unfamiliar conspecifics (pigs: de Groot et al., 2001; Rutherford et al., 2014)

F. Other stresses

    i. Audiovisual stress (Turner et al., 2002a)

    ii. Anesthesia (pigs: Kostopanagiotou et al., 2010)

    iii. Epidural blockade (Apple et al., 1995).

    v. Exercise on treadmill (D'Allaire and DeRoth, 1986)

    vi. Feed deprivation followed by feed being out-of-reach (Velie et al., 2012)

Table 1 cont....

...Table 1 cont.

   vii. Halothane (pigs: Neubert et al., 1996)
  viii. Hypoxia (fetal lambs: Ducsay et al., 2009)
   ix. Hypotension following phenylephrine (O'Connor et al., 2005)
   xi. Military and police devices that incapacitate by electromuscular disruption/ incapacitation (Werner et al., 2012)

Table 2: Effect of glucocorticoids on growth in livestock and poultry.

Species	Change	Reference
Cattle	+	+ Brethour, 1972; Gottardo et al., 2008
Pig	-/+?	- Chapple et al., 1989a,b; Gaines et al., 2004; Lopes et al., 2004 + Carroll, 2001; Seaman-Bridges et al., 2003
Sheep	-/+?	- Spurlock and Clegg, 1962; Jobe et al., 1998 + Spurlock and Clegg, 1962
Horse	- -	- Glade et al., 1981
Chicken	- -	- Donker and Beuving, 1989; Post et al., 2003; Zulkifli et al., 2014

cells *in vitro* (Carsia et al., 1986). In addition, CRH increases the expression of the precursor for ACTH, pro-opiomelanocortin (POMC), in corticotropes (chicken: Kang and Kuenzel, 2014).

CRH is synthesized in cell bodies in the paraventricular nucleus (PVN) and released into hypophyseal portal blood vessels in the median eminence. This concept is supported, for instance, by the report that release of ACTH is suppressed following PVN lesioning (sheep fetus: Bell et al., 2005). There are also inputs from the hippocampus (reviewed Massart et al., 2012) with CRH expressed in the hippocampus and expression increased following central administration of CRH (pig: Vellucci and Parrott, 2000a,b).

Biogenic Amines and CRH Release

There is noradrenergic stimulation of the release of CRH as supported by the ability of NE when administered into the 3rd ventricle to increases circulating concentrations of F (cattle: Sutoh et al., 2016). Isolation stress is followed by hypothalamic release of NE (sheep: Hasiec et al., 2014). The effect of NE is prevented by the concomitant intravenous administration of tryptophan supporting a serotonergic input (cattle: Sutoh et al., 2016). In addition, there is evidence that the HPA is inhibited by derivative of DA (1-metyl-6,7-dihydroxy-1,2,3,4-tetrahydroisoquinoline or salsolinol) with the DA metabolite reducing both ACTH and F responses to stress (sheep: Hasiec et al., 2014, 2015).

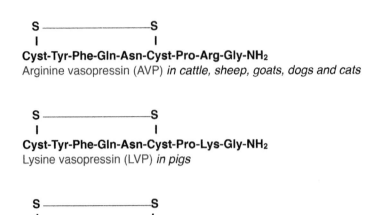

Figure 1A: Structure of neurohypophyseal hormones in domestic animals. These neuropeptides stimulate the corticotrophin cells in the anterior pituitary gland to release ACTH.

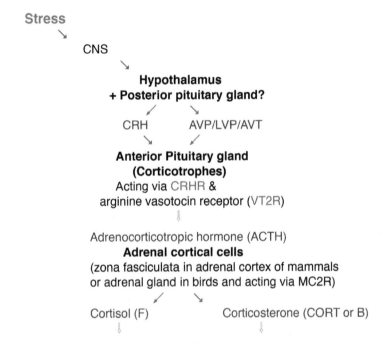

Figure 1B: The hypothalami-pituitary-adrenal axis. CRH is Corticotropin releasing hormone (CRH) or Corticoliberin. AVP is arginine vasopressin; LVP is lysine vasopressin and AVT is arginine vasotocin.

CRH Binding Protein (CRH-BP)

A CRH binding protein (CRH-BP) for CRH have been characterized (sheep: Behan et al., 1996). The activity of CRH is reduced when bound to CRH-BP (reviewed Seasholtz et al., 2002). The level of CRH-BP is higher in broiler (heavy/meat-type) than layer (light) type chickens (Khan et al., 2015).

Other Roles of CRH

There are effects of CRH on behavior and physiological processes. For instance, following administration of CRH into the brain, the pigs become excited and active (Parrott et al., 2000). Associated with early weaning in pigs, CRH has been reported to have direct stress related effects on jejunal functioning (pig: Smith et al., 2010).

Other Neuropeptides/hormones Modulating ACTH Release

Posterior Pituitary Hormones Modulating ACTH Release

Posterior pituitary hormones influence ACTH release and synthesis. In sheep, arginine vasopressin (AVP) stimulated both ACTH release and POMC expression in anterior pituitary cells *in vitro* (van de Pavert et al., 1997). The homologue of AVP in pigs is lysine vasopressin (LVP). Challenges with either CRH or LVP increases plasma concentrations of ACTH and cortisol in pigs (Janssens et al., 1995). LVP has been demonstrated to potentiate the effect of CRH on ACTH release from porcine anterior pituitary cells but has no effect *per se* on ACTH release (Abraham and Minton, 1996). The effect of LVP is via a protein kinase C mechanism (Abraham and Minton, 1996). For instance, when CRH sensitive cells are eliminated, ACTH release is stimulated by phorbol esters, activators of the protein kinase C signal transduction (van de Pavert et al., 1997). The homologue of AVP in chickens, arginine vasotocin (AVT) also influences ACTH release/synthesis by the chicken corticotropes. AVT in the presence of CRH increases expression of POMC in the chicken anterior pituitary cells (Kang and Kuenzel, 2014). The effect of AVT is mediated via the VT2 receptor (VT2R) (Sharma et al., 2009); the equivalent of mammalian vasopressin V1b receptor (V1bR).

Other Hormones Stimulating ACTH Release

Other hormones are capable of influencing ACTH release directly. Although the physiological significance of these is not established. Parathyroid hormone-related peptide (PTHrP) stimulates ACTH release (Nakayama et al., 2011a). While, calcitonin also augments release of ACTH in the presence of CRH from the chicken anterior pituitary gland (Nakayama et al., 2011b). There is evidence that somatostatin (SRIF) exerts a negative effect directly or indirectly on CORT secretion with administration of antisera to SRIF increasing CORT secretion in young chickens (Cheung et al., 1988a). Moreover, growth hormone increases circulating concentrations of CORT in chickens (Cheng et al., 1988b). It might be presumed that this effect is mediated at the corticotropes with SRIF inhibiting ACTH release. This has not been examined.

There are also neuropeptides/hormones that influence ACTH secretion likely acting at the hypothalamic level. For instance, intracerebroventricular (icv) administration of urotensin II increases circulating concentrations of ACTH (Watson et al., 2003). Opioid receptors are involved in the ACTH responses to stress based on the greater ACTH response in the presence of naloxone (pigs: Ciepielewski et al., 2013); naloxone increasing circulating concentrations of β endorphin, ACTH and cortisol in pigs (Ciepielewski et al., 2013). It would appear that opioid receptor ligands (possibly endorphin) influence ACTH release presumably acting at the hypothalamic level.

Adrenocorticotropic Hormone (ACTH)

Pro-opiomelanocortin (POMC)

POMC is the precursor of ACTH, α-, β- and γ-(melanocyte stimulating hormone) MSH and β-endorphin (see Figure 2). The structure of POMC has been established [e.g., cattle (Nakanishi et al., 1979) and chickens (Takeuchi et al., 1999)]. POMC is processed by prohormone convertase (PC) (Figure 2)

POMC mRNA

Pre-pro-opiomelanocorticotropin (POMC)
Signal peptide ~ pro-opiomelanocorticotropin (POMC)

Signal peptide + Pro-opiomelanocorticotropin (POMC)
16 KDa protein ~ Adrenocorticotropin (ACTH) ~ β lipotropin

16 KDa protein + Adrenocorticotropin (ACTH) + β lipotropin
(39 amino-acid residues)
acting via MCR2

γ MSH   α MSH + CLIP   β MSH   β Endorphin
                  (1-13)
                  acting via MCR4

Figure 2: Pro-opiomelanocorticotropin (POMC) is processed to Adrenocorticotropin (ACTH), MSH (Melanocyte stimulating hormone) and β Endorphin.

with both PC1 and 2 found in the anterior pituitary gland (pig: Seidah et al., 1992; sheep: Bell et al., 1998; Holloway et al., 2001; chicken: Ling et al., 2004).

POMC is expressed in the anterior pituitary gland (chicken: Sharma et al., 2009; Kang and Kuenzel, 2014). POMC is expressed in other tissues. For instance, in chickens, POMC but also in the adipose tissue, adrenals, brain (Takeuchi et al., 1999), feather follicles (Yoshihara et al., 2011), gonads (Takeuchi et al., 1999), and immune tissues-bursa Fabricius and thymus (Franchini and Ottaviana, 1999).

Melanocortin Receptors (MCR)

There are at least five melanocortin receptors (MCR) genes. The receptor in the adrenal cortex that binds ACTH is MC2R (EST characterized - cattle: NCBI Reference Sequence: XM_005892355.1; chicken: Barlock et al., 2014; pig: NCBI Reference Sequence: NM_001123137.1).

Stress Related Effects of ACTH not Mediated by Glucocorticoids

There are effects of ACTH not mediated via F or CORT. ACTH increases the adherence of *Escherichia coli* O157:H7 to pig colonic mucosa. This is thought to be via binding to MCR on the enteric nerves (Schreiber and Brown, 2005).

Negative Feedback and the HPA Axis

Glucocorticoids exert a negative feedback effect on the hypothalamus and corticotropes in the anterior pituitary gland decreasing release of respectively CRH and ACTH. This is supported by the precipitous decline in circulating concentrations of ACTH together with circulating and urinary concentrations of F in pigs receiving dexamethasone administration (Hay et al., 2000; Lopes et al., 2004). CORT decreases expression of POMC in the presence of CRH together with AVT by the chicken anterior pituitary cells (Kang and Kuenzel, 2014). The glucocorticoid dexamethasone decreased the expression of VP but not CRH in the hypothalamus (pig: Vellucci and Parrott, 2000a).

Glucocorticoids act at the levels of both the hypothalamus and anterior pituitary gland in a negative feedback manner to reduce CRH and POMC expression. CORT reduces CRH expression in chickens *in vivo* but did not influence POMC expression (Vandenborne et al., 2005). In contrast, CORT reduces POMC expression in the chicken pituitary gland *in vitro* (Vandenborne et al., 2005; Kang and Kuenzel, 2014). Moreover, CORT did not influence CRF-R1 expression (Vandenborne et al., 2005). There are positive effects of glucocorticoids at the hypothalamic and pituitary levels. Corticosterone reverses the decrease in expression of VT2R and VT4R in the presence of CRH and AVT *in vitro* (chicken: Kang and Kuenzel, 2014). Corticosterone at high concentrations increases the expression of CRH-2R and reduces that of CRH-1R in the presence of CRH and AVT *in vitro* (chicken: Kang and Kuenzel, 2014). Expression of CRH-1R is reduced that of CRH-2R is increased in the presence of CRH and AVT *in vitro* (chicken: Kang and Kuenzel, 2014).

There is also evidence for negative feedback from ACTH. Circulating concentrations of F are somewhat depressed in pigs immunized against ACTH conjugated to ovalbumin (Lee et al., 2005). In contrast, circulating concentrations of beta-endorphin, another product of POMC, were elevated (Lee et al., 2005) suggesting a homeostatic mechanism with increased release of ACTH.

Cortical Hormones

The adrenal cortex synthesizes F and CORT (discussed below) together with multiple peptides. For instance, adrenocortical cells produce opioid precursors—proenkephalin, POMC and prodynorphin (pig: Krazinski et al., 2011). The expression of prodynorphin by porcine adrenocortical cells is increased by angiotensin-II and decreased by ACTH (Krazinski et al., 2011). Angiotensin-II decreases expression of POMC in porcine adrenocortical cells (Krazinski et al., 2011).

Glucocorticoids in Livestock and Poultry

The major adrenal glucocorticoid in livestock is cortisol (F). Surprisingly given that the major glucocorticoid in horses is F, circulating concentrations of CORT are reported to be elevated by transportation stress in horses (Oikawa et al., 2004). The major adrenal glucocorticoid in birds is CORT with some F being produced albeit, at a very low levels except in the embryo (deRoos and deRoos, 1964; Hall and Koritz, 1966; Kalliecharan and Hall, 1974, 1977; Tanabe et al., 1986; reviewed: Carsia, 2015). This is supported by for instance, chicken adrenal cells releasing 115 fold more CORT than F (Carsia et al., 1987a).

Synthesis of Glucocorticoids in Livestock and Poultry

The synthesis of glucocorticoids in adrenal cortical cells is summarized in Figure 3. In mammals, the major site of synthesis of F is the *zona fasciculata* within the adrenal cortex (Robinson et al., 1983; Perry et al., 1992). The avian adrenal differs from that of mammals with chromaffin and cortical cells intermingled (reviewed: Carsia, 2015). Sub-populations of cortical cells have been separated in the turkey. These do not seem to be analogous

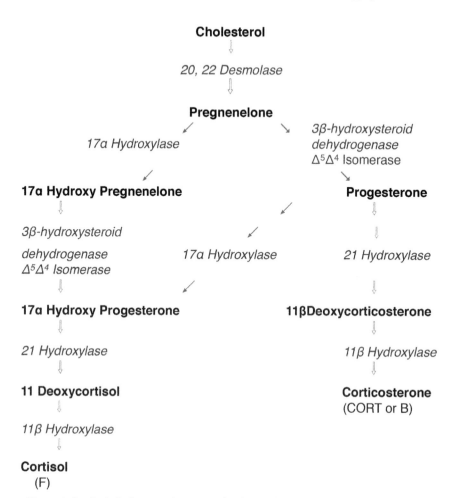

Figure 3: Synthetic Pathway and enzymes for the synthesis of cortisol or corticosterone.

functionally to the zones in the mammalian adrenal cortex as they produce both CORT and aldosterone, respond to ACTH and exhibit only minor differences in steroidogenesis and its control (Kocsis et al., 1995; Carsia and Weber, 2000). While the adrenal cortical cells are the major sites for the synthesis of glucocorticoids, there is some production in other tissues. For instance, there is 11β-hydroxylase (P450c11β) (see Figure 3) activity in chicken bursa, thymus and adrenal gland (Lechner et al., 2001).

Control of Glucocorticoid Synthesis

Glucocorticoid synthesis is controlled by ACTH binding to the MC2R receptor (e.g., chicken: Barlock et al., 2014). ACTH binds to the MC2R activating adenylate cyclase increasing cAMP. The effects of ACTH are mediated at least in part via adenylate cyclase and cAMP; ACTH elevating F or CORT production and cAMP or its analog, 8 bromo-cAMP, increasing F or CORT production by adrenocortical cells *in vitro* (pig: Huang et al., 2000; chicken: Carsia et al., 1987a). There are changes in adrenal ACTH receptors with physiological state. For instance, there are shifts in chickens on a low protein in the diet (8%) (chickens: Carsia and Weber, 1988). In turn, cAMP activates the cAMP response element binding protein (CREB) leading to phosphorylated CREB (pCREB) and increased steroidogenic acute regulatory protein (StAR) increasing cholesterol transport (Li et al., 2008). Interestingly, there is considerably less cortisol released from adrenocortical cells from halothane sensitive pigs *in vitro* (Huang et al., 2000).

There are other putative physiological stimulators of glucocorticoid production by adrenocortical cells. These include both opioid agonists and prostaglandins. Opioids increases both basal and ACTH stimulated F production by porcine adrenocortical cells; these effects being mediated via mu or kappa opioid receptors (basal) and mu receptors (ACTH stimulated) (Krazinski et al., 2011). Synthesis of CORT by chicken adrenal cortical cells is stimulated other chemical messengers including by E and NE (Mazzocchi et al., 1997) and prostaglandins (PGE_2 > PGE_1 > PGA_1 > PGB_2 > PGB_1 > $PGF_2\alpha$) (Kocsis et al., 1997). Moreover, CORT inhibits net production of CORT from chicken adrenal cells with the effect being reversed by prolactin (Carsia et al., 1987b).

Glucocorticoid Receptor

The glucocorticoids act via binding to the glucocorticoid receptor (GR) also known as the Nuclear receptor subfamily 3, group C, member 1 (NR3C1). The GR has been characterized (cattle: NCBI Reference Sequence: NM_001206634.1; pig: GenBank: AJ296022.1; horse: NCBI Reference Sequence: NM_001195191.1 and chicken: GenBank: DQ227738.1). Injections of dexamethasone increased expression of glucocorticoid receptors in the liver of young calves (Cantiello et al., 2009).

Determination of Glucocorticoids

There has been a shift from initially competitive binding protein assays (CBP) (to transcortin/corticosteroid binding protein) to radioimmunoassay (RIA) to more recently to enzyme linked immune-assay (ELISA).

Determination of Circulating Concentrations of Glucocorticoids in Livestock

The most frequently used method to determine circulating concentrations of F in livestock is RIA (e.g., pigs: McGlone et al., 1993; cattle: Tolleson et al., 2012; sheep: Kongsted et al., 2013). Other techniques employed include luminescence immunoassay (Lansade et al., 2014).

Determination of Circulating Concentrations of Glucocorticoids in Poultry

In poultry, CORT has been predominantly determined by CBP and RIA. It may be questioned whether the techniques employed today have been validated for CORT from chicken plasma and cross-reactivities fully established. For instance, Kim and colleagues (2015) recently reported on high plasma concentrations of F by ELISA; these being markedly elevated in chickens raised on a diet excluding antibiotics compared to one with antibiotics present. If the high levels of F prove to be the case, there needs to be a re-evaluation of the role of F in poultry. It might be noted that some reports include very high levels of CORT. For instance in a study of transportation stress, circulating concentrations of CORT were reported as over 30 ng ml^{-1} in control birds (by RIA) (Zhang et al., 2009; Yue et al., 2010) and about 40 ng ml^{-1} (by CBP) again in control birds (Edens and Siegel, 1975). This is much greater than the normal resting physiological range of 2–6 ng ml^{-1} (Satterlee and Gildersleeve, 1983; Kannan et al., 1997a,b; reviewed: Carsia, 2015). Again this argues for close examination of the validity, accuracy and precision of techniques used to measure CORT.

Determination of Concentrations of Glucocorticoids in Saliva and other Secretions

Determination of F in salivary is a useful approach. It has been employed in livestock (sheep: e.g., Fell et al., 1985; cattle: e.g., Negrão et al., 2004 and horses: e.g., Peeters et al., 2013). Both serum and salivary concentrations of F are increased by ACTH in sheep and horses (Yates et al., 2010; Peeters et al., 2011). Salivary concentrations of F are used as an indicator of stress (horses: Peeters et al., 2013) and are thought to reflect free F (Mormède et al., 2007) (for discussion of free versus bound glucocorticoids—see below). Concentrations of F have also been reported in tears (horses: Monk et al., 2014) and feces (elevated by transportation stress in horses: Schmidt et al., 2010).

Bioavailability of Glucocorticoids and Corticosteroid Binding Globulin (CBG)

Corticosteroid binding globulin (CBG or transcortin) binds both CORT and F in respectively poultry and livestock blood. For instance in birds, CORT is found predominantly bound to proteins with a one high affinity protein (CBG) and at least one low affinity high capacity protein (birds: Wingfield et al., 1984). It is assumed that only free CORT is biological activity (Figure 1). Evidence that bound F is less or completely inactive comes from the ability of CBG to reduce the inhibition of CRH by F stimulated ACTH release from fetal pituitary cells (Berdusco et al., 1995).

The gene encoding CBG is serpin peptidase inhibitor, clade A (alpha-1 antiproteinase, antitrypsin), member (*SERPINA66*). The gene *SERPINA66* is found on chromosome 7 in pigs (Gene ID: 396736). There is a missense mutation Gly307Arg in *SERPINA6* in some pigs which influences both the ability of CBG to bind F and circulating concentrations of F (Guyonnet-Dupérat et al., 2006). There are also multiple non-synonymous single nucleotide polymorphisms (SNPs) of the porcine *SERPINA6* gene (Görres et al., 2015).

In livestock, circulating concentrations of CBG are influenced by physiological state, for instance, being decreased with heat stress in pigs (Heo et al., 2005) and increased in late fetal development (Berdusco et al., 1995). Moreover, an index of free F that has been used in livestock is the ratio of circulating concentrations of CBG:F (Heo et al., 2005; Adcock et al., 2007). There has been little attention to characterization and determining changes in the circulating concentrations of CBG with physiological state (e.g., stress) in poultry for thirty years. It has been partially characterized (Gould and Siegel, 1978). In chickens, ACTH depresses circulating concentrations of CBG (Gould and Siegel, 1985). Moreover, CBG concentrations are elevated in obese (hypothyroid) chickens (Fässler et al., 1988).

The primary site of CBG synthesis is the liver based on the expression of the *SERPINA66* gene in sheep (Berdusco et al., 1994, 1995) and pigs (Heo et al., 2003). Fetal liver expression of CBG is increased by glucocorticoid administration (fetal liver: Sloboda et al., 2002; pigs: Adcock et al., 2007). In contrast, administration of ACTH to pregnant pigs decreased the plasma concentration of CBG (Kanitz et al., 2006). There is some expression of CBG in the fetal sheep anterior pituitary gland (Berdusco et al., 1995).

Cross Talk between the HPA Axis and Other Endocrine/metabolic Systems

Adipose Tissue

There are differences in the stress responses (increases in circulating concentrations of ACTH and F) to isolation restraint in ovariectomized sheep with increased adiposity (Tilbrook et al., 2008).

Adrenomedullin (ADM)

There are influences of adrenomedullin (ADM) on the HPA axis. Administration (ICV) of ADM depresses circulating concentrations of ACTH and F in sheep in addition to its vasodilatory effects (Parkes and May, 1995).

Hypothalamo-pituitary Gonadal Axis (HPG)

There appears to be some influence of the hypothalamo-pituitary gonadal axis (HPG) on the stress responses in livestock (reviewed, e.g., Turner and Tilbrook, 2006). For instance, testosterone attenuates both ACTH and F responses to either metabolic and psychological stress in castrated sheep (Dawood et al., 2005). Isolation/restraint stress increases circulating concentrations of F with a greater response in ewes than rams (Turner et al., 2002b). Within the hypothalamus, there is cross-talk between the HPA and the hypothalamo-pituitary-gonadal axis with CRH decreasing hypothalamic expression of gonadotropin releasing hormone (GnRH) (Ciechanowska et al., 2011).

Pancreas

There are effects of pancreatic hormones on the HPA axis. Insulin administration and/or the concomitant hypoglycaemia is followed by increased circulating concentrations of CORT or F (chickens: Scanes et al., 1980; sheep: Turner et al., 2002b).

Stress and the HPA Axis

Challenge with ACTH increases circulating concentrations of F in sheep (Mellor and Murray, 1989a). ACTH also influences the expression of 134 genes including MC2R and StAR (steroidogenic acute regulatory protein) transport protein for cholesterol in the chicken adrenal (Bureau et al., 2009). Similarly, there is up-regulation of the StAR gene expression in the adrenal of with pigs following ACTH treatment (Hazard et al., 2008). There are decreases in adrenocortical response to ACTH but not to cAMP by adrenocortical cells *in vitro* during growth (Carsia et al., 1985). Effects of stressor and other stimulators of release/synthesis of ACTH and F in livestock is discussed below.

Stress and the HPA Axis in Pigs

Stress influences multiple levels of the HPA predominantly by elevating circulating concentrations of ACTH and F (see below). In addition, stress increases expression of CRH in the brain (maternal deprivation: Schwerin et al., 2005) and POMC (e.g., transportation stress: McGlone et al., 1993). Moreover, while stress increases the expression of CRH1R and CRH2R in the amygdala and decreases expression of CRH1R in the pituitary gland (Schwerin et al., 2005) in pigs. Stress influences circulating concentrations of ACTH. Among the stresses demonstrated to influence release of ACTH in pigs are the following:

- Anesthesia (Kostopanagiotou et al., 2010)
- Exercise (Zhang et al., 1992)
- Naloxone (Ciepielewski et al., 2013)
- Restraint (Ciepielewski et al., 2013)
- Transportation (McGlone et al., 1993)
- Vibrations simulating transportation (Perremans et al., 2001)

Stresses such as the following also increase release of another product of POMC, beta-endorphin:

- Anesthesia (Kostopanagiotou et al., 2010)
- Naloxone (Ciepielewski et al., 2013)
- Restraint (Ciepielewski et al., 2013)
- Snaring (Roozen et al., 1995)

Multiple stresses increase circulating concentrations of F in pigs:

- Anesthesia and surgery (Dalin et al., 1993b).
- Exercise (Zhang et al., 1992)
- Feed restriction (Metges et al., 2015)
- Relocating and groups (Dalin et al., 1993b).
- Immobilization stress in a prone position (Rosochacki et al., 2000)

- Insulin challenge (Zhang et al., 1992; but not in fetal pigs: Spencer et al., 1983)
- Lipopolysaccharide (LPS) (Collier et al., 2011)
- Naloxone (Ciepielewski et al., 2013)
- Overcrowding (Verbrugghe et al., 2011)
- Restraint (Collier et al., 2011; Ciepielewski et al., 2013; Rosochacki et al., 2000)
- Simulated loading (Spencer, 1994)
- Snaring restraint (Farmer et al., 1991; Roozen et al., 1995)
- Transportation (Dalin et al., 1993a; McGlone et al., 1993)
- Treadmill exercise (D'Allaire and DeRoth, 1986)
- Vibration increased (Perremans et al., 2001)

The F response to treadmill exercise is higher in stress susceptible pigs (D'Allaire and DeRoth, 1986). There was no effect of chronic stress (tethered housing) on the response of pigs to challenges with either CRH or ACTH (Janssens et al., 1995). Moreover, social stress (mixing unfamiliar pigs) increases salivary concentrations of F (de Groot et al., 2001).

Stress and the HPA Axis in Sheep and Cattle

There are marked effects of a variety of stressors on the HPA axis. There have been relatively few studies reporting effects of stress on ACTH secretion. Circulating concentrations of ACTH have been reported to be increased in sheep by social isolation and weaning (Hasiec et al., 2014). Moreover, hypoxia in fetal sheep increases circulating concentrations of ACTH (Braems et al., 1996; Ducsay et al., 2009). There was no discernible effect of milking on circulating concentrations of ACTH (cattle: Tancin et al., 2000). In contrast, there are numerous reports on the effects of stressors on circulating concentrations of F in cattle and sheep. For instance, multiple stresses increase circulating concentrations of F in cattle:

- Dehorning (Laden et al., 1985; Wohlt et al., 1994)
- Freeze branding (Lay et al., 1992a)
- Hot-iron branding (Lay et al., 1992a)
- Lipopolysaccharide (LPS) (Zebeli et al., 2013)
- Low plane of nutrition (Tolleson et al., 2012)
- Milking (cattle: Tancin et al., 2000)
- Tick infestation (Tolleson et al., 2012)
- Transportation (Zavy et al., 1992)
- Weaning (Zavy et al., 1992)

It might be noted that morphine suppresses the milking induced increase in circulating concentrations of F (Tancin et al., 2000).

Similarly to the situation in cattle, multiple stresses increase circulating concentrations of F in sheep:

- Alpha-1 adrenergic agonist, phenylephrine, and concomitant hypotension (O'Connor et al., 2005)
- Alpha 2 adrenergic agonist, medetomidine (Ranheim et al., 2000)
- Audiovisual stress (Turner et al., 2002a)
- Castration (Mellor and Murray, 1989b; Paull et al., 2008)
- Epidural blockade (Apple et al., 1995)
- Hypoxia (Braems et al., 1996)
- Isolation stress (Apple et al., 1995)
- Milking (Yardimci et al., 2013)
- Restraint (Apple et al., 1995)
- Shearing (Yardimci et al., 2013)
- Tail-docking (Mellor and Murray, 1989b)
- Weaning (Hasiec et al., 2014).

In contrast, there is no effect of fasting on circulating concentrations of F (Kongsted et al., 2013) and suckling reduces the ACTH and F responses to isolation stress (Hasiec et al., 2014, 2015).

Stress and the HPA Axis in Horses

Circulating concentrations of F are increased by stresses in horses including acute or chronic disease, acute abdomen syndrome, castration and laminitis as are those of epinephrine (Ayala et al., 2012). Transportation stress is accompanied by increases the circulating concentrations of CORT and ACTH in horses (Oikawa et al., 2004). In young horses in training, there increases in the circulating concentrations of beta-endorphin but not ACTH (McCarthy et al., 1991). Circulating concentrations of F are increased by exercise/physical activity (Kędzierski et al., 2014). Following challenge with the alpha-1 adrenergic agonist, phenylephrine, there are decreased in blood pressure and increases in circulating concentrations of ACTH and F in anesthetized young horses (O'Connor et al., 2005).

More recent studies on the effects of stress in horses have employed salivary concentrations of F. There is a tight relationship between concentrations of F in saliva and plasma ($R^2 = 0.80$) of horses receiving ACTH challenges (Peeters et al., 2011). Salivary or circulating concentrations of F are increased by the following:

- Breeding season in stallions (Aurich et al., 2015)
- Environmental enrichment (Lansade et al., 2014)
- Fasting (Glade et al., 1984)
- Hot-branding (Erber et al., 2012)
- Following micro-chip implantation (Erber et al., 2012)
- Loading prior to transportation (Schmidt et al., 2010)
- Lunging or hyperflexion (Becker-Birck et al., 2013; Christensen et al., 2014)
- Performance (von Lewinski et al., 2013)
- Physical activity/exercise (Kędzierski et al., 2014)
- Rehearsal (von Lewinski et al., 2013)
- Transportation (Schmidt et al., 2010)
- Transrectal ultrasound examination (Schönbom et al., 2015)

Stress and the HPA Axis in Poultry

Plasma concentrations of CORT are elevated by multiple stressors such as the following:

- Anesthesia by pentobarbitone (Cheung et al., 1988a)
- Cold (Beuving and Vonder, 1978)
- Cooping stress (Satterlee et al., 1994)
- Diethyl ether (Scanes et al., 1980)
- *E. coli* endotoxin (Curtis et al., 1980; Scanes et al., 1980)
- Epinephrine administration (Harvey and Scanes, 1978)
- Fasting (Harvey et al., 1983)
- Feed restriction (de Jong et al., 2002; Zulkifli et al., 2011)
- Handling (Kannan et al., 1997a)
- Heat (Edens and Siegel, 1975; Beuving and Vonder, 1978)
- Immobilization (Beuving and Vonder, 1978; Kang and Kuenzel, 2014)
- Induced molt (Davis et al., 2000)
- Insulin challenge (Harvey et al., 1978)
- Insulin induced hypoglycemia (Scanes et al., 1980)
- Protein deprivation (very low protein diet) (Carsia et al., 1988)
- Shackling (Kannan et al., 1997a; Bedanova et al., 2007)
- Skip a pay feeding regimen (de Beer et al., 2008)
- Stress and the HPA Axis in Poultry

There are increases in circulating concentration with nutritional restriction such as fasting (Harvey et al., 1983) and a severe protein restriction (Carsia et al., 1988). Restoration of feed to fasted chicks is followed by decreases in circulating concentrations of CORT (Harvey et al., 1983). Chronic protein deprivation is associated elevated plasma concentrations of CORT but surprisingly depressed circulating concentrations of ACTH (Carsia et al., 1988). There are changes in adrenocortical functioning with diet in young chickens with increased basal and ACTH stimulated steroidogenesis *in vitro* (Carsia et al., 1988).

In a production setting, shackling broiler chickens is accompanied by increases in plasma concentrations of CORT rising from 6.5 ng ml^{-1} to about 14 ng ml^{-1} after 4 minutes (Kannan et al., 1997a). It should also be noted that in the same series of studies, handling resulted in an increase in plasma concentrations of CORT to 11 ng ml^{-1} (Kannan et al., 1997a).

In contrast to the observed effects of stressors on circulating concentrations of CORT, there is no effect of ammonia on plasma concentrations of CORT (Olanrewaju et al., 2008) or of crating with or in the absence of transportation stress (Kannan et al., 1997b). There are other circumstances that would be expected to elicit a HPA stress response and where none is found. For instance, there were little or no effects of crating alone or with transportation on circulating concentrations of CORT (Kannan et al., 1997a,b; Zhang et al., 2009; Yue et al., 2010). There is little effect of space or position in social hierarchy or 'peck-order' on circulating concentrations of CORT (Cunningham et al., 1988). Moreover, there are situations where circulating concentrations of CORT are paradoxically increased when it might be expected to be decreased. Perhaps surprisingly, caged hens have been reported to produce more eggs and have lower plasma concentrations of CORT compared with chickens housed in a group environment (Mashaly et al., 1984). Moreover, both basal and stressed circulating concentrations of CORT are higher in a line of chickens with low feather pecking than in a high feather pecking line (Korte et al., 1997). Similarly, circulating concentrations of CORT are increased when laying hens are transferred to an environment with perches available (Yan et al., 2014).

Diurnal Shifts in Glucocorticoid Release

Studies on the effect of stressors may be confounded by the diurnal/ circadian changes in plasma concentrations of CORT/F. For instance, there are diurnal shifts in circulating and urinary concentrations of F (pigs: Hay et al., 2000). Moreover, there is a pronounced circadian rhythm in plasma concentrations of CORT in broiler chickens being higher during the day than at night and peaking at 11.00 a.m. (de Jong et al., 2001) with an analogous circadian rhythm in plasma concentrations of CORT in 4–5 week old females of a laying strain (Wilson et al., 1984). However, as females mature to hens, the circadian rhythm is no long evident (Wilson et al., 1984) but once egg laying begins, there is a rhythm of plasma concentrations of CORT aligned with the ovulatory cycle (Wilson and Cunningham, 1981).

Actions of Glucocorticoids

The effects of glucocorticoids on growth and metabolism have been extensively investigated in poultry and livestock. This has predominantly involved *in vivo* studies in which CORT (or another glucocorticoid) or ACTH is administered in the diet, water, by infusion and by daily injection. The effects of glucocorticoids on growth in livestock and poultry are summarized in Table 2.

Glucocorticoids and Growth

Based on studies in rodent models, glucocorticoids inhibit growth with reduced muscle weight and increased adiposity. An identical situation exists in poultry (see below).

Glucocorticoids and Growth in Cattle

Surprisingly with the widespread catabolic actions of glucocorticoids, dexamethasone treatment either has no effect or, in fact, increases growth rate (ADG) in cattle (Brethour, 1972; Gottardo et al., 2008) with multiple shifts in gene expression (Carraro et al., 2009). Indeed, dexamethasone has been used illegally to improve growth rate in cattle (reviewed: Carraro et al., 2009) in view of the increased intramuscular fat (Brethour, 1972). It is not clear the extent that glucocorticoids stimulate growth via the HPA axis over via estrogen or androgen receptors or via effects on sex-steroid binding globulin or corticosteroid binding globulin.

Glucocorticoids and Growth in Horses

There is one report of a glucocorticoid, dexamethasone, having the following catabolic effects on growth in horses (Glade et al., 1981):

- Inhibiting growth,
- Suppressing nitrogen retention,
- Increasing nitrogen excretion.

Glucocorticoids and Growth in Pigs

There are reports of glucocorticoids 1. inhibiting, 2. without effect and 3. stimulating growth in pigs. Supporting the former, Administration of cortisol results in markedly decreased reduced growth rates in young pigs (Chapple et al., 1989a,b). Moreover, treatment of neonatal pigs with the glucocorticoid, isoflupredone, resulted in reduced growth up to weaning

(Gaines et al., 2004). Moreover, chronic administration of dexamethasone reduced nitrogen retention in growing pigs with depressed circulating concentrations of the growth related hormones, insulin-like growth factor I (IGF-I) (Lopes et al., 2004). Moreover, there is increased urinary concentrations of nitrogen in growing pigs treated with dexamethasone (Lopes et al., 2004). In another study, administration of glucocorticoid, betamethasone, had little effect on the growth of pre-term pigs but there were catabolic effects including decreased proliferation and increased apoptosis in the heart ventricle muscle cells (Kim et al., 2014).

In contrast, there are several reports that glucocorticoids increase growth rate. Dexamethasone has been reported to increase growth together with the circulating concentrations of two growth-related hormones/factors/binding proteins, IGF-I and IGH-BP3, between the neonatal stage to market weight age in pigs (Carroll, 2001; Seaman-Bridges et al., 2003).

Glucocorticoids and Growth in Sheep

There is evidence of both catabolic and growth promoting effects of glucocorticoids in sheep. Supporting the former contention, administration of glucocorticoid, betamethasone, to fetal lambs decreased body weight (Jobe et al., 1998). In contrast, administration of the glucocorticoid, cortisone, results in increased growth (carcass weight) in sheep due to increased feed intake and deposition of fat (Spurlock and Clegg, 1962). However, consistent with the catabolic effects of glucocorticoids, cortisone treatment reduces protein accretion (Spurlock and Clegg, 1962).

Glucocorticoids and Growth in Poultry

There is abundant evidence that glucocorticoids inhibit growth in poultry. CORT or exogenous glucocorticoids reduce growth in chickens (e.g., Donker and Beuving, 1989; Post et al., 2003; Zulkifli et al., 2014) (Table 1). The reduction in growth is accompanied by decreased skeletal muscle weight (Gross et al., 1980) but increased weights of the liver, intestine and adipose tissue (abdominal fat pad) (e.g., Bartov et al., 1980; Gross et al., 1980; Bartov, 1985; Hamano, 2006) (Table 1). Similarly, increasing physiological concentrations of CORT by implanting osmotic pumps releasing ACTH (Puvadolpirod and Thaxton, 2000a,b,c,d) results in the following:

- Decreased growth rate
- Increased nitrogen in excreta
- Reduced nitrogen in absorption.

However, daily injections of neither ACTH nor the glucocorticoid, cortisone influenced circulating concentrations of growth hormone (GH) (Davison et al., 1979; Harvey and Scanes, 1979).

Glucocorticoids and Metabolism

The release of F or CORT, following stressors (Table 1), would be expected to influence metabolism with increases in the gluconeogenesis (hepatic), glycogenesis (hepatic and skeletal muscle) and lipogenesis (hepatic) and decreases in net protein accretion (skeletal muscle) and glucose utilization (skeletal muscle). There are concomitant increases in liver and adipose tissue weights and decreases in skeletal muscle weight together with increases in the circulating concentrations of glucose and triglyceride. The actions of glucocorticoids on metabolism in poultry are discussed elsewhere in this volume (Scanes, 2016) so the effects of F on metabolism focus in the situation in livestock.

It is reasonable to assume that elevated concentrations of F or CORT increase both protein degradation in skeletal muscles and hepatic gluconeogenesis in livestock. The fetal adrenal gland exerts an important role in the control of gluconeogenesis with adrenalectomy preventing fasting induced gluconeogenesis in fetal lambs (Fowden and Forhead, 2011). Administration of dexamethasone is accompanied by increased circulating concentrations of glucose in pigs (Lopes et al., 2004) and neonatal calves (Scheuer et al., 2006). Moreover, administration of glucocorticoids increase circulating concentrations of insulin in neonatal calves (Scheuer et al., 2006).

CORT influences hypothalamic expression of neuromodulator/receptor genes related to appetite control such as agouti-related protein (AGRP), cocaine-and amphetamine-regulated transcript (CART), corticotropin-releasing hormone (CRH), ghrelin, neuropeptide Y (NPY), POMC together with the leptin and MCR1 receptors (Byerly et al., 2009; Liu et al., 2012).

Glucocorticoids and Mineral Metabolism

There is evidence links between CORT and mineral metabolism. Administration of CORT depresses plasma concentrations of zinc but increases those of copper (Klasing et al., 1987), ceruloplasmin and ovotransferin (Zulkifli et al., 2014). Similarly, *E. coli* administration decreases serum concentrations of zinc and iron (Klasing et al., 1987). Inhibitory effects of lipopolysaccharide endotoxin on growth and feed intake in chickens is overcome by increasing dietary copper oxide (Koh et al., 1996).

Glucocorticoids and Stress and Behavior

The tonic immobility test can be used to assess fear in chickens. There appears to be a link between fear and growth. Chickens with a short duration of immobility (low fear) have elevated growth rates while those with a long duration of immobility (high fear) have depressed growth rates (Wang et al., 2013). There were no differences in the circulating concentrations of CORT between the two groups (Wang et al., 2013). In parallel, following the stress related release of CORT, there may to be shifts in fear and the associated behaviors. For instance, CORT administration has been reported to increase tonic immobility (Jones et al., 1988; El-Lethey et al., 2001). Moreover, CORT administration increases feather pecking (El-Lethey et al., 2001).

Glucocorticoids and Immune Function

The immune system is inhibited by the HPA axis. Glucocorticoids including dexamethasone or prednisolone, have been reported to induce of the thymus involution with marked apoptosis in the medulla in cattle (Cannizzo et al., 2010, 2011; Vascellari et al., 2012). In chickens administered with glucocorticoids, there are large reductions in the weights of lymphoid tissues (spleen, bursa Fabricius and thymus) (summarized in Table 3). Moreover, ACTH infusion is followed by decreases in relative weights of the thymus, bursa and spleen in chickens (Puvadolpirod and Thaxton, 2000a).

There are other immunosuppressive effects of glucocorticoids in livestock and poultry. For instance, dexamethasone increases the number of *Salmonella* Typhimurium (ST) bacteria recovered from ileum, colon and cecum following ST challenge (pigs: Verbrugghe et al., 2011). Glucocorticoids also enhance the rate of proliferation of ST in pig macrophage (Verbrugghe et al., 2011).

There are positive effects of glucocorticoids on the immune system. Expression of Toll-like 2 receptor and TNF-alpha expression is increased in leukocytes from calves treated with dexamethasone (Eicher et al., 2004). In addition, serum concentrations of the acute phase protein, α1-acid glycoprotein, together with interleukin-6 and brain heat shock protein 70 (HSP 70) reelevated in chicks receiving administration of CORT as intramuscular injections (Zulkifli et al., 2014). CORT also increases expression of cytokines by chicken lymphocytes [interleukin (IL)-1 β, IL-6, IL-18 and transforming growth factor (TGF)-β4 (Shini and Kaiser, 2009) and a series of interleukins (IL-1β, IL-6, IL-10, IL-12α and IL-18) by chicken heterophils (Shini et al., 2010).

Table 3: Effect of glucocorticoids on immune parameters in poultry.

Parameter	Change	Reference
Mortality in response to *E. coli*	↑↑	Gross et al., 1980
Weight of Lymphoid Organs		
Bursa Fabricius	↓↓	Gross et al., 1980; Davison et al., 1985; Donker and Beuving, 1989
Spleen	↓↓	Gross et al., 1980; Donker and Beuving, 1989
Thymus	↓↓	Gross et al., 1980; Davison et al., 1985; Donker and Beuving, 1989
Leukocytes in Blood		
Heterophil numbers	↑↑↑	Gross et al., 1980; Gross and Siegel, 1983
Lymphocyte numbers	↓↓↓	Gross et al., 1980; Gross and Siegel, 1983
Heterophil:Lymphocyte ratio	↑↑↑	Gross et al., 1980; Gross and Siegel, 1983
Cytokines		
Plasma concentration of interleukin-6	↑↑	Zulkifli et al., 2014
Lymphocyte expression of IL-1 β	↑↑	Shini and Kaiser, 2009
Heterophil expression of IL-1 β, IL-6	↑↑	Shini et al., 2010

Leukocytes and Stress

Leukocyte Numbers

There was a higher blood concentrations of leukocytes in commercial chickens than unselected indigenous chickens and jungle fowl (Scanes and Christensen, 2014). Leukocyte concentrations were increased following tail docking in lambs (Wohlt et al., 1982) and in pigs following transportation (Dalin et al., 1993a).

Neutrophil:lymphocyte (N:L) Ratio

It is well recognized that stress evokes marked shifts the neutrophil: lymphocyte ratio (N:L) in livestock mammals (see below) and the heterophil to lymphocyte ratio (H:L) in birds (Table 3). The increase in the H:L ratio can be attributed, at least partially, to effects of stress induced CORT release. Administration of CORT is followed by dramatic increases in the H:L ratio in chickens rising from less than 0.4 to greater than 8.0 (Gross et al., 1980; Gross and Siegel, 1983). There are reports that there are increases in N:L in stressed pigs, sheep and poultry.

Stress and the Neutrophil:Lymphocyte (N:L) Ratio in Pigs

Rearing pigs in small pens was accompanied by increased N:L ratio but no change in growth rate (Yen and Pond, 1987). Pigs raised in an enhanced environment have reduced N:L ratios (Reimert et al., 2014). There is a lack of correlation between circulating concentrations of F and the N:L ratio in growing pigs (Stull et al., 1999).

Stress and the Neutrophil:Lymphocyte (N:L) Ratio in Sheep

Shifts in N:L ratio in stressed sheep are consistent with N:L ratio being a good indicator of stress. Stressing sheep by a short period of water restriction is accompanied by increases in the N:L ratio along with the concentration of F in wool (Ghassemi Nejad et al., 2014). Both circulating concentrations of F and N:L ratio lower in sheep with shade than no shade (Liu et al., 2012). Sheep transported on unpaved roads have higher circulating concentrations of F and N:L ratios compared to those moved on paved roads (Miranda-de la Lama et al., 2011).

Stress and the Heterophil:Lymphocyte (H:L) Ratio in Poultry

Maxwell (1993) concluded that H:L ratio was more reliable indicator of mild or moderate stress in poultry than circulating concentrations of CORT. Moreover, it should be noted that the H:L ratio is heritable (Al-Murrani et al., 1997) and has been proposed as a marker for resistance to stress (Al-Murrani et al., 2006). Moreover, is followed by increases in both the H:L ratio and circulating concentrations of CORT in chickens. In chickens, the H:L ratio is elevated by other stressors including:

- Fasting (Gross and Siegel, 1983)
- *E. coli* endotoxin challenge (Gross and Siegel, 1983)
- Shackling (Bedanova et al., 2006a)
- Social stress (Gross and Siegel, 1983)

together with heat, challenge with *Salmonella typhimurium*, transportation stress (reviewed: Maxwell, 1993; also see Al-Murrani et al., 1997). Similarly, either transportation and dexamethasone increases both the H:L ratios and mortality following *E. coli* challenge in turkeys (Huff et al., 2005). Most but not all of these stressors also increases circulating concentrations of CORT (see above). For instance, transportation stress does not apparently influence circulating concentrations of CORT or the H:L ratio (Kannan et al., 1997a; Zhang et al., 2009). Moreover, high fear chickens exhibit an elevated H:L ratios compared to low fear chickens but there was no differences in the adrenal CORT response to adrenocorticotropic hormone (ACTH) (Beuving et al., 1989). Furthermore, malathion decreases the H:L ratio (Gross and Siegel, 1983) despite the increases adrenocortical activity (Goyal et al., 1988). There are also elevated circulating concentrations of CORT in birds treated with parathion (Rattner et al., 1982). These studies would suggest that H:L ratio can be shifted both by CORT and in a glucocorticoid independent manner. There are no differences the H:L between commercial chickens and unselected indigenous chickens or with the red jungle fowl (Scanes and Christensen, 2014) suggesting a lack of stress effect of domestication or modern production practices.

Lymphocyte (N:L) Ratio as a Measure of Stress and its Impact

The H:L ratio has been proposed as a very useful measure of stress in poultry (reviewed for instance: Maxwell, 1993) and in wild birds (Davis et al., 2008). The N:L ratio is reported to be a useful clinical parameter in human medicine being *"a useful marker to predict subsequent mortality in patients admitted for ST segment elevation myocardial infarction"* (Núñez et al., 2008), together with after percutaneous coronary intervention (Duffy et al., 2006) and following hepatic re-section for colorectal liver metastases (Halazun et al., 2008).

Catecholamines: Epinephrine (E) and Norepinephrine (Ne) and Stress

Chromaffin Cells

Bovine medullary cells have been used extensively to study the release of E and NE. Bovine medullary cells consist of chromaffin cells that produce either E or NE (Marley and Livett, 1987a). Acetyl choline or nicotinic agents stimulate the release of catecholamines together with opioid peptides (Wilson et al., 1982; Marley and Livett, 1987b). Based on studies with antagonists, it was concluded that acetyl choline acts via either the $\alpha3\beta4$ or $\alpha3\beta4\alpha5$ nicotinic cholinergic receptors (Broxton et al., 1999). There are additional controls. Histamine increases release of both E and NE via H_1 receptors (based on the ability of mepyramine to block) from bovine chromaffin cells (Livett and Marley, 1986). Substance P inhibits release of both E and NE from their respective chromaffin cells (Krause et al., 1997). Moreover, opioids influence de-sensitization to nicotine (Marley and Livett, 1987b).

Medullary Peptides

Porcine chromaffin cells also produce a hypotensive peptide, adrenomedullin (ADM) (Nussdorfer et al., 1997).

Overview of Catecholamines and Stress

Stresses increase release of catecholamines from the adrenal medulla and sympathetic nerves. Predominantly, studies on the effects of various stresses determine circulating concentrations of E and NE in the blood. It has been proposed that the α-adrenergic receptor 2C (ADRA2C) gene, regulating NE release at pre-synaptic and chromaffin cell levels, is important to the domestication of chickens (Elfwing et al., 2014). There are also studies in which the concentrations of E and NE are determined in urine (pigs: Peeters et al., 2005; cattle: Ndlovu et al., 2008). Based on urinary determination, there are diurnal changes in urinary concentrations of E and NE (pigs: Hay et al., 2000).

E increases gastric acid secretion, heart rate and respiratory rate in pigs (Marolf et al., 1993). Other effects of catecholamines include NE increasing the adherence of *Escherichia coli* O157:H7 to pig colonic mucosa *in vitro* (Green et al., 2004).

Stress and Catecholamines in Pigs

Marked effects of stressors on E and NE have been reported in pigs. Circulating concentrations of catecholamines are influenced by stresses in pigs including the following:

- Aggressive behaviors following exposure to different pigs E ↑ and NE ↑ (Fernandez et al., 1994)
- Epidural blockage E ↑ (Apple et al., 1995)
- Feed deprivation followed by feed being out-of-reach NE ↑ (Velie et al., 2012)
- High temperature NE down but no change in E (Barrand et al., 1981)
- Immobilization stress in a prone position, E ↑, NE ↑ and DA ↑ (Rosochacki et al., 2000)
- Lipopolysaccharide (LPS) E ↑ and NE ↑ (Collier et al., 2011)
- Low temperature NE ↑ but no change in E (Barrand et al., 1981)
- Exercise E ↑ and NE ↑ (Barrand et al., 1981)
- Restraint E ↑ and NE ↑ (Rosochacki et al., 2000)
- Snaring E ↑ and NE ↑ (Roozen et al., 1995; Velie et al., 2012)

- Social interaction E ↑ and NE ↑ (Fernandez et al., 1995)
- Transportation E ↑ but no change in NE (McGlone et al., 1993).

There is evidence that nutritional stress does not interact with social interaction stress. Fed and fasted pigs exhibit similar increases in circulating concentrations of E and NE in pigs exposed to encounters with other pigs (Fernandez et al., 1995). In addition, social stress (mixing unfamiliar pigs) is followed by increases urinary concentrations of catecholamines (de Groot et al., 2001). Stressing the pregnant dam, influences the E and NE responses to LPS in the offspring (Collier et al., 2011).

Porcine Stress Syndrome (PSS)

Stress-susceptible pigs have problems of meat quality [pale soft exudative muscle (PSE)]. In addition, they exhibit tremors, hyperthermia, shock and sudden death. Halothane can induces these symptoms. Halothane challenges has allowed differentiating between stress susceptible and stress resistant pigs. PSS is due to a mutation in a single gene, the ryanodine receptor (a calcium channel protein expressed in muscle sarcoplasmic reticulum). In stress susceptible pigs, halothane challenge results in increases in E and NE (Neubert et al., 1996). In stress-susceptible pigs, there are lower concentration of the neurotransmitters in the hypothalamus (5-HT, DA, NE and DA), caudate nucleus (5-HT, DA, NE and DA) and hippocampus (5-HT and NE) (Adeola et al., 1993). Circulating concentrations of E, NE and DA are greater in stress susceptible Pietrain than Duroc pigs (Rosochacki et al., 2000). Urinary DA (24 hours) are lower in stress susceptible pigs as are caudate nuclei DA concentrations but no differences in NE (Altrogge et al., 1980).

Stress and Catecholamines in Cattle and Sheep

Stresses influence the circulating concentrations of catecholamines. Examples of the stressors influencing E and/or NE are the following:

- Epidural blockade E ↑ NE unaffected (sheep: Apple et al., 1995)
- Hot-iron branding E ↑ (calves: Lay et al., 1992b)
- Isolation stress E ↑ NE unaffected (sheep: Apple et al., 1995)
- Restraint E ↑ NE unaffected (sheep: Apple et al., 1995)
- Separation from calf from dam E ↑ and NE ↑ (cattle - dam: Lefcourt and Elsasser, 1995)
- Urotensin II administered ICV E ↑ (Watson et al., 2003).
- Weaning/separation of calf from dam E and NE up (calf: Lefcourt and Elsasser, 1995)

Branding had no consistent effect on circulating concentrations of E or NE in cows (Lay et al., 1992a). The effects of stress on catecholamines are also seen in the cerebrospinal fluid. There are increases in the concentrations of NE in cerebrospinal fluid of stressed sheep (Turner et al., 2002b).

Stress and Catecholamines in Horses

There is limited information on the effects of stressors on circulating concentrations of catecholamines in horses. Circulating concentrations of E are increased by diseases including: acute abdomen syndrome, castration and laminitis (horses: Ayala et al., 2012). In addition, circulating concentrations of NE is elevated in anesthetized foals following administration of the alpha-1 adrenergic agonist, phenylephrine, and the concomitant hypotension (O'Connor et al., 2005).

Stress and Catecholamines in Poultry

The available evidence suggests that some stresses increase circulating concentrations of NE and E in chickens. Manual restraint is accompanied by increased circulating concentrations of E and NE (Beuving and Blokhuis, 1997). Acute heat stress is followed rapidly by increases in circulating concentrations of NE and E in chickens (Bottje and Harrison, 1986). These decline during recovery (Bottje and Harrison, 1986). During stress, hens from a high feather pecking line had higher circulating concentrations of NE than a low feathering hens but there were no differences in circulating concentrations of E (Korte et al., 1997). Social stress, however, had no effect on circulating concentrations of E and NE in chickens but there was an increase in E concentrations in Raphe nucleus and a decrease in the concentration of 5-hydroxyindoleacetic acid in the hypothalamus (Cheng and Fahey, 2009). Moreover, there were no effects of crating alone or with transportation on circulating concentrations of either NE or E (Kannan et al., 1997b). There was, similarly, no effect of the availability of perches on circulating concentrations of E and NE in either pullets and hens (Yan et al., 2014).

Other Physiological Responses to Stress

There are effects of stressors on other hormones in livestock and poultry.

Growth Hormone, Prolactin, Thyroid Hormones and Stress

Stress is associated with increases in the circulating concentrations of growth hormone (GH) in pigs but suppressed circulating concentrations of GH in poultry. For instance, restraint of pigs increases circulating concentrations of GH (Ciepielewski et al., 2013). Similarly, snare restraint in pigs is followed by increased circulating concentrations of GH (Farmer et al., 1991). Insulin increases circulating concentrations of GH in fetal pigs but, interestingly, there is no change in the circulating concentrations of F (Spencer et al., 1983). In chickens, circulating concentrations of GH secretion are decreased by insulin challenge (Harvey et al., 1978), epinephrine administration (Harvey and Scanes, 1978) and heat stress (Liu et al., 2014).

There is some association between stress and circulating concentrations of prolactin. Restraint of pigs increases circulating concentrations of prolactin (Ciepielewski et al., 2013). There is no effect of isolation stress on plasma concentrations of prolactin despite the increased circulating concentrations of F (Hasiec et al., 2014). Moreover, immobilization (stress) is followed by decreases in the plasma concentrations of triiodothyronine (T_3) (Wodzicka-Tomaszewska et al., 1982). Stress also influences circulating concentrations of thyroid hormones in pigs (Spencer, 1994).

Stress and Cytokines

Stresses influence cytokines. Restraint for 4 hours is followed within 15 minutes with an increase in plasma concentrations of interleukin (IL)-1β, IL-6, IL-10 and tumor necrosis factor (TGF)-β together with a transient elevations in plasma concentrations of IL-12 and in natural killer cell cytotoxicity (NKCC) (Ciepielewski et al., 2013). Lipopolysaccharide (LPS) increases circulating concentrations of IL-6 and TNF-alpha (Collier et al., 2011).

Stress and Erythrocytes

There is evidence that at least some stressors can influence erythrocytes. There were marked differences in the hematocrit/Packed Cell Volume (PCV) between commercial chickens, indigenous/random-bred chickens and jungle fowl in a meta-analysis of multiple studies (Scanes and Christensen, 2014) with lower hematocrit/PCV in indigenous/random-bred chickens than jungle fowl and an even lower hematocrit/PCV in commercial chickens. There were no differences in blood concentrations of hemoglobin. There are decreases in circulating concentrations of hemoglobin with the stress of shackling which also results in elevated circulating concentrations of CORT (Bedanova et al., 2006a). In stress susceptible pigs, halothane challenge results in increases in hematocrit together with plasma concentrations of glucose, lactate and potassium and F (Neubert et al., 1996). However, chickens reared on reduced floor space also have increases in blood concentration of hemoglobin (Bedánová et al., 2006b).

Body Temperature

Some stressors influence body temperature. For instance, acute heat stress is followed by increases in body temperature in chickens (Bottje and Harrison, 1986). It is questioned whether the increases in body temperature reported during growth in modern young meat chickens (Christensen et al., 2012) are due to a failure to sufficiently dissipate heat and thereby stress the bird. The relationship between stress and body temperature is supported by the ability of CORT to depress body temperature in chickens (Klasing et al., 1987).

Circulating Concentrations of Metabolites and Electrolytes

Stressors can influence circulating concentrations metabolites and electrolytes. In stress susceptible pigs, halothane challenge results in increases in plasma concentrations of glucose, lactate and potassium and F (Neubert et al., 1996). Stress also influences circulating concentrations of lactate in pigs (Spencer, 1994). Transportation decreases circulating concentrations of glucose in chickens (Zhang et al., 2009; Yue et al., 2010).

Heart Weight and Rate

Stressors influence heart rate and heart weight. For instance, both freeze and hot-iron branding are followed by increased heart rate (Lay et al., 1992a,b). Lines of chickens with respectively high and low levels of feather pecking have been developed by divergent genetic selection (Kjaer and Jørgensen, 2011). These have differences in EKGs reflecting autonomous nervous responses to stress situations (Kjaer and Jørgensen, 2011). There are some stress responses in chickens related to social status. For instance, heart weights are reported to be increased in hens at the bottom of the peck-order (lowest dominance) (Cunningham et al., 1988). Heart rate is increased in acute heat stress in chickens (Bottje and Harrison, 1986).

Long Lasting or "Programming" Effects of Short Term Stress

What has been missing from the discussion are long-term or long lasting effects of stress. Early stress in chickens is followed by changes in social dominance (more dominant males and fewer dominant females) and changes in gene expression (Elfwing et al., 2015). Similarly, handling of neonatal pigs has some long-term effects on the stress responses/HPA axis in boars. These include: increased circulating concentrations of CBG and locomotory behavior together with decreased circulating concentrations of F (Weaver et al., 2000). However, there were no effects on other attributes of the conventional stress responses, namely the increases in circulating concentrations of ACTH and F (Weaver et al., 2000). Gestational social stress also results shifts in the HPA axis in the offspring pigs with increased CRH mRNA in PVN (within the hypothalamus) and amygdala (Jarvis et al., 2006) and reduced hypothalamic CRH (Haussmann et al., 2000). Moreover, there are differences in the levels of amygdala CRHR1 and 2 in 10 weeks old pigs together with subsequent maternal behavior in female offspring of pregnant pigs subject to social stress during pregnancy (Rutherford et al., 2014).

There are effects of maternal nutrition on the fetus and subsequent neonate. For instance in sheep, circulating concentrations of F are elevated in neonatal lambs from nutritional restricted dams (Kongsted et al., 2013; Vonnahme et al., 2013). Moreover, fasting depresses circulating concentrations of F in neonatal lambs from nutritionally restricted but not nutritionally replete dams (Kongsted et al., 2013). Preimplantation under-nutrition in the dam is followed by elevated fetal ACTH and F in sheep (Williams-Wyss et al., 2014).

Stress in Livestock and Poultry as Biomedical Models

Stress responses in livestock have been valuable as biomedical models. For instance, military and police devices that incapacitate by electromuscular disruption/incapacitation have been tested in pigs (Werner et al., 2012). In anesthetized pigs, there is not a strong F response (Werner et al., 2012). Moreover, epidural blockage is followed by increases in circulating concentrations of F and E (pigs: Apple et al., 1995).

Conclusions

There is not always close relationships between the effects of stressors on different indices of stress. For instance, despite, circulating concentrations of F, ACTH, GH and prolactin being increased in pigs subjected to nose snare restraint, circulating concentrations of F were not related to those of GH and prolactin (Rushen et al., 1993). Similarly, the concentration of F in wool, along with N:L ratios, is elevated by short period of water restriction but there is no effect on plasma concentrations of F (Ghassemi Nejad et al., 2014). Rosochacki and colleagues (2000) employed the circulating concentrations of F, E, NE and DA as indicator of stresses.

There are multiple physiological responses to stress. While it would be nice but unrealistic if any one response such as plasma concentrations of F or CORT (or any other hormone) or the N:L/H:L ratio or a behavioral response was the "gold standard" for stress, this is not uniformly the case. Instead, it is argued that stress might be defined as in Equation 3.

Equation 3
Stress = Constant A x Integrated plasma concentrations of glucocorticoids
    over time
    + constant B x Integrated plasma concentrations of catecholamines
    + Constant C x down-stream indicators of glucocorticoids such as
    N:L/H:L ratio + Constant D x Other indices of stress

In view of there being effects of stressors on indices of stress without impacting production, the negative effects of stress on production are expressed in Equation 4.

Equation 4
Negative effect of Stress on Production =
    Constant A x Integrated plasma concentrations of glucocorticoids
    over time
    + constant B x Integrated plasma concentrations of catecholamines
    + Constant C x N:L ratio + Constant D x Other indices of stress
    minus Constant E

It is argued that stress studies should include multiple time points and multiple physiological indices of stress. The indices should include F or CORT, circulating concentrations of E and NE, N:L ratio.

Keywords: Stress, hypothalamo-pituitary-adrenocortical axis, cortisol and growth, corticosterone and growth, glucocorticoids and carbohydrate metabolism, glucocorticoids and lipid metabolism, glucocorticoids and protein metabolism, and stress and immune function

References

Abraham, E.J., and J.E. Minton. 1996. Effects of corticotropin-releasing hormone, lysine vasopressin, oxytocin, and angiotensin II on adrenocorticotropin secretion from porcine anterior pituitary cells. Domest. Anim. Endocrinol. 13: 259–268.

Adcock, R.J., H.G. Kattesh, M.P. Roberts, J.A. Carroll, A.M. Saxton, and C.J. Kojima. 2007. Temporal relationships between plasma cortisol, corticosteroid-binding globulin (CBG), and the free cortisol index (FCI) in pigs in response to adrenal stimulation or suppression. Stress 10: 305–310.

Adeola, O., R.O. Ball, J.D. House, and P.J. O'Brien. 1993. Regional brain neurotransmitter concentrations in stress-susceptible pigs. J. Anim. Sci. 71: 968–974.

Al-Murrani, W.K., A. Kassab, H.Z. al-Sam, and A.M. al-Athari. 1997. Heterophil/lymphocyte ratio as a selection criterion for heat resistance in domestic fowls. Br. Poult. Sci. 38: 159–163.

Al-Murrani, W.K., A.J. Al-Rawi, M.F. Al-Hadithi, and B. Al-Tikriti. 2006. Association between heterophil/lymphocyte ratio, a marker of 'resistance' to stress, and some production and fitness traits in chickens. Br. Poult. Sci. 47: 443–448.

Altrogge, D.M., D.G. Topel, M.A. Cooper, J.W. Hallberg, and D.D. Draper. 1980. Urinary and caudate nuclei catecholamine levels in stress-susceptible and normal swine. J. Anim. Sci. 51: 74–77.

Apple, J.K., M.E. Dikeman, J.E. Minton, R.M. McMurphy, M.R. Fedde, D.E. Leith, and J.A. Unruh. 1995. Effects of restraint and isolation stress and epidural blockade on endocrine and blood metabolite status, muscle glycogen metabolism, and incidence of dark-cutting longissimus muscle of sheep. J. Anim. Sci. 73: 2295–2307.

Aurich, J., M. Wulf, N. Ille, R. Erber, M. von Lewinski, R. Palme, and C. Aurich. 2015. Effects of season, age, sex, and housing on salivary cortisol concentrations in horses. Domest. Anim. Endocrinol. 52: 11–16.

Ayala, I., N.F. Martos, G. Silvan, C. Gutierrez-Panizo, J.G. Clavel, and Illera, J.C. 2012. Cortisol, adrenocorticotropic hormone, serotonin, adrenaline and noradrenaline serum concentrations in relation to disease and stress in the horse. Res. Vet. Sci. 93: 103–107.

Barlock, T.K., D.T. Gehr, and R.M. Dores. 2014. Analysis of the pharmacological properties of chicken melanocortin-2 receptor (cMC2R) and chicken melanocortin-2 accessory protein 1 (cMRAP1). Gen. Comp. Endocrinol. 205: 260–267.

Barrand, M.A., M.J. Dauncey, and D.L. Ingram. 1981. Changes in plasma noradrenaline and adrenaline associated with central and peripheral thermal stimuli in the pigs. J. Physiol. 279: 139–151.

Bartov, I., L.S. Jensen, and J.R. Veltmann, Jr. 1980. Effect of corticosterone and prolactin on fattening in broiler chicks. Poult. Sci. 59: 1328–1334.

Bartov, I. 1985. Effects of dietary protein concentration and corticosterone injections on energy and nitrogen balances and fat deposition in broiler chicks. Br. Poult. Sci. 26: 311–324.

Becker-Birck, M., A. Schmidt, M. Wulf, J. Aurich, A. von der Wense, E. Möstl, R. Berz, and C. Aurich. 2013. Cortisol release, heart rate and heart rate variability, and superficial body temperature, in horses lunged either with hyperflexion of the neck or with an extended head and neck position. J. Anim. Physiol. Anim. Nutr. (Berl.) 97: 322–330.

Bedánová, I., E. Voslarova, P.V. Vecerek, V. Pistěková, and P. Chloupek. 2006a. Stress in broilers resulting from shackling. Acta Vet. Brno. 76: 129–135.

Bedánová, I., E. Voslárová, V. Vecerek, V. Pistěková, and P. Chloupek. 2006b. Effects of reduction in floor space during crating on haematological indices in broilers. Berl. Munch. Tierarztl. Wochenschr. 119: 17–21.

Bedánová, I., E. Voslarova, P. Chloupek, V. Pistekova, P. Suchy, J. Blahova, R. Dobsikova, and V. Vecerek. 2007. Stress in broilers resulting from shackling. Poult. Sci. 86: 1065–1069.

Behan, D.P., D. Cepoi, W.H. Fischer, M. Park, S. Sutton, P.J. Lowry, and W.W. Vale. 1996. Characterization of a sheep brain corticotropin-releasing factor-binding protein. Brain Res. 709: 265–274.

Bell, M.E., T.R. Myers, and D.A. Myers. 1998. Expression of proopiomelanocortin and prohormone convertase-1 and -2 in the late gestation fetal sheep pituitary. Endocrinology 139: 5135–5143.

Bell, M.E., T.J. McDonald, and D.A. Myers. 2005. Proopiomelanocortin processing in the anterior pituitary of the ovine fetus after lesion of the hypothalamic paraventricular nucleus. Endocrinology 146: 2665–2673.

Berdusco, E.T., W.K. Milne, and J.R. Challis. 1994. Low-dose cortisol infusion increases plasma corticosteroid-binding globulin (CBG) and the amount of hepatic CBG mRNA in fetal sheep on day 100 of gestation. J. Endocrinol. 140: 425–430.

Berdusco, E.T., K. Yang, G.L. Hammond, and J.R. Challis. 1995. Corticosteroid-binding globulin (CBG) production by hepatic and extra-hepatic sites in the ovine fetus; Effects of CBG on glucocorticoid negative feedback on pituitary cells *in vitro*. J. Endocrinol. 146: 121–130.

Beuving, G., and G.M.A. Vonder. 1978. Effect of stressing factors on CORT levels in the plasma of laying hens. Gen. Comp. Endocrinol. 35: 153–159.

Beuving, G., R.B. Jones, and H.J. Blokhius. 1989. Adrenocortical and heterophil/lymphocyte responses to challenge in hens showing short or long tonic immobility reactions. Br. Poult. Sci. 30: 175–184.

Beuving, G., and H.J. Blokhius. 1997. Effect of novelty and restraint on catecholamines in plasma of laying hens. Br. Poult. Sci. 38: 297–300.

Bottje, W.G., and P.C. Harrison. 1986. Alpha adrenergic regulation of celiac blood flow and plasma catecholamine response during acute heat stress in fed cockerels. Poult Sci. 65: 1598–1605.

Braems, G.A., S.G. Matthews, and J.R. Challis. 1996. Differential regulation of proopiomelanocortin messenger ribonucleic acid in the pars distalis and pars intermedia of the pituitary gland after prolonged hypoxemia in fetal sheep. Endocrinology 137: 2731–2738.

Brethour, J.R. 1972. Effects of acute injections of dexamethasone on selective deposition of bovine intramuscular fat. J. Anim. Sci. 35: 351–356.

Broxton, N.M., J.G. Down, J. Gehrmann, P.F. Alewood, D.G. Satchell, and B.G. Livett. 1999. Alpha-conotoxin ImI inhibits the alpha-bungarotoxin-resistant nicotinic response in bovine adrenal chromaffin cells. J. Neurochem. 72: 1656–1662.

Bureau, C., C. Hennequet-Antier, M. Couty, and D. Guémené. 2009. Gene array analysis of adrenal glands in broiler chickens following ACTH treatment. BMC Genomics 10: 430.

Byerly, M.S., J. Simon, E. Lebihan-Duval, M.J. Duclos, L.A. Cogburn, and T.E. Porter. 2009. Effects of BDNF, T3, and CORT on expression of the hypothalamic obesity gene network *in vivo* and *in vitro*. Am. J. Physiol. 296: R1180–R1189.

Cannizzo, F.T., F. Spada, R. Benevelli, C. Nebbia, P. Giorgi, N. Brina, E. Bollo, and B. Biolatti. 2010. Thymus atrophy and regeneration following dexamethasone administration to beef cattle. Vet. Rec. 167: 338–343.

Cannizzo, F.T., P. Capra, S. Divari, V. Ciccotelli, B. Biolatti, and M. Vincenti. 2011. Effects of low-dose dexamethasone and prednisolone long term administration in beef calf: chemical and morphological investigation. Anal. Chim. Acta 700: 95–104.

Cantiello, M., M. Giantin, C. Monica, R.M. Lopparelli, F. Capolongo, F. Lasserre, E. Bollo, C. Nebbia, P.G.P. Martin, T. Pineau, and M. Dacasto. 2009. Effects of dexamethasone, administered for growth promoting purposes, upon the hepatic cytochrome P450 3A expression in the veal calf. Biochem. Pharmacol. 77: 451–463.

Carraro, L., S. Ferraresso, B. Cardazzo, C. Romualdi, C. Montesissa, F. Gottardo, T. Patarnello, M. Castagnaro, and L. Bargelloni. 2009. Expression profiling of skeletal muscle in young bulls treated with steroidal growth promoters. Physiol. Genomics 38: 138–148.

Carroll, J.A. 2001. Dexamethasone treatment at birth enhances neonatal growth in swine. Domest. Anim. Endocrinol. 21: 97–109.

Carsia, R.V., C.G. Scanes, and S. Malamed. 1985. Loss of sensitivity to ACTH of adrenocortical cells isolated from maturing domestic fowl. Proc. Soc. Exp. Biol. Med. 179: 279–282.

Carsia, R.V., H. Weber, and F.M. Perez. 1986. Corticotropin-releasing factor stimulates the release of adrenocorticotropin from domestic fowl pituitary cells. Endocrinology 118: 143–148.

Carsia, R.V., M.E. Morin, H.D. Rosen, and H. Weber. 1987a. Ontogenic corticosteroidogenesis of the domestic fowl: response of isolated adrenocortical cells. Proc. Soc. Exp. Biol. Med. 184: 436–445.

Carsia, R.V., C.G. Scanes, and S. Malamed. 1987b. Polyhormonal regulation of avian and mammalian corticosteroidogenesis *in vitro*. Comp. Biochem. Physiol. A 88: 131–140.

Carsia, R.V., and H. Weber. 1988. Protein malnutrition in the domestic fowl induces alterations in adrenocortical cell adrenocorticotropin receptors. Endocrinology 122: 681–688.

Carsia, R.V., H. Weber, and T.J. Lauterio. 1988. Protein malnutrition in the domestic fowl induces alterations in adrenocortical function. Endocrinology 122: 673–680.

Carsia, R.V., and H. Weber. 2000. Remodeling of turkey adrenal steroidogenic tissue induced by dietary protein restriction: the potential role of cell death. Gen. Comp. Endocrinol. 118: 471–479.

Carsia, R.V. 2015. Adrenals. pp. 577–611. In: C.G. Scanes (ed.). Sturkie's Avian Physiology, 6th edition, Academic Press, New York.

Chapple, R.P., J.A. Cuaron, and R.A. Eater. 1989a. Effect of glucocorticoids and limiting nursing on the carbohydrate digestive capacity and growth rate in pigs. J. Anim. Sci. 67: 2956–2973.

Chapple, R.P., J.A. Cuaron, and R.A. Eater. 1989b. Response of digestive carbohydrates and growth to graded does and administration frequency of hydrocortisone and adrenocorticotropic hormone in nursing piglets. J. Anim. Sci. 67: 2974–2984.

Cheng, H.W., and A. Fahey. 2009. Effects of group size and repeated social disruption on the serotonergic and dopaminergic systems in two genetic lines of White Leghorn laying hens. Poult. Sci. 88: 2018–2025.

Cheung, A., S. Harvey, T.R. Hall, S.K. Lam, and G.S. Spencer. 1988a. Effects of passive immunization with antisomatostatin serum on plasma corticosterone concentrations in young domestic cockerels. J. Endocrinol. 116: 179–183.

Cheung, A., T.R. Hall, and S. Harvey. 1988b. Stimulation of corticosterone release in the fowl by recombinant DNA-derived chicken growth hormone. Gen. Comp. Endocrinol. 69: 128–32.

Chobotská, K., M. Arnold, P. Werner, and V. Pliska. 1998. A rapid assay for tyrosine hydroxylase activity, an indicator of chronic stress in laboratory and domestic animals. Biol. Chem. 379: 59–63.

Christensen, J.W., M. Beekmans, M. van Dalum, and M. VanDierendonck. 2014. Effects of hyperflexion on acute stress responses in ridden dressage horses. Physiol. Behav. 128: 39–45.

Christensen, K., Y. Vizzier Thaxton, J.P. Thaxton, and C.G. Scanes. 2012. Changes in body temperature during growth and in response to fasting in growing modern meat type chickens. Br. Poult. Sci. 53: 531–537.

Ciechanowska, M., M. Łapot, T. Malewski, K. Mateusiak, T. Misztal, and F. Przekop. 2011. Effects of corticotropin-releasing hormone and its antagonist on the gene expression of gonadotrophin-releasing hormone (GnRH) and GnRH receptor in the hypothalamus and anterior pituitary gland of follicular phase ewes. Reprod. Fertil. Dev. 23: 780–787.

Ciepielewski, Z.M., W. Stojek, A. Borman, D. Myślińska, W. Glac and M. Kamyczek. 2013. Natural killer cell cytotoxicity, cytokine and neuroendocrine responses to opioid receptor blockade during prolonged restraint in pigs. Res. Vet. Sci. 95: 975–985.

Collier, C.T., P.N. Williams, J.A. Carroll, T.H. Welsh, and J.C. Laurenz. 2011. Effect of maternal restraint stress during gestation on temporal lipopolysaccharide-induced neuroendocrine and immune responses of progeny. Domest. Anim. Endocrinol. 40: 40–50.

Cunningham, D.L., A. van Tienhoven, and G. Gvaryahu. 1988. Population size, cage area, and dominance rank effects on productivity and well-being of laying hens. Poult. Sci. 67: 399–406.

Curtis, M.J., I.H. Flack, and S. Harvey. 1980. The effect of *Escherichia coli* endotoxins on the concentrations of CORT and growth hormone in the plasma of the domestic fowl. Res. Vet. Sci. 28: 123–127.

Dalin, A.M., U. Magnusson, J. Häggendal, and L. Nyberg. 1993a. The effect of transport stress on plasma level of catecholamines, cortisol, corticosteroid-binding globulin, blood cell count, and lymphocyte proliferation in pigs. Acta Vet. Scand. 34: 59–68.

Dalin, A.M., U. Magnusson, J. Häggendal, and L. Nyberg. 1993b. The effect of thiopentone-sodium anesthesia and surgery, relocation, grouping, and hydrocortisone treatment of the blood levels of cortisol, corticosteroid-binding globulin, and catecholamines in pigs. J. Anim. Sci. 71: 1902–1909.

D'Allaire, S., and L. DeRoth. 1986. Physiological responses to treadmill exercise and ambient temperature in normal and malignant hyperthermia susceptible pigs. Can. J. Vet. Res. 50: 78–83.

Davis, A.K., D.L. Maney, and J.C. Maerz. 2008. The use of leukocyte profiles to measure stress in vertebrates: A review for ecologists. Funct. Ecol. 22: 760–772.

Davis, G.S., K.E. Anderson, and A.S. Carroll. 2000. The effects of long-term caging and molt of Single Comb White Leghorn hens on heterophil to lymphocyte ratios, CORT and thyroid hormones. Poult. Sci. 79: 514–518.

Davison, T.F., C.G. Scanes, I.H. Flack, and S. Harvey. 1979. Effect of daily injections of ACTH on growth and on the adrenal and lymphoid tissues of two strains of immature fowls. Br. Poult. Sci. 20: 575–585.

Davison, T.F., B.M. Freeman, and J. Rea. 1985. Effects of continuous treatment with synthetic ACTH or corticosterone on immature *Gallus Domesticus*. Gen. Comp. Endocrinol. 59: 416–423.

Dawood, T., M.R. Williams, M.J. Fullerton, K. Myles, J. Schuijers, J.W. Funder, K. Sudhir, and P.A. Komesaroff. 2005. Glucocorticoid responses to stress in castrate and testosterone-replaced rams. Regul. Pept. 125: 47–53.

de Beer, M., J.P. McMurtry, D.M. Brocht, and C.N. Coon. 2008. An examination of the role of feeding regimens in regulating metabolism during the broiler breeder grower period. 2. Plasma hormones and metabolites. Poult. Sci. 87: 264–275.

de Groot, J., M.A. Ruis, J.W. Scholten, J.M. Koolhaas, and W.J. Boersma. 2001. Long-term effects of social stress on antiviral immunity in pigs. Physiol. Behav. 73: 145–158.

de Jong, I.C., A.S. van Voorst, J.H. Erkens, D.A. Ehlhardt, and H.J. Blokhuis. 2001. Determination of the circadian rhythm in plasma CORT and catecholamine concentrations in growing broiler breeders using intravenous cannulation. Physiol. Behav. 74: 299–304.

de Jong, I.C., S. van Voorst, D.A. Ehlhardt, and H.J. Blokhuis. 2002. Effects of restricted feeding on physiological stress parameters in growing broiler breeders. Br. Poult. Sci. 43: 157–168.

deRoos, R., and C. deRoos. 1964. Effects of mammalian corticotropin and chicken adenohypophysial extracts on steroidogenesis by chicken adrenal tissue *in vitro*. Gen. Comp. Endocrinol. 4: 602–607.

Donker, R.A., and G. Beuving. 1989. Effect of corticosterone infusion on plasma corticosterone concentration, antibody production, circulating leukocytes and growth in chicken lines selected for humoral immune responsiveness. Br. Poult. Sci. 30: 361–369.

Ducsay, C.A., M. Mlynarczyk, K.M. Kaushal, K. Hyatt, K. Hanson, and D.A. Myers. 2009. Long-term hypoxia enhances ACTH response to arginine vasopressin but not corticotropin-releasing hormone in the near-term ovine fetus. Am. J. Physiol. 297: R892–R899.

Duffy, B.K., H.S Gurm, V. Rajagopal, R. Gupta, S.G. Ellis, and D.L. Bhatt. 2006. Usefulness of an elevated neutrophil to lymphocyte ratio in predicting long-term mortality after percutaneous coronary intervention. Am. J. Cardiol. 97: 993–996.

Edens, F.W., and H.S. Siegel. 1975. Adrenal responses in high and low ACTH response lines of chickens during acute heat stress. Gen. Comp. Endocrinol. 25: 64–73.

Elfwing, M., A. Fallahshahroudi, I. Lindgren, P. Jensen, and J. Altimiras. 2014. The strong selective sweep candidate gene ADRA2C does not explain domestication related changes in the stress response of chickens. PLoS One. 9: e103218.

Elfwing, M., D. Nätt, V.C. Goerlich-Jansson, M. Persson, J. Hjelm, and P. Jensen. 2015. Early stress causes sex-specific, life-long changes in behaviour, levels of gonadal hormones, and gene expression in chickens. PLoS One 10: e0125808.

Eicher, S.D., K.A. McMunn, H.M. Hammon, and S.S. Donkin. 2004. Toll-like receptors 2 and 4, and acute phase cytokine gene expression in dexamethasone and growth hormone treated dairy calves. Vet. Immunol. Immunopathol. 98: 115–125.

El-Lethey, H., T.W. Jungi, and B. Huber-Eicher. 2001. Effects of feeding CORT and housing conditions on feather pecking in laying hens (*Gallus gallus domesticus*). Physiol. Behav. 73: 243–251.

Erber, R., M. Wulf, M. Becker-Birck, S. Kaps, J.E. Aurich, E. Möstl, and C. Aurich. 2012. Physiological and behavioural responses of young horses to hot iron branding and microchip implantation. Vet. J. 191: 171–175.

Farmer, C., P. Dubreuil, Y. Couture, P. Brazeau, and D. Petitclerc. 1991. Hormonal changes following an acute stress in control and somatostatin-immunized pigs. Domest. Anim. Endocrinol. 8: 527–536.

Fässler, R., H. Dietrich, G. Krömer, G. Böck, H.P. Brezinschek, and G. Wick. 1988. The role of testosterone in spontaneous autoimmune thyroiditis of Obese strain (OS) chickens. J. Autoimmun. 1: 97–108.

Fell, L.R., D.A. Shutt, and C.J. Bentley. 1985. Development of a salivary cortisol method for detecting changes in plasma "free" cortisol arising from acute stress in sheep. Aust. Vet. J. 62: 403–406.

Fernandez, X., M.C. Meunier-Salaün, and P. Mormede. 1994. Agonistic behavior, plasma stress hormones, and metabolites in response to dyadic encounters in domestic pigs: interrelationships and effect of dominance status. Physiol. Behav. 56: 841–847.

Fernandez, X., M.C. Meunier-Salaun, P. Ecolan, and P. Mormède. 1995. Interactive effect of food deprivation and agonistic behavior on blood parameters and muscle glycogen in pigs. Physiol. Behav. 58: 337–345.

Fowden, A.L., and A.J. Forhead. 2011. Adrenal glands are essential for activation of glucogenesis during undernutrition in fetal sheep near term. Am. J. Physiol. 300: E94–E102.

Franchini, A., and E. Ottaviani. 1999. Immunoreactive POMC-derived peptides and cytokines in the chicken thymus and bursa of Fabricius microenvironments: age-related changes. J. Neuroendocrinol. 11: 685–692.

Gaines, A.M., J.A. Carroll, and G.L. Allee. 2004. Evaluation of exogenous glucocorticoid injection on preweaning growth performance of neonatal pigs under commercial conditions. J. Anim. Sci. 82: 1241–1245.

Ghassemi Nejad, J., J.D. Lohakare, J.K. Son, E.G. Kwon, J.W. West, and K.I. Sung. 2014. Wool cortisol is a better indicator of stress than blood cortisol in ewes exposed to heat stress and water restriction. Animal 8: 128–132.

Glade, M.J., L. Krook, H.F. Schryver, and H.F. Hintz. 1981. Growth inhibition induced by chronic dexamethasone treatment of foals. J. Equine Vet. Sci. 1: 198–201.

Glade, M.J., S. Gupta, and T.J. Reimers. 1984. Hormonal responses to high and low planes of nutrition in weanling thoroughbreds. J. Anim. Sci. 59: 658–665.

Görres, A., S. Ponsuksili, K. Wimmers, and E. Muráni. 2015. Analysis of non-synonymous SNPs of the porcine SERPINA6 gene as potential causal variants for a QTL affecting plasma cortisol levels on SSC7. Anim. Genet. 46: 239–246.

Gottardo, F., M. Brscic, G. Pozza, C. Ossensi, B. Contiero, A. Marin, and G. Cozzi. 2008. Administration of dexamethasone *per os* in finishing bulls. I. Effects on productive traits, meat quality and cattle behaviour as indicator of welfare. Animal 2: 1073–1079.

Gould, N.R., and H.S. Siegel. 1978. Partial purification and characterization of chicken corticosteroid-binding globulin. Poult. Sci. 57: 1733–1739.

Gould, N.R., and H.S. Siegel. 1985. Effects of corticotropin and heat on corticosteroid-binding capacity and serum corticosteroid in White Rock chickens. Poult. Sci. 64: 144–148.

Goyal, B.S., S.K. Garg, and B.D. Garg. 1988. Malathion-induced hyper-adrenal activity in WLH chicks. Current Sci. 55: 526–528.

Grandin, T. 1997. Assessment of stress during handling and transport. J. Anim. Sci. 75: 249–257.

Green, B.T., M. Lyte, C. Chen, Y. Xie, M.A. Casey, A. Kulkarni-Narla, L. Vulchanova, and D.R. Brown. 2004. Adrenergic modulation of *Escherichia coli* O157:H7 adherence to the colonic mucosa. Am. J. Physiol. 287: G1238–G1246.

Gross, W.B., P.B. Siegel, and R.T. DuBose. 1980. Some effects of feeding CORT to chickens. Poult. Sci. 59: 516–522.

Gross, W.B., and H.S. Siegel. 1983. Evaluation of the heterophil/lymphocyte ratio as a measure of stress in chickens. Avian Dis. 27: 972–979.

Guyonnet-Dupérat, V., N. Geverink, G.S. Plastow, G. Evans, O. Ousova, C. Croisetière, A. Foury, E. Richard, P. Mormède, and M.P. Moisan. 2006. Functional implication of an Arg307Gly substitution in corticosteroid-binding globulin, a candidate gene for a quantitative trait locus associated with cortisol variability and obesity in pig. Genetics 173: 2143–2149.

Halazun, K.J., A. Aldoori, H.Z. Malik, A. Al-Mukhtar, K.R. Prasad, G.J. Toogood, and J.P. Lodge. 2008. Elevated preoperative neutrophil to lymphocyte ratio predicts survival following hepatic resection for colorectal liver metastases. Eur. J. Surg. Oncol. 34: 55–60.

Hall, P.F., and S.B. Koritz. 1966. Action of ACTH upon steroidogenesis in the chicken-adrenal gland. Endocrinology 79: 652–654.

Hamano, Y. 2006. Effects of dietary lipoic acid on plasma lipid, *in vivo* insulin sensitivity, metabolic response to corticosterone and *in vitro* lipolysis in broiler chickens. Br. J. Nutr. 95: 1094–1101.

Harvey, S., and C.G. Scanes. 1978. Effect of adrenaline and adrenergic active drugs on growth hormone secretion in immature cockerels. Experientia 34: 1096–1097.

Harvey, S., C.G. Scanes, A. Chadwick, and N.J. Bolton. 1978. Influence of fasting, glucose and insulin on the levels in the plasma of growth hormone and prolactin of the domestic fowl. J. Endocrinol. 76: 501–506.

Harvey, S., and C.G. Scanes. 1979. Plasma growth hormone levels in growth retarded cortisone treated chicks. Br. Poult. Sci. 20: 331–335.

Harvey, S., H. Klandorf, and Y. Pinchasov. 1983. Visual and metabolic stimuli cause adrenocortical suppression in fasted chickens during refeeding. Neuroendocrinology 37: 59–63.

Hasiec, M., D. Tomaszewska-Zaremba, and T. Misztal. 2014. Suckling and salsolinol attenuate responsiveness of the hypothalamic-pituitary-adrenal axis to stress: Focus on catecholamines, corticotrophin-releasing hormone, adrenocorticotrophic hormone, cortisol and prolactin secretion in lactating sheep. J. Neuroendocrinol. 26: 844–852.

Hasiec, M., A.P. Herman, and T. Misztal. 2015. Salsolinol: A potential modulator of the activity of the hypothalamic-pituitary-adrenal axis in nursing and postweaning sheep. Domest. Anim. Endocrinol. 53: 26–34.

Haussmann, M.F., J.A. Carroll, G.D. Weesner, M.J. Daniels, R.L. Matteri, and D.C. Lay. 2000. Administration of ACTH to restrained, pregnant sows alters their pigs' hypothalamic-pituitary-adrenal (HPA) axis. J. Anim. Sci. 78: 2399–2411.

Hay, M., and P. Mormède. 1998. Urinary excretion of catecholamines, cortisol and their metabolites in Meishan and large white sows: validation as a non-invasive and integrative assessment of adrenocortical and sympathoadrenal axis activity. Vet. Res. 29: 119–128.

Hay, M., M.C. Meunier-Salaün, F. Brulaud, M. Monnier, and P. Mormède. 2000. Assessment of hypothalamic-pituitary-adrenal axis and sympathetic nervous system activity in pregnant sows through the measurement of glucocorticoids and catecholamines in urine. J. Anim. Sci. 78: 420–428.

Hazard, D., Liaubet, L., Sancristobal, M., and P. Mormède. 2008. Gene array and real time PCR analysis of the adrenal sensitivity to adrenocorticotropic hormone in pig. BMC Genomics 9: 101.

Heo, J., H.G. Kattesh, R.L. Matteri, J.D. Grizzle, and J.D. Godkin. 2003. Partial nucleotide sequence of the porcine corticosteroid-binding globulin (CBG) cDNA and specification of CBG expression sites in postnatal pigs. Domest. Anim. Endocrinol. 24: 257–264.

Heo, J., H.G. Kattesh, M.P. Roberts, J.L. Morrow, D.J. Wailey, and A.M. Saxton. 2005. Hepatic corticosteroid-binding globulin (CBG) messenger RNA expression and plasma CBG concentrations in young pigs in response to heat and social stress. J. Anim. Sci. 83: 208–215.

Holloway, A.C., W.L. Whittle, and J.R. Challis. 2001. Effects of cortisol and estradiol on pituitary expression of proopiomelanocortin, prohormone convertase-1, prohormone convertase-2, and glucocorticoid receptor mRNA in fetal sheep. Endocrine 14: 343–348.

Huang, H.S., M.C. Wu, and P.H. Li. 2000. Expression of steroidogenic enzyme messenger ribonucleic acid and cortisol production in adrenocortical cells isolated from halothane-sensitive and halothane-resistant pigs. J. Cell. Biochem. 79: 58–70.

Huff, G.R., W.E. Huff, J.M. Balog, N.C. Rath, N.B. Anthony, and K.E. Nestor. 2005. Stress response differences and disease susceptibility reflected by heterophil to lymphocyte ratio in turkeys selected for increased body weight. Poult. Sci. 84: 709–717.

Janssens, C.J., F.A. Helmond, and V.M. Wiegant. 1995. Chronic stress and pituitary-adrenocortical responses to corticotropin-releasing hormone and vasopressin in female pigs. Eur. J. Endocrinol. 132: 479–486.

Jarvis, S., C. Moinard, S.K. Robson, E. Baxter, E. Ormandy, A.J. Douglas, J.R. Seckl, J.A. Russell, and A.B. Lawrence. 2006. Programming the offspring of the pig by prenatal social stress: neuroendocrine activity and behaviour. Horm. Behav. 9: 68–80.

Jobe, A.H., N. Wada, L.M. Berry, M. Ikegami, and M.G. Ervin. 1998. Single and repetitive maternal glucocorticoid exposures reduce fetal growth in sheep. Am. J. Obstet. Gynecol. 178: 880–885.

Johnson, R.W., E.H. von Borell, L.L. Anderson, L.D. Kojic, and J.E. Cunnick. 1994. Intracerebroventricular injection of corticotropin-releasing hormone in the pig: acute effects on behavior, adrenocorticotropin secretion, and immune suppression. Endocrinology 135: 642–648.

Jones, R.B., G. Beuving, and H.J. Blokhuis. 1988. Tonic immobility and heterophil/lymphocyte responses of the domestic fowl to corticosterone infusion. Physiol. Behav. 42: 249–253.

Kalliecharan, R., and B.K. Hall. 1974. A developmental study of the levels of progesterone, CORT, cortisol, and cortisone circulating in plasma of chick embryos. Gen. Comp. Endocrinol. 24: 364–372.

Kalliecharan, R., and B.K. Hall. 1977. The *in vitro* biosynthesis of steroids from pregnenolone and cholesterol and the effects of bovine ACTH on corticoid production by adrenal glands of embryonic chicks. Gen. Comp. Endocrinol. 33: 147–159.

Kang, S.W., and W.J. Kuenzel. 2014. Regulation of gene expression of vasotocin and corticotropin-releasing hormone receptors in the avian anterior pituitary by CORT. Gen. Comp. Endocrinol. 204: 25–32.

Kanitz, E., Otten, W., and M. Tuchscherer. 2006. Changes in endocrine and neurochemical profiles in neonatal pigs prenatally exposed to increased maternal cortisol. J. Endocrinol. 191: 207–220.

Kannan, G., J.L. Heath, C.J. Wabeck, and J.A. Mench. 1997a. Shackling of broilers: Effects on stress responses and breast meat quality. Br. Poult. Sci. 38: 323–332.

Kannan, G., J.L. Heath, C.J. Wabeck, M.C. Souza, J.C. Howe and J.A. Mench. 1997b. Effects of crating and transport on stress and meat quality characteristics in broilers. Poult. Sci. 76: 523–529.

Kędzierski, W., A. Cywińska, K. Strzelec, and S. Kowalik. 2014. Changes in salivary and plasma cortisol levels in Purebred Arabian horses during race training session. Anim. Sci. J. 85: 313–317.

Khafipour, E., P.M. Munyaka, C.M. Nyachoti, D.O. Krause, and J.C. Rodriguez-Lecompte. 2014. Effect of crowding stress and *Escherichia coli* K88+ challenge in nursery pigs supplemented with anti-*Escherichia coli* K88+ probiotics. J. Anim. Sci. 92: 2017–2029.

Khan, M.S., C. Shigeoka, Y. Takahara, S. Matsuda, and T. Tachibana. 2015. Ontogeny of the corticotrophin-releasing hormone system in slow- and fast-growing chicks (*Gallus gallus*). Physiol. Behav. 151: 38–45.

Kim, D.W., M.M. Mushtaq, R.H.K. Kang, J.H. Kim, J.C. Na, J. Hwangbo, J.D. Kim, C.B. Yang, B.J. Park, and H.C. Choi. 2015. Various levels and forms of dietary α-lipoic acid in broiler chickens: Impact on blood biochemistry, stress response, liver enzymes, and antibody titers. Poult. Sci. Epub ahead of print.

Kim, M.Y., Y.A. Eiby, E.R. Lumbers, L.L. Wright, K.J. Gibson, A.C. Barnett, and B.E. Lingwood. 2014. Effects of glucocorticoid exposure on growth and structural maturation of the heart of the preterm piglet. PLoS One 9: e93407.

Kjaer, J.B., and H. Jørgensen. 2011. Heart rate variability in domestic chicken lines genetically selected on feather pecking behavior. Genes Brain Behav. 10: 747–755.

Klasing, K.C., D.E. Laurin, R.K. Peng, and D.M. Fry. 1987. Immunologically mediated growth depression in chicks: Influence of feed intake, corticosterone and interleukin-1. J. Nutr. 117: 1629–37.

Kocsis, J.F., E.T. Lamm, P.J. McIlroy, C.G. Scanes, and R.V. Carsia. 1995. Evidence for functionally distinct subpopulations of steroidogenic cells in the domestic turkey (*Meleagris gallopavo*) adrenal gland. Gen. Comp. Endocrinol. 98: 57–72.

Kocsis, J.F., N.E. Rinkardt, D.G. Satterlee, H. Weber, and R.V. Carsia. 1997. Concentration-dependent, biphasic effect of prostaglandins on avian corticosteroidogenesis *in vitro*. Gen. Comp. Endocrinol. 115: 132–142.

Koh, T.S., R.K. Peng, and K.C. Klasing. 1996. Dietary copper level affects copper metabolism during lipopolysaccharide-induced immunological stress in chicks. Poult. Sci. 75: 867–872.

Kongsted, A.H., S.V. Husted, M.P. Thygesen, V.G. Christensen, D. Blache, A. Tolver, T. Larsen, B. Quistorff, and M.O. Nielsen. 2013. Pre- and postnatal nutrition in sheep affects β-cell secretion and hypothalamic control. J. Endocrinol. 219: 159–171.

Korte, S.M., G. Beuving, W. Ruesink, and H.J. Blokhuis. 1997. Plasma catecholamine and corticosterone levels during manual restraint in chicks from a high and low feather pecking line of laying hens. Physiol. Behav. 62: 437–441.

Kostopanagiotou, G., K. Kalimeris, K. Christodoulaki, C. Nastos, N. Papoutsidakis, C. Dima, C. Chrelias, A. Pandazi, I. Mourouzis, and C. Pantos. 2010. The differential impact of volatile and intravenous anaesthetics on stress response in the swine. Hormones (Athens) 9: 67–75.

Krause, W., Michael, N., Lübke, C., Livett, B.G., and P. Oehme. 1997. Substance P and epibatidine-evoked catecholamine release from fractionated chromaffin cells. Eur. J. Pharmacol. 328: 249–254.

Krazinski, B.E., M. Koziorowski, P. Brzuzan, and S. Okras. 2011. The expression of genes encoding opioid precursors and the influence of opioid receptor agonists on steroidogenesis in porcine adrenocortical cells *in vitro*. J. Physiol. Pharmacol. 62: 461–468.

Laden, S.A., J.E. Wohlt, P.K. Zajac, and R.V. Carsia. 1985. Effects of stress from electrical dehorning on feed intake, growth, and blood constituents of Holstein heifer calves. J. Dairy. Sci. 68: 3062–3066.

Lansade, L., M. Valenchon, A. Foury, C. Neveux, S.W. Cole, S. Layé, B. Cardinaud, F. Lévy, and M.P. Moisan. 2014. Behavioral and transcriptomic fingerprints of an enriched environment in horses (*Equus caballus*). PLoS One 9: e114384.

Lay, D.C., Jr., T.H. Friend, C.L. Bowers, K.K. Grissom, and O.C. Jenkins. 1992a. A comparative physiological and behavioral study of freeze and hot-iron branding using dairy cows. J. Anim. Sci. 70: 1121–1125.

Lay, D.C., Jr., T.H. Friend, R.D. Randel, C.L. Bowers, K.K. Grissom, and O.C. Jenkins. 1992b. Behavioral and physiological effects of freeze or hot-iron branding on crossbred cattle. J. Anim. Sci. 70: 330–336.

Lechner, O., H. Dietrich, G.J. Wiegers, M. Vacchio, and G. Wick. 2001. Glucocorticoid production in the chicken bursa and thymus. Int. Immunol. 13: 769–776.

Lee, C., L.R. Giles, W.L. Bryden, J.A. Downing, C.D. Collins, and P.C. Wynn. 2005. The effect of active immunization against adrenocorticotropic hormone on cortisol, beta-endorphin, vocalization, and growth in pigs. J. Anim. Sci. 83: 2372–2379.

Lefcourt, A.M., and T.H. Elsasser. 1995. Adrenal responses of Angus x Hereford cattle to the stress of weaning. J. Anim. Sci. 73: 2669–2676.

Li, L.A., D. Xia, S. Wei, J. Hartung, and R.Q. Zhao. 2008. Characterization of adrenal ACTH signaling pathway and steroidogenic enzymes in Erhualian and Pietrain pigs with different plasma cortisol levels. Steroids 73: 806–814.

Ling, M.K., E. Hotta, Z. Kilianova, T. Haitina, A. Ringholm, L. Johansson, N. Gallo-Payet, S. Takeuchi, and H.B. Schiöth. 2004. The melanocortin receptor subtypes in chicken have high preference to ACTH-derived peptides. Br. J. Pharmacol. 143: 626–637.

Liu, H.W., Y. Cao, and D.W. Zhou. 2012. Effects of shade on welfare and meat quality of grazing sheep under high ambient temperature. J. Anim. Sci. 90: 4764–4770.

Liu, L.L., J.H. He, H.B. Xie, Y.S. Yang, J.C. Li, and Y. Zou. 2014. Resveratrol induces antioxidant and heat shock protein mRNA expression in response to heat stress in black-boned chickens. Poult. Sci. 93: 54–62.

Livett, B.G., and P.D. Marley. 1986. Effects of opioid peptides and morphine on histamine-induced catecholamine secretion from cultured, bovine adrenal chromaffin cells. Br. J. Pharmacol. 89: 327–334.

Lopes, S.O., R. Claus, M. Lacorn, A. Wagner, and R. Mosenthin. 2004. Effects of dexamethasone application in growing pigs on hormones, N-retention and other metabolic parameters. J. Vet. Med. A Physiol. Pathol. Clin. Med. 51: 97–105.

Lü, F., K. Yang, and J.R. Challis. 1994. Regulation of ovine fetal pituitary function by corticotrophin-releasing hormone, arginine vasopressin and cortisol *in vitro*. J. Endocrinol. 143: 199–208.

Marley, P.D., and B.G. Livett. 1987a. Differences between the mechanisms of adrenaline and noradrenaline secretion from isolated, bovine, adrenal chromaffin cells. Neurosci. Lett. 77: 81–86.

Marley, P.D., and B.G. Livett. 1987b. Effects of opioid compounds on desensitization of the nicotinic response of isolated bovine adrenal chromaffin cells. Biochem. Pharmacol. 36: 2937–2944.

Marolf, C.J., B.D. Schultz, and E.T. Clemens. 1993. Epinephrine effects on gastrin and gastric secretions in normal and stress-susceptible pigs and in dogs. Comp. Biochem. Physiol. C. 106: 367–370.

Mashaly, M.M., M.L. Webb, S.L. Youtz, W.B. Roush, and H.B. Graves. 1984. Changes in serum corticosterone concentration of laying hens as a response to increased population density. Poult. Sci. 63: 2271–2274.

Massart, R., R. Mongeau, and L. Lanfumey. 2012. Beyond the monoaminergic hypothesis: neuroplasticity and epigenetic changes in a transgenic mouse model of depression. Philos. Trans. R. Soc. Lond. B. Biol. Sci. 367: 2485–9244.

Maxwell, M.H. 1993. Avian blood leukocyte responses to stress. World's Poult. Sci. J. 49: 34–43.

Mazzocchi, G., G. Gottardo, and G.G. Nussdorfer. 1997. Catecholamines stimulate steroid secretion of dispersed fowl adrenocortical cells, acting through the beta-receptor subtype. Horm. Metab. Res. 29: 190–192.

McCarthy, R.N., L.B. Jeffcott, J.W. Funder, M. Fullerton, and I.J. Clarke. 1991. Plasma beta-endorphin and adrenocorticotrophin in young horses in training and adrenocorticotrophin in young horses in training. Aust. Vet. J. 68: 359–361.

McGlone, J.J., J.L. Salak, E.A. Lumpkin, R.I. Nicholson, M. Gibson, and R.L. Norman. 1993. Shipping stress and social status effects on pig performance, plasma cortisol, natural killer cell activity, and leukocyte numbers. J. Anim. Sci. 71: 888–896.

Mellor, D.J., and L. Murray. 1989a. Changes in the cortisol responses of lambs to tail docking, castration and ACTH injection during the first seven days after birth. Res. Vet. Sci. 46: 392–395.

Mellor, D.J., and L. Murray. 1989b. Effects of tail docking and castration on behaviour and plasma cortisol concentrations in young lambs. Res. Vet. Sci. 46: 387–391.

Metges, C.C., S. Görs, K. Martens, R. Krueger, B.U. Metzler-Zebeli, C. Nebendahl, W. Otten, E. Kanitz, A. Zeyner, H.M. Hammon, R. Pfuhl, and G. Nürnberg. 2015. Body composition and plasma lipid and stress hormone levels during 3 weeks of feed restriction and refeeding in low birth weight female pigs. J. Anim. Sci. 93: 999–1014.

Miranda-de la Lama, G.C., P. Monge, M. Villarroel, J.L. Olleta, S. García-Belenguer, and G.A. María. 2011. Effects of road type during transport on lamb welfare and meat quality in dry hot climates. Trop. Anim. Health Prod. 43: 915–922.

Monk, C.S., K.A. Hart, R.D. Berghaus, N.A. Norton, P.A. Moore, and K.E. Myrna. 2014. Detection of endogenous cortisol in equine tears and blood at rest and after simulated stress. Vet. Ophthalmol. 17 Suppl. 1: 53–60.

Mormède, P., S. Andanson, B. Aupérin, B. Beerda, D. Guémené, J. Malmkvist, X. Manteca, G. Manteuffel, P. Prunet, C.G. van Reenen, S. Richard, and I. Veissier. 2007. Exploration of the hypothalamic–pituitary–adrenal function as a tool to evaluate animal welfare. Physiol. Behav. 92: 317–339.

Nakanishi, S., A. Inoue, T. Kita, M. Nakamura, A.C. Chang, S.N. Cohen, and S. Numa. 1979. Nucleotide sequence of cloned cDNA for bovine corticotropin-beta-lipotropin precursor. Nature 278: 423–427.

Nakayama, H., T. Takahashi, Y. Oomatsu, K. Nakagawa-Mizuyachi, and M. Kawashima. 2011a. Parathyroid hormone-related peptide directly increases adrenocorticotropic hormone secretion from the anterior pituitary in hens. Poult. Sci. 90: 175–180.

Nakayama, H., T. Takahashi, K. Nakagawa-Mizuyachi, and M. Kawashima. 2011b. Effect of calcitonin on adrenocorticotropic hormone secretion stimulated by corticotropin-releasing hormone in the hen anterior pituitary. Anim. Sci. J. 82: 475–480.

Naqvi, S.M.K., and A.K. Rai. 1990. Effect of nutritional stress on wool yield, characteristics and efficiency of feed conversion to wool. Livestock Res. Rur. Dev. 2. http://www.lrrd.org/lrrd2/2/naqvi.htm Accessed September 7, 2015.

Ndlovu, T., M. Chimonyo, A.I. Okoh, and V. Muchenje. 2008. A comparison of stress hormone concentrations at slaughter in Nguni, Bonsmara and Angus steers. Afr. J. Agric. Res. 3: 96–100.

Negrão, J.A., M.A. Porcionato, A.M. de Passillé, and J. Rushen. 2004. Cortisol in saliva and plasma of cattle after ACTH administration and milking. J. Dairy Sci. 87: 1713–1718.

Neubert, E., H. Gürtler, and G. Vallentin. 1996. Effect of acute stress on plasma levels of catecholamines and cortisol in addition to metabolites in stress-susceptible growing swine. Berl. Munch. Tierarztl. Wochenschr. 109: 381–384.

Núñez, J., E. Núñez, V. Bodí, J. Sanchis, G. Miñana, L. Mainar, E. Santas, P. Merlos, E. Rumiz, H. Darmofal, A.M. Heatta, and A. Llàcer. 2008. Usefulness of the neutrophil to lymphocyte ratio in predicting long-term mortality in ST segment elevation myocardial infarction. Am. J. Cardiol. 101: 747–752.

Nussdorfer, G.G., G.P. Rossi, and G. Mazzocchi. 1997. Role of adrenomedullin and related peptides in the regulation of the hypothalamo-pituitary-adrenal axis. Peptides 18: 1079–1089.

O'Connor, S.J., D.S. Gardner, J.C.N. Ousey Holdstock, P. Rossdale, C.M. Edwards, A.L. Fowden, and D.A. Giussani. 2005. Development of baroreflex and endocrine responses to hypotensive stress in newborn foals and lambs. Pflugers Arch. 450: 298–306.

Oikawa, D., S. Takagi, and K. Yashiki. 2004. Some aspects of the stress responses to road transport in thoroughbred horses with special reference to shipping fever. J. Equine Sci. 15: 99–102.

Olanrewaju, H.A., J.P. Thaxton, W.A. Dozier, J. Purswell, S.D. Collier, and S.L. Branton. 2008. Interactive effects of ammonia and light intensity on hematochemical variables in broiler chickens. Poult. Sci. 87: 1407–1414.

Oliver, W.J., and J.J. Oliver. 2010. The effect of feeding stress on wool production of strong and fine wool Merino sheep. Grootfontein Agricultural Development Institute. http://gadi.agric.za/Agric/Vol7No1_2007/Olivier-effect-of-feeding-stress.php Accessed September 7, 2015.

Online Etymology Dictionary, 2015 http://www.etymonline.com/index.php?term=stress Accessed September 7, 2015.

Ott, S., L. Soler, C.P. Moons, M.A. Kashiha, C. Bahr, J. Vandermeulen, S. Janssens, A.M. Gutiérrez, D. Escribano, J.J. Cerón, D. Berckmans, F.A. Tuyttens, and T.A. Niewold. 2014. Different stressors elicit different responses in the salivary biomarkers cortisol, haptoglobin, and chromogranin A in pigs. Res. Vet. Sci. 97: 124–128.

Parkes, D.G., and C.N. May. 1995. ACTH-suppressive and vasodilator actions of adrenomedullin in conscious sheep. J. Neuroendocrinol. 7: 923–929.

Parrott, R.F., S.V. Vellucci, and J.A. Goode. 2000. Behavioral and hormonal effects of centrally injected "anxiogenic" neuropeptides in growing pigs. Pharmacol. Biochem. Behav. 65: 123–129.

Patthy, M., J. Horvath, M. Mason-Garcia, B. Szoke, D.H. Schlesinger, and A.V. Schally. 1985. Isolation and amino acid sequence of corticotropin-releasing factor from pig hypothalami. Proc. Natl. Acad. Sci. U.S.A. 82: 8762–8766.

Paull, D.R., C. Lee, S.J. Atkinson, and A.D. Fisher. 2008. Effects of meloxicam or tolfenamic acid administration on the pain and stress responses of Merino lambs to muleing. Aust. Vet. J. 86: 303–311.

Peeters, E., A. Neyt, F. Beckers, S. De Smet, A.E. Aubert, and R. Geers. 2005. Influence of supplemental magnesium, tryptophan, vitamin C, and vitamin E on stress responses of pigs to vibration. J. Anim. Sci. 83: 1568–1580.

Peeters, M., J. Sulon, J.F. Beckers, D. Ledoux, and M. Vandenheede. 2011. Comparison between blood serum and salivary cortisol concentrations in horses using an adrenocorticotropic hormone challenge. Equine Vet. J. 43: 487–493.

Peeters, M., C. Closson, J.-F. Beckers, and M. Vandenheede. 2013. Rider and horse salivary cortisol levels during competition and impact on performance. J. Equine Vet. Sci. 33: 155–160.

Perremans, S., J.M. Randall, G. Rombouts, E. Decuypere, and R. Geers. 2001. Effect of whole-body vibration in the vertical axis on cortisol and adrenocorticotropic hormone levels in piglets. J. Anim. Sci. 79: 975–981.

Perry, R.A., K. Tangalakis, and E.M. Wintour. 1992. Cytological maturity of zona fasciculata cells in the fetal sheep adrenal following ACTH infusion: an electron microscope study. Acta Endocrinol. (Copenh.) 127: 536–541.

Post, J., J.M. Rebel, and A.A. ter Huurne. 2003. Physiological effects of elevated plasma corticosterone concentrations in broiler chickens. An alternative means by which to assess the physiological effects of stress. Poult. Sci. 82: 1313–1318.

Puvadolpirod, S., and J.P. Thaxton. 2000a. Model of physiological stress in chickens 1. Response parameters. Poult. Sci. 79: 363–369.

Puvadolpirod, S., and J.P. Thaxton. 2000b. Model of physiological stress in chickens 2. Dosimetry of adrenocorticotropin. Poult. Sci. 79: 370–376.

Puvadolpirod, S., and J.P. Thaxton. 2000c. Model of physiological stress in chickens 3. Temporal patterns of response. Poult. Sci. 79: 377–382.

Puvadolpirod, S., and J.P. Thaxton. 2000d. Model of physiological stress in chickens 4. Digestion and metabolism. Poult. Sci. 79: 383–390.

Ranheim, B., T.E. Horsberg, N.E. Søli, K.A. Ryeng, and J.M. Arnemo. 2000. The effects of medetomidine and its reversal with atipamezole on plasma glucose, cortisol and noradrenaline in cattle and sheep. J. Vet. Pharmacol. Ther. 23: 379–387.

Rattner, B.A., L. Sileo, and C.G. Scanes. 1982. Hormonal responses and tolerance to cold of female quail following parathion ingestion. Pest. Biochem. Physiol. 18: 132–138.

Reimert, I., T.B. Rodenburg, W.W. Ursinus, B. Kemp, and J.E. Bolhuis. 2014. Selection based on indirect genetic effects for growth, environmental enrichment and coping style affect the immune status of pigs. PLoS One 9: e108700.

Robinson, P.M., R.S. Comline, A.L. Fowden, and M. Silver. 1983. Adrenal cortex of fetal lamb: changes after hypophysectomy and effects of Synacthen on cytoarchitecture and secretory activity. Q. J. Exp. Physiol. 68: 15–27.

Roozen, A.W., V.T. Tsuma, and U. Magnusson. 1995. Effects of short-term restraint stress on plasma concentrations of catecholamines, beta-endorphin, and cortisol in gilts. Am. J. Vet. Res. 56: 1225–1227.

Rosochacki, S.J., A.B. Piekarzewska, J. Połoszynowicz, and T. Sakowski. 2000. The influence of restraint immobilization stress on the concentration of bioamines and cortisol in plasma of Pietrain and Duroc pigs. J. Vet. Med. A Physiol. Pathol. Clin. Med. 47: 231–242.

Rushen, J., N. Schwarze, J. Ladewig, and G. Foxcroft. 1993. Opioid modulation of the effects of repeated stress on ACTH, cortisol, prolactin, and growth hormone in pigs. Physiol. Behav. 53: 923–928.

Rutherford, K.M., A. Piastowska-Ciesielska, R.D. Donald, S.K. Robson, S.H. Ison, S. Jarvis, P.J. Brunton, J.A. Russell, and A.B. Lawrence. 2014. Prenatal stress produces anxiety prone female offspring and impaired maternal behaviour in the domestic pig. Physiol. Behav. 129: 255–264.

Sapolsky, R.M., L.M. Romero, and A.U. Munck. 2000. How do glucocorticoids influence stress responses? Integrating permissive, suppressive, stimulatory, and preparative actions. Endocr. Rev. 21: 55–89.

Satterlee, D.G., and R.P. Gildersleeve. 1983. Factors affecting broiler processing parameters and plasma corticosterone. Poult. Sci. 62: 785–792.

Satterlee, D.G., R.B. Jones, and F.H. Ryder. 1994. Effects of ascorbyl-2-polyphosphate on adrenocortical activation and fear-related behavior in broiler chickens. Poult. Sci. 73: 194–201.

Scanes, C.G., G.F. Merrill, R. Ford, P. Mauser, and C. Horowitz. 1980. Effects of stress (Hypoglycaemia, endotoxin, and ether) on the peripheral circulating concentration of corticosterone in the Domestic Fowl (*Gallus domesticus*). Comp. Biochem. Physiol. 66C: 183–186.

Scanes, C.G., and K. Christensen. 2014. Comparison of meta-analysis of the hematological parameters of commercial and indigenous poultry to wild birds: Implications to domestication and development of commercial breeds/lines. J. Vet. Sci. Anim. Health. 1: 1.

Scanes, C.G. 2016. Control of metabolism in poultry. *In*: Biology of Domestic Animals (in press).

Scheuer, B.H., Y. Zbinden, P. Schneiter, L. Tappy, J.W. Blum, and H.M. Hammon. 2006. Effects of colostrum feeding and glucocorticoid administration on insulin-dependent glucose metabolism in neonatal calves. Domest. Anim. Endocrinol. 31: 227–245.

Schmidt, A., S. Biau, E. Möstl, M. Becker-Birck, B. Morillon, J. Aurich, J.M. Faure, and C. Aurich. 2010. Changes in cortisol release and heart rate variability in sport horses during long-distance road transport. Domest. Anim. Endocrinol. 38: 179–189.

Schönbom, H., A. Kassens, C. Hopster-Iversen, J. Klewitz, M. Piechotta, G. Martinsson, A. Kißler, D. Burger, and H. Sieme. 2015. Influence of transrectal and transabdominal ultrasound examination on salivary cortisol, heart rate, and heart rate variability in mares. Theriogenology 83: 749–756.

Schreiber, K.L., and D.R. Brown. 2005. Adrenocorticotrophic hormone modulates *Escherichia coli* O157:H7 adherence to porcine colonic mucosa. Stress 8: 185–190.

Schwerin, M., E. Kanitz, M. Tuchscherer, K.P. Brüssow, G. Nürnberg, and W. Otten. 2005. Stress-related gene expression in brain and adrenal gland of porcine fetuses and neonates. Theriogenology 63: 1220–1234.

Seaman-Bridges, J.S., J.A. Carroll, T.J. Safranski, and E.P. Berg. 2003. Short- and long-term influence of perinatal dexamethasone treatment on swine growth. Domest. Anim. Endocrinol. 24: 193–208.

Seasholtz, A.F., R.A. Valverde, and R.J. Denver. 2002. Corticotropin-releasing hormone-binding protein: biochemistry and function from fishes to mammals. J. Endocrinol. 175: 89–97.

Seidah, N.G., H. Fournier, G. Boileau, S. Benjannet, N. Rondeau, and M. Chrétien. 1992. The cDNA structure of the porcine pro-hormone convertase PC2 and the comparative processing by PC1 and PC2 of the N-terminal glycopeptide segment of porcine POMC. FEBS Lett. 310: 235–239.

Selye, H. 1936. A syndrome produced by diverse nocuous agents. Nature 138: 32.

Sharma, D., L.E. Cornett, and C.M. Chaturvedi. 2009. Corticosterone- or metapyrone-induced alterations in adrenal function and expression of the arginine vasotocin receptor VT2 in the pituitary gland of domestic fowl, *Gallus gallus*. Gen. Comp. Endocrinol. 161: 208–215.

Shini, S., and P. Kaiser. 2009. Effects of stress, mimicked by administration of corticosterone in drinking water, on the expression of chicken cytokine and chemokine genes in lymphocytes. Stress 12: 388–399.

Shini, S., A. Shini, and P. Kaiser. 2010. Cytokine and chemokine gene expression profiles in heterophils from chickens treated with corticosterone. Stress 13: 185–194.

Sloboda, D.M., J.P. Newnham, and J.R. Challis. 2002. Repeated maternal glucocorticoid administration and the developing liver in fetal sheep. J. Endocrinol. 175: 535–543.

Smith, F., J.E. Clark, B.L. Overman, C.C. Tozel, J.H. Huang, J.E. Rivier, A.T. Blikslager, and A.J. Moeser. 2010. Early weaning stress impairs development of mucosal barrier function in the porcine intestine. Am. J. Physiol. 298: G352–G363.

Spencer, G.S., G.J. Garssen, B. Colenbrander, A.A. Macdonald, and M.M. Bevers. 1983. Glucose, growth hormone, somatomedin, cortisol and ACTH changes in the plasma of unanaesthetised pig foetuses following intravenous insulin administration *in utero*. Acta Endocrinol. (Copenh). 104: 240–245.

Spencer, G.S. 1994. Hormone and metabolite changes with stress in stress-susceptible Pietrain pigs. Endocr. Regul. 28: 73–78.

Spencer, J.D., R.D. Boyd, R. Cabrera, and G.L. Allee. 2003. Early weaning to reduce tissue mobilization in lactating sows and milk supplementation to enhance pig weaning weight during extreme heat stress. J. Anim. Sci. 81: 2041–2052.

Spurlock, G.M., and M.T. Clegg. 1962. Effect of cortisone acetate on carcass composition and wool characteristics of weaned lambs. J. Anim. Sci. 21: 494–500.

Stull, C.L., C.J. Kachulis, J.L. Farley, and G.J. Koenig. 1999. The effect of age and teat order on alpha1-acid glycoprotein, neutrophil-to-lymphocyte ratio, cortisol, and average daily gain in commercial growing pigs. J. Anim. Sci. 77: 70–74.

Sutherland, M.A., D.J. Mellor, K.J. Stafford, N.G. Gregory, R.A. Bruce, and R.N. Ward. 2002. Cortisol responses to dehorning of calves given a 5-h local anaesthetic regimen plus phenylbutazone, ketoprofen, or adrenocorticotropic hormone prior to dehorning. Res. Vet. Sci. 73: 115–123.

Sutoh, M., E. Kasuya, K.I. Yayou, F. Ohtani, and Y. Kobayashi. 2016. Intravenous tryptophan administration attenuates cortisol secretion induced by intracerebroventricular injection of noradrenaline. Anim. Sci. J. Aug 11. 87: 266–270.

Takeuchi, S., K. Teshigawara, and S. Takahashi. 1999. Molecular cloning and characterization of the chicken pro-opiomelanocortin (POMC) gene. Biochim. Biophys. Acta 1450: 452–459.

Tanabe, Y., N. Saito, and T. Nakamura. 1986. Ontogenetic steroidogenesis by testes, ovary, and adrenals of embryonic and postembryonic chickens (*Gallus domesticus*). Gen. Comp. Endocrinol. 63: 456–463.

Tancin, V., D. Schams, and W.D. Kraetzl. 2000. Cortisol and ACTH release in dairy cows in response to machine milking after pretreatment with morphine and naloxone. J. Dairy Res. 67: 467–474.

Thwaites, C.J. 1968. Heat stress and wool growth in sheep. Proc. Aust. Soc. Anim. Prod. 7: 259–263.

Tilbrook, A.J., E.A. Rivalland, A.I. Turner, G.W. Lambert, and I.J. Clarke. 2008. Responses of the hypothalamopituitary adrenal axis and the sympathoadrenal system to isolation/restraint stress in sheep of different adiposity. Neuroendocrinology 87: 193–205.

Tolleson, D.R., G.E. Carstens, T.H. Welsh, Jr., P.D. Teel, O.F. Strey, M.T. Longnecker, S.D. Prince, and K.K. Banik. 2012. Plane of nutrition by tick-burden interaction in cattle: Effect on growth and metabolism. J. Anim. Sci. 90: 3442–3450.

Turner, A.I., E.T. Rivalland, I.J. Clarke, G.W. Lambert, M.J. Morris, and A.J. Tilbrook. 2002a. Noradrenaline, but not neuropeptide Y, is elevated in cerebrospinal fluid from the third cerebral ventricle following audiovisual stress in gonadectomised rams and ewes. Neuroendocrinology 76: 373–380.

Turner, A.I., B.J. Canny, R.J. Hobbs, J.D. Bond, I.J. Clarke, and A.J. Tilbrook. 2002b. Influence of sex and gonadal status of sheep on cortisol secretion in response to ACTH and on cortisol and LH secretion in response to stress: importance of different stressors. J. Endocrinol. 173: 113–122.

Turner, A.I., and A.J. Tilbrook. 2006. Stress, cortisol and reproduction in female pigs. Soc. Reprod. Fertil. Suppl. 62: 191–203.

van de Pavert, S.A., I.J. Clarke, A. Rao, K.E. Vrana, and J. Schwartz. 1997. Effects of vasopressin and elimination of corticotropin-releasing hormone-target cells on pro-opiomelanocortin mRNA levels and adrenocorticotropin secretion in ovine anterior pituitary cells. J. Endocrinol. 154: 139–147.

Vandenborne, K., B. De Groef, S.M. Geelissen, E.R. Kühn, V.M. Darras, and S. Van der Geyten. 2005. Corticosterone-induced negative feedback mechanisms within the hypothalamo-pituitary-adrenal axis of the chicken. J. Endocrinol. 185: 383–391.

Vascellari, M., K. Capello, A. Stefani, G. Biancotto, L. Moro, R. Stella, G. Pozza, and F. Mutinelli. 2012. Evaluation of thymus morphology and serum cortisol concentration as indirect biomarkers to detect low-dose dexamethasone illegal treatment in beef cattle. BMC Vet. Res. 8: 129.

Velie, B.D., J.P. Cassady, and C.S. Whisnant. 2012. Endocrine response to acute stress in pigs with differing backtest scores. Livestock Sci. 145: 140–144.

Vellucci, S.V., and R.F. Parrott. 2000a. Gene expression in the forebrain of dexamethasone-treated pigs: Effects on stress neuropeptides in the hypothalamus and hippocampus and glutamate receptor subunits in the hippocampus. Res. Vet. Sci. 69: 25–31.

Vellucci, S.V., and R.F. Parrott. 2000b. Hippocampal gene expression in the pig: Upregulation of corticotrophin releasing hormone mRNA following central administration of the peptide. Neuropeptides 34: 221–228.

Verbrugghe, E., F. Boyen, A. Van Parys, K. Van Deun, S. Croubels, A. Thompson, N. Shearer, B. Leyman, F. Haesebrouck and F. Pasmans. 2011. Stress induced *Salmonella Typhimurium* recrudescence in pigs coincides with cortisol induced increased intracellular proliferation in macrophages. Vet Res. 42: 118.

von Lewinski, M., S. Biau, R. Erber, N. Ille, J. Aurich, J.M. Faure, E. Möstl, and C. Aurich. 2013. Cortisol release, heart rate and heart rate variability in the horse and its rider: different responses to training and performance. Vet. J. 97: 229–232.

Vonnahme, K.A., T.L. Neville, L.A. Lekatz, L.P. Reynolds, C.J. Hammer, D.A. Redmer, and J.S. Caton. 2013. Thyroid hormones and cortisol concentrations in offspring are influenced by maternal supranutritional selenium and nutritional plane in sheep. Nutr. Metab. Insights 6: 11–21.

Wang, S., Y. Ni., F. Guo, W. Fu, R. Grossmann, and R. Zhao. 2013. Effect of corticosterone on growth and welfare of broiler chickens showing long or short tonic immobility. Comp. Biochem. Physiol. 164A: 537–543.

Watson, A.M., G.W. Lambert, K.J. Smith, and C.N. May. 2003. Urotensin II acts centrally to increase epinephrine and ACTH release and cause potent inotropic and chronotropic actions. Hypertension 42: 373–379.

Weaver, S.A., F.X. Aherne, M.J. Meaney, A.L. Schaefer, and W.T. Dixon. 2000. Neonatal handling permanently alters hypothalamic-pituitary-adrenal axis function, behaviour, and body weight in boars. J. Endocrinol. 164: 349–359.

Werner, J.R., D.M. Jenkins, W.B. Murray, E.L. Hughes, D.A. Bienus, and M.J. Kennett. 2012. Human electromuscular incapacitation devices characterization: A comparative study on stress and the physiological effects on swine. J. Strength Cond. Res. 26: 804–810.

Williams-Wyss, O., S. Zhang, S.M. MacLaughlin, D. Kleemann, S.K. Walker, C.M. Suter, J.E. Cropley, J.L. Morrison, C.T. Roberts, and I.C. McMillen. 2014. Embryo number and periconceptional undernutrition in the sheep have differential effects on adrenal epigenotype, growth, and development. Am. J. Physiol. 307: E141–E150.

Wilson, S.C., and F.C. Cunningham. 1981. Effect of photoperiod on the concentrations of corticosterone and luteinizing hormone in the plasma of the domestic hen. J. Endocrinol. 91: 135–143.

Wilson, S.C., R.C. Jennings, and F.J. Cunningham. 1984. Developmental changes in the diurnal rhythm of secretion of corticosterone and LH in the domestic hen. J. Endocrinol. 101: 299–304.

Wilson, S.P., K.J. Chang, and O.H. Viveros. 1982. Proportional secretion of opioid peptides and catecholamines from adrenal chromaffin cells in culture. J. Neurosci. 2: 1150–1156.

Wingfield, J.C., K.S. Matt, and D.S. Farner. 1984. Physiologic properties of steroid hormone-binding proteins in avian blood. Gen. Comp. Endocrinol. 53: 281–292.

Wodzicka-Tomaszewska, M., T. Stelmasiak, and R.B. Cumming. 1982. Stress by immobilization, with food and water deprivation, causes changes in plasma concentration of triiodothyronine, thyroxine and CORT in poultry. Aust. J. Biol. Sci. 35: 393–401.

Wohlt, J.E., T.D. Wright, V.S. Sirois, D.M. Kniffen, and L. Lelkes. 1982. Effect of docking on health, blood cells and metabolites and growth of Dorset lambs. J. Anim. Sci. 54: 23–28.

Wohlt, J.E., M.E. Allyn, P.K. Zajac, and L.S. Katz. 1994. Cortisol increases in plasma of Holstein heifer calves from handling and method of electrical dehorning. J. Dairy Sci. 77: 3725–3729.

Yan, F.F., P.Y. Hester, and H.W. Cheng. 2014. The effect of perch access during pullet rearing and egg laying on physiological measures of stress in White Leghorns at 71 weeks of age. Poult. Sci. 93: 1318–1326.

Yardimci, M., E.H. Sahin, I.S. Cetingul, I. Bayram, R. Aslan, and E. Sengor. 2013. Stress responses to comparative handling procedures in sheep. Animal 7: 143–150.

Yates, D.T., T.T. Ross, D.M. Hallford, L.J. Yates, and R.L. Wesley. 2010. Comparison of salivary and serum cortisol concentrations after adrenocorticotropic hormone challenge in ewes. J. Anim. Sci. 88: 599–603.

Yen, J.T., and W.G. Pond. 1987. Effect of dietary supplementation with vitamin C or carbadox on weanling pigs subjected to crowding stress. J. Anim. Sci. 64: 1672–1681.

Yoshihara, C., Y. Tashiro, S. Taniuchi, H. Katayama, S. Takahashi, and S. Takeuchi. 2011. Feather follicles express two classes of pro-opiomelanocortin (POMC) mRNA using alternative promoters in chickens. Gen. Comp. Endocrinol. 171: 46–51.

Yue, H.Y., L. Zhang, S.G. Wu, L. Xu, H.J. Zhang, and G.H. Qi. 2010. Effects of transport stress on blood metabolism, glycolytic potential, and meat quality in meat-type yellow-feathered chickens. Poult. Sci. 89: 413–419.

Zavy, M.T., P.E. Juniewicz, W.A. Phillips, and D.L. Von Tungeln. 1992. Effects of initial restraint, weaning and transport stress on baseline and ACTH stimulated cortisol responses in beef calves of different genotypes. Am. J. Vet. Res. 53: 551–557.

Zebeli, Q., S. Sivaraman, S.M. Dunn, and B.N. Ametaj. 2013. Intermittently-induced endotoxaemia has no effect on post-challenge plasma metabolites, but increases body temperature and cortisol concentrations in periparturient dairy cows. Res. Vet. Sci. 95: 1155–1162.

Zhang, L., H.Y. Yue, H.J. Zhang, L. Xu, S.G. Wu, H.J. Yan, Y.S. Gong, and G.H. Qi. 2009. Transport stress in broilers: I. Blood metabolism, glycolytic potential, and meat quality. Poult. Sci. 88: 2033–2041.

Zhang, S.H., D.P. Hennessy, P.D. Cranwell, D.E. Noonan, and H.J. Francis. 1992. Physiological responses to exercise and hypoglycaemia stress in pigs of differing adrenal responsiveness. Comp. Biochem. Physiol. 103: 695–703.

Zulkifli, I., A.F. Soleimani, M. Khalil, A.R. Omar, and A.R. Raha. 2011. Inhibition of adrenal steroidogenesis and heat shock protein 70 induction in neonatally feed restricted broiler chickens under heat stress condition. Arch. Geflügelk. 75: 246–252.

Zulkifli, I., P. Najafi, A.J. Nurfarahin, A.F. Soleimani, S. Kumari, A.A. Aryani, E.L. O'Reilly, and P.D. Eckersall. 2014. Acute phase proteins, interleukin 6, and heat shock protein 70 in broiler chickens administered with corticosterone. Poult. Sci. 93: 3112–3118.

Section E
Future Directions

CHAPTER-11

Nutrient Transporter Gene Expression in Poultry, Livestock and Fish

Eric A. Wong,[1,*] *Elizabeth R. Gilbert*[1] and
Katarzyna B. Miska[2]

INTRODUCTION

The transport of nutrients is mediated by membrane transporters that are members of the Solute Carrier (SLC) gene series. For the human genome, this series includes 52 families and 395 transporter genes (Hediger et al., 2013). A number of excellent reviews that discuss the structure and function of the various SLC genes in humans and rodents have been published as a series with an introduction by Hediger et al. (2013). This review will focus on the mRNA abundance of nutrient transporter genes in poultry, livestock, and fish, with an emphasis on the embryonic and post-hatch period in chickens, turkeys and pigeons and the perinatal period in pigs, sheep, cattle and fish. Most of these studies utilize real time PCR for measurement of mRNA abundance of nutrient transporters using relative quantification based on a reference gene or absolute quantification based on a standard curve. The mRNA abundance, rather than protein abundance, is mainly reported due to the paucity of reagents to detect these proteins in non-human, non-rodent species. Nevertheless, mRNA abundance provides some insight into the regulation and expression of these genes. In addition, although *in vitro* studies are essential to understand the molecular mechanisms underlying transporter expression and function, we have focused this review on whole animal studies.

[1] Department of Animal and Poultry Sciences, Virginia Tech, Blacksburg, VA 24061.
[2] Animal Bioscience and Biotechnology Laboratory, Henry A Wallace Beltsville Agricultural Research Center, Beltsville, MD 20705.
* Corresponding author

Nutrient transporters are membrane proteins that mediate the import of nutrients into the cell or efflux of nutrients out of the cell. These transporters can be divided into active and passive transporters (Hediger et al., 2013). Active transporters utilize energy-coupling mechanisms to transport substrates and typically involve ATP hydrolysis to power ion (e.g., Na^+, K^+, H^+) pumps. Whereas, passive transporters, which are also known as facilitated transporters, mediate the transport of substrates down a concentration gradient. Some transporters act as exchangers, i.e., exchanging the influx of one amino acid coupled with the efflux of another amino acid.

Examples of active and facilitated transport in intestinal epithelial cells are shown in Figures 1 and 2. The enterocyte is a polarized epithelial cell lining the villi of the small intestine. More than 90% of cells lining the villi are enterocytes, with a brush border membrane (also known as microvilli) facing the lumen in contact with end-products of digestion. The brush border membrane houses a vast array of transporters that are specific for different nutrients. Inside the cell, nutrients such as amino acids and monosaccharides

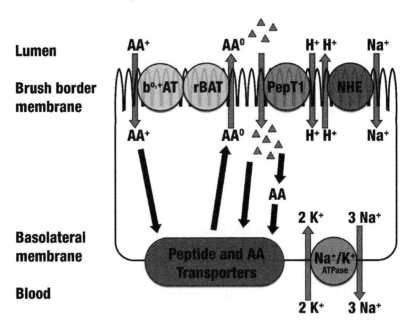

Mechanisms of amino acid uptake at the brush border membrane: active transport of di- and tripeptides by PepT1 and free amino acid exchange by the $b^{o,+}AT/rBAT$ heterodimer

Figure 1: Mechanism of amino acid uptake by PepT1 and $b^{o,+}AT/rBAT$. The peptide transporter PepT1 (SLC15A1) transports di- and tripeptides (▲) with a H^+ ion, which is coupled with the action of NHE3 (Na^+/H^+ exchanger, SLC9A3) and the Na^+/K^+ ATPase. The heterodimeric amino acid transporter $b^{o,+}AT/rBAT$ (SLC7A9/SLC3A1) exchanges extracellular cationic amino acids (AA^+) for intracellular neutral amino acids (AA^o).

are subject to further metabolism, or are transported out of the cell via the basolateral membrane into venules that empty into the hepatic portal vein that drains into the liver. Nutrients may also be transported into the enterocyte via the basolateral membrane from the bloodstream. The peptide transporter PepT1 actively co-transports inward di- and tri-peptides with a H$^+$ ion (Figure 1). The proton gradient is maintained by the efflux of H$^+$ by the Na$^+$-H$^+$ exchanger (NHE3), which is ultimately coupled to the Na$^+$-K$^+$ ATPase (Gilbert et al., 2008a; Smith et al., 2013). The amino acid transporter b$^{o,+}$AT/rBAT is a heterodimeric protein that preferentially exchanges extracellular cationic amino acids for intracellular neutral amino acids (Fotiadis et al., 2013). For the monosaccharide transporters, SGLT1 actively co-transports inward glucose and galactose with a Na$^+$ ion (Figure 2). GLUT5 and GLUT2 are both facilitated transporters, which mediate the uptake of fructose and export of glucose, galactose, and fructose, respectively. The amino acid, peptide, monosaccharide, and mineral transporters that have been profiled in the studies reviewed are listed in Table 1.

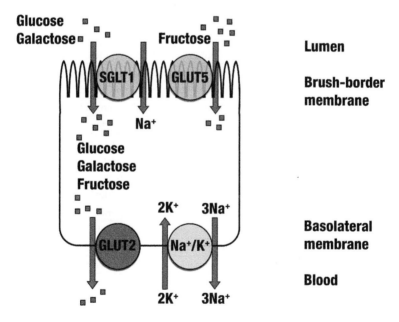

Active transport of monosaccharides via SGLT1 and facilitated fiffusion via GLUT5 and GLUT2

Figure 2: Active transport of monosaccharides via SGLT1 and facilitated diffusion via GLUT5 and GLUT2. SGLT1 (sodium glucose transporter 1, SLC5A1) mediates the active cotransport of glucose (■) and galactose (●) with a Na$^+$ ion, which is coupled to the action of the Na$^+$/K$^+$ ATPase. The facilitated transporter GLUT5 (glucose transporter 5, SLC2A5) transports fructose (◆) and the facilitated transporter GLUT2 (glucose transporter 2, SLC2A2) transports fructose, glucose, and galactose.

Table 1: SLC genes and digestive enzymes that have been profiled in poultry, livestock and fish.

SLC name[1]	Gene name	Function
SLC1A1	EAAT3	Glu, Asp transporter
SLC1A4	ASCT1	Ala, Ser, Cys, Thr transporter
SLC1A5	ASCT2	Ala, Ser, Cys, Thr, Gln, Asn transporter
SLC2A1	GLUT1	Glucose, galactose, mannose, glucosamine transporter
SLC2A2	GLUT2	Glucose, galactose, fructose, mannose transporter
SLC2A4	GLUT4	Glucose, glucosamine transporter
SLC2A5	GLUT5	Fructose transporter
SLC3A1	rBAT	Heterodimerizes with $b^{o,+}AT$
SLC3A2	4F2hc	Heterodimerizes with LAT1, y^+LAT1, y^+LAT2
SLC5A1	SGLT1	Glucose and galactose transporter
SLC5A9	SGLT4	Mannose, fructose and glucose transporter
SLC6A14	ATB^{0+}	Neutral and cationic amino acid transporter
SLC6A19	B^0AT	Neutral amino acid transporter
SLC7A1	CAT1	Cationic amino acid transporter
SLC7A2	CAT2	Cationic amino acid transporter
SLC7A3	CAT3	Cationic amino acid transporter
SLC7A5	LAT1	Large, neutral amino acid exchanger
SLC7A6	y^+LAT2	Cationic/large neutral amino acid exchanger
SLC7A7	y^+LAT1	Cationic/large neutral amino acid exchanger
SLC7A8	LAT2	Neutral amino acid transporter
SLC7A9	$b^{o,+}AT$	Cationic and large amino acid exchanger
SLC9A2	NHE2	Sodium-hydrogen exchanger
SLC9A3	NHE3	Sodium-hydrogen exchanger
SLC15A1	PepT1	Di- and tri-peptide transporter
SLC15A2	PepT2	Di- and tri-peptide transporter
SLC15A3	PHT1	Di- and tri-peptide and histidine transporter
SLC30A1	ZnT1	Zinc transporter
SLC34A2	NPT2b	Sodium phosphate transporter, type IIb
SLC38A1	SNAT1	Sodium dependent amino acid (System A) transporter
SLC38A2	SNAT2	Sodium dependent amino acid (System A) transporter
SLC43A2	LAT4	Branched chain amino acid transporter
	TRPV6	Calcium channel
	APN	Aminopeptidase N
	SI	Sucrase-isomaltase

[1]Reviews of SLC1 (Kanai et al., 2013), SLC2 (Mueckler and Thorens, 2013), SLC3 (Fotiadis et al., 2013), SLC5 (Wright, 2013), SLC6 (Pramod et al., 2013); SLC7 (Fotiadis et al., 2013), SLC9 (Donowitz et al., 2013), SLC15 (Smith et al., 2013), SLC30 (Huang and Tepaamorndech, 2013), SLC34 (Forster et al., 2013), SLC38 (Schioth et al., 2013), and SLC43 (Bodoy et al., 2013) gene families.

The majority of studies involving the regulation of intestinal nutrient transporter (particularly monosaccharide, amino acid, and peptide) gene expression in animals have been conducted with rodents or chickens. There are relatively fewer studies on the regulation of gut transporter expression in cattle, pigs, and sheep, with most of these studies involving the initial cloning and characterization of PepT1. For more details, see review by Gilbert et al. (2008a). This review will thus focus on the development of the chick intestine during embryonic development and during the post-hatch period, the distribution of transporters in the small intestine, and various factors regulating the gene expression of intestinal nutrient transporters including: development, diet composition, environmental stress, and disease. Discussion of mammalian nutrient transporters focuses on species of agricultural relevance and is mainly focused on ontogenetic and dietary regulation aspects. Studies of fish are focused on those involving species of most economic relevance, with other species included to demonstrate the diverse mechanisms (e.g., differences in pH dependence). To discuss all transporters and studies involving their regulation is beyond the scope of this review. Instead, we will focus on studies involving peptide, amino acid, and monosaccharide transporters, and where appropriate the accompanying study of some related intestinal digestive enzymes.

Development of the Embryonic Chick Intestine and Yolk Sac

The modern commercial broiler spends one third of its development as an embryo during a 21 day incubation period and two thirds (42 days) of its development post-hatch; thus growth and development during the embryonic stage is as important as the post-hatch stage. During the transition from pre-hatch to post-hatch, the main nutrient source for the chick changes from yolk, which is high in fat, to feed, which is high in protein and carbohydrates. Thus, the expression of various nutrient transporters is expected to change during this transition.

The embryonic chick relies on the yolk as its source of nutrients. The development of the intestine and the yolk sac covering the yolk has been described by Patten (1971). During the first few days of incubation, the extra-embryonic splanchnopleure spreads over the yolk forming the yolk sac; whereas the intra-embryonic splanchnopleure undergoes a number of changes that results in the development of the intestine. The embryonic intestine and the yolk sac form a contiguous membrane and thus share many physiological and anatomical features involved in nutrient uptake. The various stages of nutrition for the developing chick embryo have been reviewed by Moran (2007). During embryogenesis, nutrients are absorbed from the yolk by transporters located on the apical side of the epithelial cells lining the yolk sac and then are either metabolized by the cell or transported out of the cell into the extra-embryonic blood supplying nutrients to the embryo. Late in development, the seroamniotic membrane ruptures and albumen enters the amniotic sac. This composite fluid is swallowed by the embryo and enters the gastrointestinal tract, where nutrients can be absorbed by intestinal epithelia. Prior to hatch, the yolk sac is internalized and uptake from the yolk begins to diminish as nutrients in the yolk are depleted. There is a concomitant shift to nutrient uptake in the intestine from ingested feed. Because the yolk sac and embryonic intestine form a contiguous structure and share nutrient uptake functions, expression of nutrient transporters in both tissues is likely coordinated.

Nutrient Transporters in the Yolk Sac and Embryonic Intestine of Chickens, Turkeys and Pigeons

During embryogenesis the yolk sac functions as the intestine by expressing a number of nutrient transporters, while the intestine undergoes development and maturation. The expression of selected transporters in the yolk sac has been reported for chickens from embryo day 11 (e11) to day of hatch (doh) and pigeons from e12 to doh. In chickens, the digestive enzyme aminopeptidase N (APN), the amino acid transporters B°AT and cationic amino acid transporter 1 (CAT1), the peptide transporter PepT1, and the monosaccharide transporter GLUT5 showed greater mRNA abundance during early embryogenesis that declined towards hatch (Yadgary et al., 2011; Speier et al., 2012). In contrast, EAAT3, SGLT1 and NPT2b (Sodium phosphate transporter) increased towards hatch. Sucrase-isomaltase (SI), the calcium channel (TRPV6) and the zinc transporter (ZnT1) did not change from e11 to doh. In pigeons, similar decreasing expression profiles towards hatch were observed for APN and PepT1, while expression of SGLT1, GLUT2 and SI rose from e12 to e16 and then declined towards hatch (Dong et al., 2012b). A transcriptome analysis of the chicken yolk sac has been completed for e13 to e21, which profiled all of the expressed SLC genes (Yadgary et al., 2014). Comparable increased expression towards hatch of EAAT3 and SGLT1 and decreased expression towards hatch of GLUT5, B°AT and PepT1 was observed along with the expression profiles of many other members of the SLC gene family.

In the embryonic intestine, the expression profiles of a large number of nutrient transporters have been reported for chicken, turkey and pigeon. In the intestinal epithelia, brush border membrane transporters generally increase while basolateral membrane transporters decrease with age. This phenomenon can be explained by the changing source of nutrients. During early embryogenesis, nutrients are supplied to the embryo via the extra-embryonic blood system, which supplies nutrients to the intestinal epithelial cells through the basolateral membrane. As the embryo develops less nutrients are supplied via the yolk and there is a transition to nutrient uptake through the brush border membrane, thus explaining the decrease in basolateral membrane transporters and an increase in brush border membrane transporters. In general, the expression profiles of the brush border membrane and basolateral membrane transporters are similar for chickens, turkeys and pigeons.

At the brush border membrane most transporters increase in expression, which reflects the maturation of the intestine during development of the functional capacity of epithelial cells to absorb nutrients. The digestive enzymes APN and SI, the amino acid/peptide transporters $b^{o,+}AT/rBAT$, $B^{o}AT$, PepT1, EAAT3, and ATB^{o+}, and the monosaccharide transporters SGLT1, GLUT2 and GLUT5 showed increased mRNA abundance towards hatch in chickens (Gilbert et al., 2007; Speier et al., 2012; Li et al., 2013; Miska et al., 2014) (see Figure 2). This pre-feeding increase in the coordinated expression of multiple transporters may be a "hard-wired" mechanism for preparing the intestine to absorb dietary nutrients. At the basolateral membrane, mRNA abundance of CAT1, CAT2, LAT1, y^+LAT1, and y^+LAT2 decreased from e15 to doh (Speier et al., 2012; Miska et al., 2014). Zwarycz and Wong (2013) examined the mRNA abundance of the peptide transporters PepT1, PepT2 and PHT1 in various tissues (proventriculus, small intestine, ceca, large intestine, brain, heart, bursa, lung, kidney and liver) of chicks at e18 and e20. They found that PepT1 was mainly expressed in the small intestine, PepT2 in the brain, kidney and liver and PHT1 in all tissues. A comprehensive microarray analysis of all SLC genes was conducted for the chick intestine between e18 and doh, which showed similar up- or downregulation profiles for these as well as other brush border and basolateral membrane transporters (Li et al., 2008). In pigeons, results similar to those of chicken were reported (Zeng et al., 2011; Dong et al., 2012b; Chen et al., 2015). The mRNA abundances of genes for brush border membrane proteins in pigeon (APN, SI, $b^{o,+}AT/rBAT$, EAAT3, PepT1, SGLT1 and GLUT2) increased during embryonic development. The cationic amino acid transporters (CAT1, CAT2 and CAT3) showed an initial increase during early embryogenesis and then a decrease towards hatch. In contrast, y^+LAT1 and y^+LAT2 increased towards hatch in pigeon but declined in chicken. Other genes examined in pigeons showed increased expression (LAT4, NHE2 and NHE3) and decreased expression (SNAT1 and SNAT2) (Chen et al., 2015). Studies with a subset of transporters in turkeys showed similar increases in mRNA for brush border membrane transporters (PepT1, EAAT3, GLUT5, SGLT1, SGLT4) and a decrease in mRNA for basolateral membrane transporters (ASCT1, CAT1, LAT1 y^+LAT2 and GLUT2) from early embryogenesis until hatch (de Oliveira et al., 2009; Weintraut et al., 2016).

The effect of breed and sex on embryonic intestinal expression of nutrient transporters has been examined in chickens and turkeys. Zeng et al. (2011) compared Wenshi Yellow-Feathered chicks (WYFC) and White Recessive Rock chicks (WRRC). The mRNA abundance of CAT1, CAT4, rBAT, y⁺LAT1, y⁺LAT2, LAT4, SNAT1 and SNAT2 was greater in WYFC than WRRC. Speier et al. (2012) compared differences between Cobb and Leghorn embryos and reported that mRNA abundance of B°AT, CAT1, SI, SGLT1 and GLUT5 was greater in the embryonic intestine of Cobb compared to Leghorn embryos. Within the WYFC and WRRC breeds there was an effect of sex on gene expression (Zeng et al., 2011). The mRNA abundance of CAT1, CAT4, and LAT2 was greater in WYFC males than females, whereas y⁺LAT2 was greater in females than males. For WRRC, females had greater mRNA abundance of B°AT than males. In turkeys, there was no main effect of sex for any gene examined, although there was a sex by age interaction for EAAT3 with males expressing greater EAAT3 at e21 than females (Weintraut et al., 2016).

Two reports examined the expression of selected nutrient transporters in the yolk sac and embryonic intestine of the same chick or pigeon (Dong et al., 2012b; Speier et al., 2012). For chicken, the yolk sac was examined from e11 to e21/doh and the intestine from e15 to e21/doh; while for pigeon the yolk sac and intestine were examined from e12 to doh. In both chickens and pigeons, expression of APN and PepT1 mRNA showed an early peak in the yolk sac that declined towards hatch, whereas in the intestine expression increased towards hatch (Dong et al., 2012b; Speier et al., 2012). This pattern reflects the shift of the protein source and the need for a digestive enzyme and the peptide transporter to mediate entry of nutrients from the yolk sac into the intestine. Not all genes, however, showed this inverse pattern of expression between yolk sac and intestine. For example, SI and SGLT1 both increased expression towards hatch in the yolk sac and intestine of chickens and pigeons.

Nutrient Transporter Expression in Poultry Post-Hatch

Several nutrient transporters have been described in post-hatch poultry but only a handful have been described at a functional or protein level. Most of the work in post-hatch poultry has been carried out in domesticated chickens but some work has been done in other bird species such as turkey, duck, and pigeons.

The cationic amino acid transporters belong to the y⁺ cationic amino acid transport (CAT) family and have been extensively investigated in chickens. Chicken genomes appear to possess three CAT2 isoforms, CAT2-A, CAT2-B, as well as CAT2-C. However, in functional studies of the chicken CAT2 proteins expressed in mammalian cells, it appeared that only CAT2-A might be capable of transporting Lys through the plasma membrane (Kirsch and Humphrey, 2010). The CAT2-C isoform encoded a truncated protein and its mRNA appeared to be targeted for nonsense mediated mRNA decay, which is a cellular mechanism that degrades transcripts containing a premature termination codon (Humphrey et al., 2008). At the mRNA level all three CAT2 isoforms were expressed with differing levels of abundance in several tissues such as skeletal muscle and liver. Neither isoform was highly expressed in the heart; however the functional CAT2-A was highly expressed in both skeletal muscle as well as liver (Humphrey et al., 2008).

The ontogeny of CATs (1-3) as well as glucose transporters (GLUT1-3) at the mRNA level has also been investigated in chicks from hatch to 14 days post-hatch (Humphrey et al., 2004). The authors investigated expression in skeletal muscle, heart, liver spleen, bursa, and thymus. The GLUT2 isoform was the only one expressed in the liver at any of the time-points sampled, and GLUT2 expression appeared to increase following hatch to day 7 with the exception of the thymus. On the other hand most tissues expressed a CAT isoform at hatch and that expression persisted through day 7. Interestingly, neither spleen nor thymus expressed any CAT isoforms even at day 7. Overall most of the genes examined in this study were differentially expressed over time and tissue.

The expression of CAT1-3 mRNA was investigated in chicks fed a Lys deficient diet (Humphrey et al., 2006). Because Lys deficient diets were consumed in lesser amounts than Lys adequate diets the expression of CATs was also investigated in pair-fed chickens fed a Lys adequate diet in the same amounts as those consumed by a chicken fed a Lys deficient diet. The expression of CAT1-3 in Lys deficient chickens was much lower in the liver, pectoralis muscle, bursa, and thymus. Pair fed chickens had expression levels similar to chickens fed a Lys adequate diet *ad libitum*. The expression of CAT1-3 in the liver, pectoralis muscle, and bursa was actually greater than in chickens fed *ad libitum*. Interestingly, the amount of plasma Lys levels in chickens fed a Lys deficient or pair fed Lys adequate diet was very similar, while chickens fed a Lys adequate diet *ad libitum*, had much higher amounts of Lys in their plasma. Therefore, it is unlikely that plasma levels of amino acids regulate the expression of CATs.

The ontogeny of nutrient transporter expression post-hatch has been thoroughly studied in chickens, turkeys and pigeons. The first comprehensive study of nutrient transporters in the small intestine of chickens was published by Gilbert et al. (2007). In this study 10 amino acid (b$^{o,+}$AT/rBAT, ATB^o, CAT1, CAT2, LAT1, y$^+$LAT1, y$^+$LAT2, B^{o}AT, and EAAT3), a peptide (PepT1), and four monosaccharide (SGLT1, SGLT5 (later redesignated SGLT4), GLUT5, and GLUT2) transporters along with APN were examined from doh to 14 days post-hatch in two lines of chickens. Line A was selected for several generations on feed consisting of corn-soybean meal, while Line B on feed composed of wheat and amino acid concentrations that were 15–20% higher than those fed to Line A chickens. Expression of PepT1 was always greater in Line B chickens than line A and APN was consistently the most highly expressed gene of those investigated. In general, PepT1 was expressed in greatest amounts in the proximal region of the small intestine, the monosaccharide transporters in the jejunum and the amino acid transporters in the ileum. The mRNA abundance of PepT1 and the amino acid transporters reflected the relative concentrations of the substrates, i.e., greater peptide concentrations in the proximal intestine and greater free amino acid concentrations in the distal intestine. Gilbert et al. (2007) reported that PepT1, EAAT3, B°AT, APN, SGLT1, SGLT5 (SGLT4), SGLT4, GLUT5 and GLUT2 increased from doh to d14, while CAT1 and CAT2 declined. Miska et al. (2015) reported similar increases in APN, ATB^{o+}, B°AT, b$^{o,+}$AT/rBAT, EAAT3, y$^+$LAT2, and PepT1 and a decrease in CAT1 and CAT2 from doh until d21 post-hatch, although various profiles were noted for the different intestinal segments.

Pigeons and turkeys showed similar patterns of nutrient transporter gene expression as chickens. There was increased mRNA abundance of PepT1, SGLT1, GLUT2, and APN from doh until d14 in pigeons (Dong et al., 2012a). Abundance of SI mRNA showed a more complex pattern with decreased expression from doh to d3/d5 followed by an increase to d14. Dong et al. (2012a) further showed that total intestinal enzyme activities of maltase, sucrase and APN increased from doh to d14 in pigeons, which was consistent with the increase in mRNA abundance. In turkeys, there was an increase in the post-hatch expression of EAAT3, GLUT5, and SGLT1 mRNA and a decrease in ASCT1, CAT1, LAT1, y$^+$LAT2 and GLUT2 mRNA from doh to d28 post-hatch (Weintraut et al., 2016). Interestingly, EAAT3 mRNA abundance was 6-fold greater in the ileum of turkeys than chickens at d14. This result suggests that turkeys have an increased capacity to absorb anionic amino acids such as glutamate, which serves as the major energy source for intestinal epithelial cells (Brosnan and Brosnan, 2013). The mRNA expression results are consistent with the finding that glutamate is the most digestible amino acid for turkeys compared to chickens and ducks (Kluth and Rodehutscord, 2006).

The effect of genetic line and sex on gene expression was examined in chickens and turkeys. A study by Mott et al. (2008) addressed whether lines of chicks selected for high or low body weight resulted in different mRNA abundance of PepT1, EAAT3, SGLT1, and GLUT5. Greater expression of PepT1, EAAT3, and GLUT5 mRNA was observed in chicks selected for low body weight, while expression of SGLT1 was not different between any mating combinations. Mott et al. (2008) also explored the presence of sexual dimorphism in chicks from doh to d14 post-hatch. When averaged across all time points, the expression of EAAT3 and SGLT1 mRNA was greater in females than males. Sexual dimorphic gene expression has also been documented in turkeys from doh to d28 post-hatch using a larger panel of transporter genes (Weintraut et al., 2016). Nine genes (APN, b$^{o,+}$AT, EAAT3, PepT1, GLUT5, ASCT1, CAT1, LAT1, and y$^+$LAT2) showed greater mRNA abundance in females than males, underlining that females may have a greater need for amino acid or sugars or that they may be utilized less efficiently in females. Only SGLT1 mRNA abundance was greater in male than female turkeys, which is opposite to chickens where female chickens showed greater SGLT1 mRNA abundance than male chickens.

The effect of dietary protein quality, feed restriction and dietary protein composition on expression of nutrient transporter expression has been investigated. To determine the effect of protein quality and feed restriction on nutrient transporter gene expression, chickens were fed a diet containing high quality SBM (soybean meal) or lower quality CGM (corn gluten meal) as the supplemental protein source either *ad libitum* or a restricted amount of SBM (Gilbert et al., 2008b). Because feed intake of birds fed CGM was less than birds fed SBM, a set of birds fed SBM was restricted to the amounts eaten by chicks fed CGM. When feed intake was equal (CGM vs. restricted SBM) expression of PepT1 and b$^{o,+}$AT mRNA was greater in SBM than CGM, whereas EAAT3 and GLUT2 mRNA was greater in CGM than SBM. When feed intake was restricted (ad lib SBM vs. restricted SBM), PepT1 mRNA was increased during feed restriction. As seen previously, only PepT1 was influenced by genetic line with Line B expressing greater levels than line A. Expression of EAAT3 mRNA was greater in line B chicks, while APN and SGLT1 mRNA was greater in Line A chicks. The authors suggested that while genetic selection changed the expression of nutrient transporters in the gut, the changes were masked when birds were fed a balanced diet (SBM) but were accentuated when fed a deficient diet (CGM). The upregulation of PepT1 mRNA following feed restriction has been previously reported in mammals (Thamotharan et al., 1999; Ihara et al., 2000) and subsequently reported in chickens (Chen et al., 2005; Duarte et al., 2011; Madsen and Wong, 2011).

The effect of dietary protein composition on expression of nutrient transporters has been investigated by Gilbert et al. (2010). The purpose of this study was to determine if feeding a partially digested or completely digested protein would alter the expression profile of transporters in the small intestine. Line A or B chickens were fed three different diets beginning at day 8 post-hatch, in which protein sources were whey protein concentrate, a whey partial hydrolysate (WPH), or a mixture of free amino acids identical to the composition of whey. The abundance of PepT1, b$^{o,+}$AT, EAAT3, y$^+$LAT2, CAT1, and APN mRNA was greater in line B than line A chickens. For all genes except LAT1, Line B chickens that consumed the WPH diet showed greater expression of nutrient transporters, indicating that genetic selection as well as diet formulation can be used to change and perhaps improve nutrient uptake.

Nutrient transporters in post-hatch chicks are also expressed in tissues other than the small intestine. In the ceca, expression of digestive enzymes, amino acid, peptide and monosaccharide transporters was reported, although at varying levels of expression compared to the small intestine (Miska et al., 2015; Su et al., 2015). The genes examined included digestive enzymes (APN and SI), amino acid/peptide transporters (ATB^{o+}, B^{o}AT, b$^{o,+}$AT/rBAT, EAAT3, ASCT1, PepT1, CAT1, CAT2, LAT1, y$^+$LAT1, y$^+$LAT2, SNAT1 and SNAT2) and monosaccharide transporters (GLUT2, GLUT5, SGLT1 and SGLT4). For most genes, mRNA abundance in the ceca ranged from one to two orders of magnitude lower than in the small intestine. Su et al. (2015) reported that CAT2 and GLUT1 expression was greater in ceca than the small intestine; however Miska et al. (2015) did not observe increased expression of CAT2 in the ceca. These results demonstrate that the ceca have the ability to absorb nutrients, although at a lower level than the small intestine. In the liver, mRNA abundance of APN, ATB^{o}, b$^{o,+}$AT/rBAT, EAAT3, PepT1, CAT1, CAT2, y$^+$LAT1, y$^+$LAT2, SNAT1 and SNAT2 was detected, but again at a level of one to two orders of magnitude less than the small intestine (Miska et al., 2015). Zwarycz and Wong (2013) also reported low expression of PepT1 mRNA in the ceca and large intestine post-hatch. PepT2 mRNA was expressed at high levels in the brain and kidney and low levels in the proventriculus, ceca, large intestine, heart, bursa, lung and liver, while PHT1 was expressed in all tissues examined. In the oviduct, differential expression of selected SLC genes has been reported by Lim et al. (2012), where expression of ASCT1 was limited to glandular epithelium while CAT3 was expressed predominantly in the luminal epithelium of the magnum.

Nutrient Transporter Expression in Poultry in Response to Thermal Stress and Disease

Environmental factors, such as temperature, can cause stress that negatively affects growth rate. Both high and low temperatures have been shown to alter nutrient transporter gene expression. Carriga et al. (2005) showed that heat stress caused decreased body weight that can only partially be attributed to decreased food intake. Birds reared for two weeks at 30°C exhibited increased transport of glucose through the brush border membrane of the jejunum, which was associated with increased amounts of SGLT1. Combined with increased glucose uptake and SGLT1 expression, the villi of heat stressed birds were longer than those of normally reared birds, therefore these three characteristics may function in the jejunum to guarantee an energy supply under stressful conditions. Sun et al. (2015) also reported the effect of heat stress on broilers on the expression of nutrient transporters at the mRNA level. The levels of SGLT1 mRNA in the jejunum in heat stressed birds did not change while levels of GLUT2 decreased. These results are not in agreement with results reported by Carriga et al. (2005). However, it is possible that levels of mRNA abundance are not the same as protein levels. The birds used by Sun et al. (2015) were older than those used by Carriga et al. (2005) and were heat stressed for a lesser period of seven days for 10 hours per day. In addition to monosaccharide transporters Sun et al. (2015) measured the change in mRNA abundance of three amino acid transporters (rBAT, CAT1, and y+LAT1), as well as PepT1, however these were unaltered by heat stress. Barri et al. (2011) also examined the effect of egg incubation temperature on SGLT1, GLUT2, GLUT5, PepT1 and EAAT3 mRNA abundance in the intestine. They found that expression of PepT1 mRNA showed a temperature by age interaction, with greater expression of PepT1 on day 6 in chicks hatched from eggs incubated at 37.4°C compared to 39.6°C.

Low temperatures also affected nutrient transporter gene expression. Cold acclimation in ducklings resulted in increased glucose transport by sarcolemmal vesicles, which was associated with increased GLUT4 protein. This result is different from mammals where no change in skeletal muscle glucose transport was associated with cold acclimation. The authors concluded that in birds muscle is the main thermogenic site, whereas some mammals such as sheep and cattle use brown adipose as the thermogenic tissue thus highlighting basic differences between mechanisms of nutrient transport of avian species compared to some mammals. In addition, in mammals GLUT4 expression and activity was inducible by insulin; however, this induction was stronger in mammals (13-fold in rat) compared to duck (5-fold) (Thomas-Delloye et al., 1999). In chickens (and possibly other avian species) the gene encoding the GLUT4 ortholog is absent and perhaps replaced in function by GLUT12 (Seki et al., 2003; Coudert et al., 2015). However, other birds like ducks and pigeons appear to express a GLUT4-like protein. Therefore, it would be of interest to determine if GLUT4 is absent in all galliformes and if a different glucose transporter responds to insulin stimulation in these birds.

Pathogens that infect the intestine have a profound effect on nutrient uptake and expression of nutrient transporters. Coccidiosis is a major disease of poultry caused by the protozoan *Eimeria*. This parasite invades intestinal epithelial cells to complete its replication cycle and causes intestinal lesions, leading to reduced body weight gain, reduced feed efficiency and severe economic losses for the poultry industry (Dalloul and Lillehoj, 2006). The sites of lesions are species specific. For example, *E. acervulina* and *E. praecox* primarily affect the duodenum, *E. maxima* affects the jejunum, and *E. tenella* affects the cecum. These intestinal lesions would be expected to affect the function, such as nutrient uptake, of the intestinal epithelial cells by dysregulation of the nutrient transporters. A series of recent studies have examined the effect of different *Eimeria* species (*E. acervulina*, *E. maxima*, *E. praecox* and *E. tenella*) on nutrient transporter gene expression in infected chickens (Paris and Wong, 2013; Su et al., 2014; Fetterer et al., 2014; Su et al., 2015; Yin et al., 2015).

Transporter gene expression was profiled in different segments of the small intestine (duodenum, jejunum, and ileum) or cecum of *Eimeria* infected chickens. The genes examined included digestive enzymes (APN and SI), amino acid transporters (b$^{o,+}$AT/rBAT, B^{o}AT, EAAT3, ASCT1, CAT1, CAT2, y$^+$LAT1, y$^+$LAT2, LAT1), the peptide transporter PepT1, monosaccharide transporters (GLUT1, GLUT2, GLUT5, SGLT1, SGLT4), and the zinc transporter ZnT1. Paris and Wong (2013) and Fetterer et al. (2014) examined the changes in nutrient transporter gene expression in *E. maxima* infected broilers. In both studies mRNA abundance of EAAT3 and b$^{o,+}$AT/rBAT in the jejunum was downregulated, while LAT1 was upregulated. In addition, ASCT1 (Paris and Wong, 2013), CAT1, and ATB$^{o,+}$ mRNA (Fetterer et al., 2014) were upregulated. Su et al. (2014) compared changes in gene expression in *E. acervulina* challenged layers and broilers. In the target tissue (duodenum), there was downregulation of APN, b$^{o,+}$AT/rBAT, B^{o}AT, EAAT3, CAT2, SI, GLUT2 and ZnT1 mRNA in both broilers and layers. Although many genes were downregulated or upregulated in the jejunum and ileum of layers, there were no changes in the jejunum and ileum of broilers, demonstrating that layers and broilers respond to an *Eimeria* infection differently. Differential expression of transporters in the small intestine and cecum in response to different *Eimeria* species (*E. acervulina*, *E. maxima*, *E. tenella*, and *E. praecox*) has been reported (Su et al., 2015; Yin et al., 2015). Changes in gene expression were found to be tissue- and species-specific. *E. acervulina* caused downregulated expression in the duodenum and ceca. *E. maxima* and *E. tenella* caused both up- and downregulated expression in all four tissues and in the jejunum, ileum and cecum, respectively. *E. praecox* caused downregulated expression in all three intestinal segments, however cecum was not examined in this study. Although there were *Eimeria*-species specific differences in gene expression, a common set of genes was altered following infection with all four *Eimeria* species. These transporters included b$^{o,+}$AT/rBAT, EAAT3, GLUT2, GLUT5 and ZnT1 and the digestive enzymes APN and SI. A general cellular model in response to *Eimeria* infection has been proposed, which involved down-regulation of mRNA for key transporters such as EAAT3, b$^{o,+}$AT/rBAT and ZnT1 (Paris and Wong, 2013; Su et al., 2014, 2015). Decreased expression of the brush border membrane transporters b$^{o,+}$AT/rBAT and EAAT3 would result in decreased uptake of some essential amino acids and glutamate, the major energy source for intestinal epithelial cells. This would lead to a depletion of intracellular pools of these key amino acids resulting in either inhibition of *Eimeria* replication or cell death. Downregulation of the basolateral zinc transporter ZnT1, which mediates the efflux of zinc, may result in an increase in intracellular concentrations of zinc, which is important for the activity of antioxidant enzymes. The *Eimeria* induced oxidative damage reported by Georgieva et al. (2011), may be counteracted by a decrease in ZnT1 expression and the resultant increase in intracellular zinc.

Nutrient Transporter Gene Expression in Livestock

In livestock species, there are a number of studies involving the effects of various dietary factors on gastrointestinal tract expression of amino acid and monosaccharide transporters. In ruminants, the dietary nutrients are subjected to metabolism and modification by the ruminal microflora, with the majority of dietary protein hydrolyzed to amino acids and further degraded to ammonia and carbon skeletons used by the microbes for rebuilding protein. Thus, much of the amino acids available to the ruminant animal for growth are derived from microbial protein. There has been a great deal of interest in understanding the potential for absorbing amino acids in the rumen, and contribution of the various stomach compartments and small intestine to whole body amino acid nutrition. In general, there is gene expression and functional activity of PepT1, in the rumen of sheep and cattle (Gilbert et al., 2008a), suggesting that the foregut of ruminants can absorb amino acids as peptides. PepT1 mRNA in the rumen, omasum, and small intestine was not influenced by age, between 2 and 8 weeks of age, in lambs, and lambs that nursed but did not have access to a creep diet expressed more ruminal PepT1 mRNA (Poole et al., 2003). In contrast, in Angus steers, there were developmental changes in jejunal CAT1 expression, with greater quantities of mRNA in growing animals as compared to suckling, weanling, or finishing animals (Liao et al., 2008). Ruminal infusion of starch hydrolysate to Angus steers for 15 days downregulated the jejunal mRNA abundance of CAT1, rBAT, y$^+$LAT2, and 4F2hc (Liao et al., 2009). Abomasal infusion only downregulated the jejunal expression of y$^+$LAT2 and 4F2hc mRNA (Liao et al., 2009). GLUT2 and SGLT1 mRNA but not GLUT5 mRNA was increased in the ileum after abomasal starch hydrolysate infusion, whereas duodenal SGLT1 expression was increased after ruminal infusion (Liao et al., 2010). The abomasal infusions bypass the ruminal microflora, thus representing nutrients that are made directly available to the animal, thus it is interesting to note the differences in small intestinal nutrient transporter transcription in response to ruminal vs. abomasal infusion of nutrients.

Similar to proteins, ingested complex carbohydrates are hydrolyzed into monosaccharides and fermented into volatile fatty acids (and other small organic acids) in the rumen-reticulum. The capacity for immediate monosaccharide absorption in the rumen is unclear, as it has been largely assumed that most of these sugars are fermented in the rumen, with little to no monosaccharide transport in more distal areas of the gastrointestinal tract. It is well known that increased dietary supplementation of grain concentrate in a ruminant diet will lead to reduced ruminal pH and greater production of lactic acid and volatile fatty acids. Goats fed 60% barley grain for 6 weeks expressed less SGLT1 in the rumen as compared to goats that consumed 30% or no grain (Metzler-Zebeli et al., 2013).

There has been increasing interest in reducing crude protein content in the swine diet and using free amino acid supplementation to provide a more ideal balance of amino acids, combined with energy content, in order to maximize growth performance while minimizing nitrogen excretion. Consumption of a low 14% crude protein, wheat-soybean meal-based diet supplemented with l-Lys, l-Thr, dl-Met, l-Leu, l-Ile, l-Val, l-His, l-Trp, and l-Phe for 21 days was associated with greater $b^{o,+}AT$ mRNA in the duodenum as compared with crossbred pigs that ate a high 22% crude protein diet that did not contain supplemental amino acids (Morales et al., 2015). Expression was not affected in pigs that consumed the low-protein diet that was only supplemented with free glycine. In a 28-day study, pigs fed wheat and cornstarch-based diets (containing a basal level of supplemented free Lys, Thr, and dl-Met) supplemented with Leu expressed much lower quantities of $b^{o,+}AT$ mRNA than those fed the unsupplemented diet (Morales et al., 2013).

In Yucatan miniature piglets that were weaned at 4 weeks of age and sampled at 1, 2, 3, or 6 wk of age, there were age- and site-specific differences in PepT1 mRNA abundance and peptide uptake in the small intestine and colon (Nosworthy et al., 2013). Peptide uptake did not differ significantly among intestinal regions prior to weaning; however, post-weaning a distinct gradient was established whereby uptake was greatest in the distal small intestine (Nosworthy et al., 2013). There was variability in the corresponding gene expression of PepT1 with one distinct pattern of a dramatic decline in mRNA abundance after weaning in the colon (Nosworthy et al., 2013). In Tibetan piglets, expression of PepT1 mRNA was greatest in the jejunum as compared to duodenum and ileum, and expression increased from 1 to 14 days followed by a decrease from days 21 to 35 (Wang et al., 2009). Similar to chickens, expression of $b^{o,+}AT$ in pigs (at 90 days of age) was greater in the distal than proximal intestine and greater in the small intestine than colon, with ileum expression increasing from 1 day to 150 days after birth (Feng et al., 2008).

While many studies have examined the effects of diet from weaning through adulthood on regulation of intestinal nutrient transporter gene expression, few have focused on effects of maternal nutrition on intestinal development of the offspring. Weaned piglets from Landrace x Yorkshire dams that were subjected to over nutrition throughout gestation (150% of NRC recommendations) had heavier small intestines, and greater mRNA abundance of SGLT1, GLUT2, and PepT1 (more than 10-fold for PepT1) in the jejunum, whereas piglets from under nourished dams (75% of NRC recommendations) showed fewer differences from the control group (Cao et al., 2014). At birth, SGLT1 mRNA abundance was greater in the piglets from over nourished dams than control or under nourished dams. The piglets from underfed dams showed stunted intestinal development, including shorter villi, but no differences in monosaccharide or peptide transporter gene expression at birth or weaning (Cao et al., 2014). The relative birth weight of the piglet has also been shown to affect PepT1 regulation in the gut. Low birth weight piglets showed increased colonic PepT1 activity as compared to normal weight piglets (Boudry et al., 2014).

Nutrient Transporter Gene Expression in Fish

Studies of transcriptional regulation of PepT1 and other transporters has been explored quite extensively in fish species of commercial relevance that are used for aquaculture because of the potential to improve their growth and feed efficiency through dietary manipulation of transporter expression. As compared to birds and mammals, regulation of transporter activity (particularly pH-dependent activity) is quite complex and mechanisms more diverse in fish, most likely due to the greater species diversity and adaptations to specific ecological niches and genome diversity, such as whole genome duplication, frequent gene-linkage disruptions, etc. (Romano et al., 2014).

The PepT1 gene has been cloned in various species of fish such as Atlantic salmon (Ronnestad et al., 2010), sea bass (*Dicentrarchus labrax*) (Terova et al., 2009), sea bream (*Sparus aurata*) (Terova et al., 2013), zebrafish (Verri et al., 2003), Atlantic cod (Ronnestad et al., 2007), and the Asian weatherloach (Goncalves et al., 2007). Similar to birds and mammals, expression of PepT1 in fish is greatest in the gastrointestinal tract (Ronnestad et al., 2010; Liu et al., 2013). Within the tract, expression was greatest in the pyloric ceca and proximal to mid-section of the small intestine as compared to the other sections between the gastroesophageal junction and rectum (Terova et al., 2009; Ronnestad et al., 2010). Expression of SGLT1 was also greater in the intestine as compared to other tissues in the Asian weatherloach, although it was not detected in the most distal region of the hindgut in contrast to PepT1 (Goncalves et al., 2007).

Expression of PepT1 also shows developmental changes in fish. In larval Atlantic cod (*Gadus morhua*), an important commercial fish and aquaculture species, PepT1 mRNA was detected throughout the gastrointestinal tract (except the esophagus and sphincters) before first feeding at hatch, and continued to increase during the first three weeks after feeding in response to two different types of prey (zooplankton or rotifers) (Amberg et al., 2008). As observed in embryonic chicks and rodents, pre-feeding expression of PepT1 in fish is upregulated probably by a hard-wired mechanism. PepT1 expression posterior to the esophagus and more ubiquitous expression throughout the intestinal tract is different from birds and mammals where PepT1 was shown to be primarily localized to the duodenum and jejunum of the small intestine (Amberg et al., 2008).

In contrast to birds and mammals, in several fish species, such as zebrafish and Atlantic salmon, functional studies have shown that PepT1 shows greater activity at an alkaline pH. PepT1 shows less proton dependency as reflected by low expression of the Na^+-H^+ exchanger in enterocytes and is likely adapted for maximum activity in the alkaline environment of the fish intestinal lumen (Ronnestad et al., 2010). Interestingly, in *Fundulus heteroclitus*, a euryhaline species capable of adapting to a wide range of salinities, two PepT1 isoforms were cloned that were differentially expressed in freshwater and saltwater conditions (Bucking and Schulte, 2012). Only PepT1b was expressed in the intestine of saltwater-acclimated fish, whereas both isoforms were expressed in freshwater-adapted fish. Furthermore, *in vitro* studies showed that an increase in luminal pH decreased dipeptide transport in freshwater-adapted fish but increased transport in saltwater-acclimated fish (Bucking and Schulte, 2012).

There have been a variety of studies to assess the effects of fasting/ starvation and refeeding on transcriptional regulation of PepT1 and other transporters in different fish species. Zebrafish represent an important biomedical model and also serve as a model for fish physiology, in part because of their complete and well-annotated genome. Fasting for 384 h led to changes in mRNA abundance of transporters in 35-day-old mixed sex zebrafish (Tian et al., 2015). During the first 6 to 12 h of starvation, mRNA abundance of B^0AT1, PepT1, PepT2, and $ATB^{0,+}$ all decreased whereas ASCT2 increased. However, after 12 h, there was an increase in B^0AT1 and $ATB^{0,+}$ to 24 h and increase in PepT2 to 96 h, followed by decreases in expression of all genes. Expression of ASCT2 decreased after 6 hours and plateaued to lowest levels at 24 hours. Of the transporters evaluated, PepT1 was the only one that continually decreased after the onset of starvation, reaching lowest levels at 192 h post-initiation of starvation. These findings contrast a number of avian and mammalian studies that show upregulation in PepT1 expression in response to fasting, feed restriction, and other treatments that impose a nutrient restriction/deprivation. A potential limitation of this study though is that whole fish were used for RNA isolation and may not be representative of changes occurring only in the small intestine as these genes are known to have a wide tissue distribution in birds and mammals.

In an earlier study, however, where zebrafish were fasted for 1, 2 and 5 days and then refed, mRNA abundance of PepT1 in the gastrointestinal tract also decreased at 1, 2 and 5 days post-food withdrawal, but then dramatically increased at 24 and 48 h post-refeeding followed by another sharp increase to 96 h post-refeeding (Koven and Schulte, 2012). Interestingly, the expression levels at 96 hours post-refeeding were more than five times greater than quantities in the control fed zebrafish, suggesting a robust compensatory mechanism in response to the prolonged fasting. In the same study, a variety of other growth-related genes were measured, but none showed the same compensatory response. Zebrafish display a compensatory growth response following feed restriction. The authors went so far as to suggest that PepT1 may play an important role during the compensatory growth response, with its role paralleling that in humans with short bowel syndrome that show an upregulation of PepT1 in the gut to maintain amino acid absorption capacities (Koven and Schulte, 2012).

Consistent with these reports, sea bass were either fed to satiety or feed-deprived for 5 weeks and then refed for 3 weeks (Terova et al., 2009). At 4 and 35 days of fasting, PepT1 mRNA was reduced in the proximal intestine and at 4 and 10 days of refeeding expression was greater than in control fish that had been fed to satiety throughout the experiment. Expression returned to normal control levels at 21 days of refeeding (Terova et al., 2009). Similarly, 6 d of food deprivation was associated with downregulated PepT1 mRNA in the pyloric ceca of juvenile Atlantic salmon (Ronnestad et al., 2010). In *F. heteroclitus*, there was increased PepT1 mRNA in both the proximal and distal intestine after 3 days of fasting; however, expression was down-regulated by 21 days of fasting, and 7 days of refeeding upregulated PepT1 to quantities greater than in pre-feeding fish (Bucking and Schulte, 2012). Similarly, in European sea bass fasted for 3, 7, 14, and 21 days, intestinal expression of PepT1 mRNA was increased at 7 days of fasting but then was reduced to pre-fasting quantities at 14 and 21 days (Hakim et al., 2009). The authors speculated that this reflected weight loss rates, where losses peaked during the first week but were then "steady state" during the final two weeks of the study (Hakim et al., 2009). However, one month of fasting in Asian weatherloach, a species that uses the intestine for both digestion/absorption and respiration, did not alter PepT1 or SGLT1 mRNA abundance in the foregut or hindgut (Goncalves et al., 2007). The epithelial cell layer in the gut of the air-breathing weatherloach has a short diffusion distance to the blood, conducive to gas exchange, but may alter nutrient transporter functions of the intestine relative to other species (Goncalves et al., 2007). Terova et al. (2009) speculated that the difference between fish and bird/

mammal studies could be due to the relative physiological magnitude of the fasting imposed—that the fasting treatments reported for fish are far more dramatic relative to the stress imposed in other taxa.

Dietary composition has also been shown to affect transporter expression in fish. In particular, due to the expense associated with feeding fish meal, there has been intense focus on substituting fish meal with alternative protein sources, such as those of vegetative origin. There have been many studies published on understanding effects of dietary protein source and quality on regulation of PepT1 and amino acid transporter gene expression and amino acid absorption. Grass carp (*Ctenopharyngodon idella*) are a freshwater species native to eastern Asia and used extensively in China as a food fish. Grass carp displayed differences in intestinal PepT1 mRNA when fed isonitrogenous diets formulated to contain either fishmeal or soybean meal as the primary source of protein in the diet (Liu et al., 2013). During the month-long feeding trial, expression of PepT1 mRNA in the intestine was greater in fish fed the fishmeal diet than in fish fed the soybean meal. When fed isocaloric diets that differed in the amount of protein (22, 32, or 42% CP), carp that were fed the 22% CP diet expressed less PepT1 mRNA during the first 7 days of the feeding trial than either of the other two groups, but after day 14 expressed more intestinal PepT1 mRNA throughout the remainder of the trial (days 21 and 28). Expression of PepT1 mRNA in fish that were fed the 32% CP diet was lowest relative to the other two dietary groups (Liu et al., 2013). These results show that dietary protein quantity influences the regulation of PepT1 expression, consistent with reports in chickens.

In Atlantic cod that were fed either intact protein as fish meal, or in combination with a fish meal hydrolysate or a mixture of free amino acids, there were no differences in expression between the control and treated groups, but within groups there were differences in the spatial distribution of PepT1. PepT1 expression was greater in the mid-region of the intestinal tract compared to the proximal region in fish fed the free amino acid or hydrolysate mixtures, as compared to the fishmeal-fed controls (Bakke et al., 2010). In a study with Asian weatherloach, there was no difference in intestinal expression of PepT1 or SGLT1 mRNA in high-protein vs. high-carbohydrate-fed fish after one month, although growth rates were lower in the carnivorous fish that were fed the high-carbohydrate (low protein) diet (Goncalves et al., 2007). In European sea bass fed a salt-enriched diet with a relatively low proportion of fish meal (10% vs. 50%), PepT1 mRNA was upregulated in the hindgut after 6 weeks of feeding, whereas B°AT was not affected by dietary manipulations (Rimoldi et al., 2015). When green pea was substituted for 15% of fish meal in the diet of sea bream, PepT1 mRNA was downregulated in the proximal intestine relative to the fish fed the control diet, although body weights were not significantly different (Terova et al., 2013).

Amino acid supplementation was also shown to influence expression of PepT1 in fish. In the intestine of 1 month-old carp, consumption of a wheat gluten protein-based diet supplemented with Lys-Gly for 4 weeks was associated with an upregulation in PepT1 mRNA as compared to fish that consumed the same diet without dipeptide supplementation (Ostaszewska et al., 2010). Dipeptide supplementation also improved growth performance relative to non-supplemented fish.

Conclusions and Implications

In conclusion, results from these studies all show that gene expression of nutrient transporters in birds, mammals, and fish is influenced by development, environmental factors, dietary composition, as well as anatomical location in the gut. Transcriptional regulation is responsive to the diet and to the environment, with transporter activity (and hence nutrient absorption) being ultimately controlled by the amount of transporter protein produced and translocated to the cell membrane. With nutrient absorption being a known bottleneck to nutrient assimilation in the animal, mechanisms that enhance uptake are especially important. These data highlight the potential for housing and dietary management to influence nutrient availability and thereby affect overall growth performance. As more data become known about how transporter expression and function are regulated, this information can be used to provide diets that more closely match the digestive and absorptive potential of the animal so that the animal is primed to reach its full genetic potential.

Acknowledgments

We thank Christopher Pritchett for the drawings in the figures and Raymond Fetterer for careful reading of the manuscript.

Keywords: Absorption, amino acids, peptides, monosaccharides, membrane bound transporters, solute carrier gene family, gastrointestinal, diet, environmental stressors, disease, poultry, chickens, turkeys, pigeons, livestock, fish, growth performance

References

Amberg, J.J., C. Myr, Y. Kamisaka, A.E. Jordal, M.B. Rust, R.W. Hardy, R. Koedijk, and I. Ronnestad. 2008. Expression of the oligopeptide transporter, PepT1, in larval Atlantic cod (Gadus morhua). Comp. Biochem. Physiol. Part B, Biochem. Mol. Biol. 150: 177–182.
Bakke, S., A.E. Jordal, P. Gomez-Requeni, T. Verri, K. Kousoulaki, A. Aksnes, and I. Ronnestad. 2010. Dietary protein hydrolysates and free amino acids affect the spatial expression of peptide transporter PepT1 in the digestive tract of Atlantic cod (*Gadus morhua*). Comp. Biochem. Physiol. Part B, Biochem. Mol. Biol. 156: 48–55.

Barri, A., C.F. Honaker, J.R. Sottosanti, R.M. Hulet, and A.P. McElroy. 2011. Effect of incubation temperature on nutrient transporters and small intestine morphology of broiler chickens. Poult. Sci. 90: 118–125.

Bodoy, S., D. Fotiadis, C. Stoeger, Y. Kanai, and M. Palacin. 2013. The small SLC43 family: Facilitator system L amino acid transporters and the orphan EEG1. Mol. Asp. Med. 34: 638–645.

Boudry, G., V. Rome, C. Perrier, A. Jamin, G. Savary, and I. Le Huerou-Luron. 2014. A high-protein formula increases colonic peptide transporter 1 activity during neonatal life in low-birth-weight piglets and disturbs barrier function later in life. Br. J. Nutr. 112: 1073–1080.

Brosnan, J.T., and M.E. Brosnan. 2013. Glutamate: A truly functional amino acid. Amino Acids 45: 413–418.

Bucking, C., and P.M. Schulte. 2012. Environmental and nutritional regulation of expression and function of two peptide transporter (PepT1) isoforms in a euryhaline teleost. Comp. Biochem. and Physiol. Part A Mol. Int. Physiol. 161: 379–387.

Cao, M., L. Che, J. Wang, M. Yang, G. Su, Z. Fang, Y. Lin, S. Xu, and D. Wu. 2014. Effects of maternal over- and undernutrition on intestinal morphology, enzyme activity, and gene expression of nutrient transporters in newborn and weaned pigs. Nutrition 30: 1442–1447.

Carriga, C., R.R. Hunter, C. Amat, J.M. Planas, M.A. Mitchell, and M. Moreto. 2005. Heat stress increases apical glucose transport in the chicken jejunum. Am. J. Physiol. Regul. Integr. Comp. Physiol. 290: R195–R201.

Chen, H., Y.X. Pan, E.A. Wong, and K.E. Webb, Jr. 2005. Dietary protein level and stage of development of an intestinal peptide transporter (cPepT1) in chickens. J. Nutr. 135: 193–198.

Chen, M., X. Li, J. Yang, C. Gao, B. Wang, X. Wang, and H. Yan. 2015. Growth of embryo and gene expression of nutrient transporters in the small intestine of the domestic pigeon (*Columbia livia*). J. Zhejiang Univ. Sci. B. 16: 511–523.

Coudert, E., G. Pascal, J. Dupont, J. Simon, E. Cailleau-Audouin, S. Crochet, M.J. Duclos, S. Tesseraud, and S. Metayer-Coustard. 2015. Phylogenesis and biological characterization of a new glucose transporter in the chicken (*Gallus gallus*), GLUT12. PLoS One 10(10): e0139517. doi:10.1371/journal.pone.0139517.

Dalloul, R.A., and H.S. Lillehoj. 2006. Poultry coccidiosis: Recent advancements in control measures and vaccine development. Expert. Rev. Vaccines 5: 143–163.

de Oliveira, J.E., S. Druyan, Z. Uni, C.M. Ashwell, and P.R. Ferket. 2009. Prehatch intestinal maturation of turkey embryos demonstrated through gene expression patterns. Poult. Sci. 88: 2600–2609.

Dong, X.Y., Y.M. Wang, L. Dai, M.M.M. Azzam, C. Wang, and X.T. Zou. 2012a. Posthatch development of intestinal morphology and digestive enzyme activities in domestic pigeons (*Columbia livia*). Poult. Sci. 91: 1886–1892.

Dong, X.Y., Y.M. Wang, C. Yuan, and X.T. Zou. 2012b. The ontogeny of nutrient transporter and digestive enzyme gene expression in domestic pigeon (*Columba livia*) intestine and yolk sac membrane during pre- and posthatch development. Poult. Sci. 91: 1974–1982.

Donowitz, M., C.M. Tse, and D. Fuster. 2013. SLC9/NHE gene family, a plasma membrane and organeller family of Na^+/H^+ exchangers. Mol. Asp. Med. 34: 236–251.

Duarte, C.R.A., M.L.M. Vicentini-Paulino, J. Buratini, A.C.S. Castillo, and D.F. Pinheiro. 2011. Messenger ribonucleic acid abundance of intestinal enzymes and transporters in feed-restricted and refed chickens at different ages. Poult. Sci. 90: 863–868.

Feng, D.Y., X.Y. Zhou, J.J. Zuo, C.M. Zhang, Y.L. Yin, X.Q. Wang, and T. Wang. 2008. Segmental distribution and expression of two heterodimeric amino acid transporter mRNAs in the intestine of pigs during different ages. J. Sci. Food. Agr. 88: 1012–1018.

Fetterer, R.H., K.B. Miska, M.C. Jenkins, and E.A. Wong. 2014. Expression of nutrient transporters in duodenum, jejunum and ileum of *Eimeria maxima*-infected broiler chickens. Parasitol. Res. 113: 3891–3894.

Forster, I.C., N. Hernando, J. Biber, and H. Murer. 2013. Phosphate transporters of the SLC20 and SLC34 families. Mol. Asp. Med. 34: 386–395.

Fotiadis, D., Y. Kanai, and M. Palacin. 2013. The SLC3 and SLC7 families of amino acid transporters. Mol. Asp. Med. 34: 139–158.

Georgieva, N.V., M. Gabrashanska, V. Koinarski, and Z. Yaneva. 2011. Zinc supplementation against *Eimeria acervulina*-induced oxidative damage in broiler chickens. Vet. Med. Intl. 2011: 647124. doi 10.4061/2011/647124.

Gilbert, E.R., H. Li, D.A. Emmerson, K.E. Webb, and E.A. Wong. 2007. Developmental regulation of nutrient transporter and enzyme mRNA abundance in the small intestine of broilers. Poult. Sci. 86: 1739–1753.

Gilbert, E.R., E.A. Wong, and K.E. Webb, Jr. 2008a. Board-invited review: Peptide absorption and utilization: Implications for animal nutrition and health. J. Animal Sci. 86: 2135–2155.

Gilbert, E.R., H. Li, D.A. Emmerson, K.E. Webb, and E.A. Wong. 2008b. Dietary protein quality and feed restriction influence abundance of nutrient transporter mRNA in the small intestine of broiler chicks. J. Nutr. 138: 262–271.

Gilbert, E.R., H. Li, D.A. Emmerson, K.E. Webb, and E.A. Wong. 2010. Dietary protein composition influences abundance of peptide and amino acid transporter mRNA in the small intestine of two lines of broiler chicks. Poult. Sci. 89: 1663–1676.

Goncalves, A.F., L.F. Castro, C. Pereira-Wilson, J. Coimbra, and J.M. Wilson. 2007. Is there a compromise between nutrient uptake and gas exchange in the gut of *Misgurnus anguillicaudatus*, an intestinal air-breathing fish? Comp. Biochem. Physiol. Part D, Genomics Proteomics. 2: 345–355.

Hakim, Y., S. Harpaz, and Z. Uni. 2009. Expression of brush border enzymes and transporters in the intestine of European sea bass (*Dicentrarchus labrax*) following food deprivation. Aquaculture 290: 110–115.

Hediger, M.A., B. Clemencon, R.E. Burtrier, and E.A. Bruford. 2013. The ABCs of membrane transporters in health and disease (SLC) series: Introduction. Mol. Aspects Med. 34: 95–107.

Huang, L., and S. Tepaamorndech. 2013. The SLC30 family of zinc transporters-A review of current understanding of their biological and pathophysiological roles. Mol. Asp. Med. 34: 548–560.

Humphrey, B.D., C.B. Stephensen, C.C. Calvert, and K.C. Klasing. 2004. Glucose and cationic amino acid transporter expression in growing chickens (*Gallus gallus domesticus*). Comp. Biochem. Physiol. Part A. 138: 515–525.

Humphrey, B.D., C.B. Stephensen, C.C. Calvert, and K.C. Klasing. 2006. Lysine deficiency and feed restriction independently alter cationic amino acid transporter expression in chickens (*Gallus gallus domesticus*) Comp. Biochem. Physiol. Part A 143: 218–227.

Humphrey, B.D., S. Kirsch, and D. Morris. 2008. Molecular cloning and characterization of the chicken cationic amino acid transporter-2 gene. Comp. Biochem. Physiol. Part B 150: 301–311.

Ihara, T., T. Tsujikawa, Y. Fujiyama, and T. Bamba. 2000. Regulation of PepT1 peptide transporter expression in the rat small intestine under malnourished conditions. Digestion 61: 59–67.

Kanai, Y., B. Clemencon, A. Simonion, M. Leuenberger, M. Lochner, M. Weisstanner, and M.A. Hediger. 2013. The SLC1 high-affinity glutamate and neutral amino acid transporter family. Mol. Asp. Med. 34: 108–120.

Kirsch, S., and B.D. Humphrey. 2010. Functional characterization of the chicken cationic amino acid transporter-2 isoforms. Comp. Biochem. Physiol. Part B 156: 279–286.

Kluth, H., and M. Rodehutscord. 2006. Comparison of amino acid digestibility in broiler chickens, turkeys, and Pekin ducks. Poult. Sci. 85: 1953–1960.

Koven, W., and P. Schulte. 2012. The effect of fasting and refeeding on mRNA expression of PepT1 and gastrointestinal hormones regulating digestion and food intake in zebrafish (*Danio rerio*). Fish Physiol. Biochem. 38: 1565–1575.

Li, X.G., X. Chen, and X.Q. Wang. 2013. Changes in relative organ weights and intestinal transporter gene expression in embryos from white Plymouth Rock and WENS Yellow Feather chickens. Comp. Biochem. Physiol. Part A 164: 368–375.

Liao, S.F., D.L. Harmon, E.S. Vanzant, K.R. McLeod, J.A. Boling, and J.C. Matthews. 2010. The small intestinal epithelia of beef steers differentially express sugar transporter messenger ribonucleic acid in response to abomasal versus ruminal infusion of starch hydrolysate. J. Animal Sci. 88: 306–314.

Lim, C.H., W. Jeong, W. Lim, J. Kim, G. Song, and F.W. Bazer. 2012. Differential expression of select members of the SLC family of genes and regulation of expression by microRNAs in the chicken oviduct. Biol. Reprod. 87: 1–9.

Liu, Z., Y. Zhou, J. Feng, S. Lu, Q. Zhao, and J. Zhang. 2013. Characterization of oligopeptide transporter (PepT1) in grass carp (*Ctenopharyngodon idella*). Comp. Biochem. Physiol. Part B, Biochem. Mol. Biol. 164: 194–200.

Madsen, S.L., and E.A. Wong. 2011. Expression of the chicken peptide transporter 1 and the peroxisome proliferator-activated receptor α following feed restriction and subsequent refeeding. Poult. Sci. 90: 2295–2300.

Metzler-Zebeli, B.U., M. Hollmann, S. Sabitzer, L. Podstatzky-Lichtenstein, D. Klein, and Q. Zebeli. 2013. Epithelial response to high-grain diets involves alteration in nutrient transporters and Na$^+$/K$^+$-ATPase mRNA expression in rumen and colon of goats. J. Anim. Sci. 91: 4256–4266.

Miska, K.B., R.H. Fetterer, and E.A. Wong. 2014. The mRNA expression of amino acid transporters, aminopeptidase N, and the di- and tri-peptide transporter PepT1 in the embryo of the domesticated chicken (*Gallus gallus*) shows developmental regulation. Poult. Sci. 93: 2262–2270.

Miska, K.B., R.H. Fetterer, and E.A. Wong. 2015. The mRNA expression of amino acid transporters, aminopeptidase, and the di- and tri-peptide transporter PepT1 in the intestine and liver of post-hatch broiler chicks. Poult. Sci. 94: 1323–1332.

Morales, A., M.A. Barrera, A.B. Araiza, R.T. Zijlstra, H. Bernal, and M. Cervantes. 2013. Effect of excess levels of lysine and leucine in wheat-based, amino acid-fortified diets on the mRNA expression of two selected cationic amino acid transporters in pigs. J. Anim. Physiol. Anim. Nutr. (Berl) 97: 263–270.

Morales, A., L. Buenabad, G. Castillo, N. Arce, B.A. Araiza, J.K. Htoo, and M. Cervantes. 2015. Low-protein amino acid-supplemented diets for growing pigs: Effect on expression of amino acid transporters, serum concentration, performance, and carcass composition. J. Anim. Sci. 93: 2154–2164.

Moran, E.T., Jr. 2007. Nutrition of the developing embryo and hatchling. Poult. Sci. 86: 1043–1049.

Mott, C.R., P.B. Siegel, K.E. Webb, and E.A. Wong. 2008. Gene expression of nutrient transporters in the small intestine of chickens from lines divergently selected for high or low juvenile body weight. Poultry Sci. 87: 2215–2224.

Mueckler, M., and B. Thorens. 2013. The SLC2 (GLUT) family of membrane transporters. Mol. Asp. Med. 34: 121–138.

Nosworthy, M.G., R.F. Bertolo, and J.A. Brunton. 2013. Ontogeny of dipeptide uptake and peptide transporter 1 (PepT1) expression along the gastrointestinal tract in the neonatal Yucatan miniature pig. Br. J. Nutr. 110: 275–281.

Ostaszewska, T., K. Dabrowski, M. Kamaszewski, P. Grochowski, T. Verri, M. Rzepkowska, and J. Wolnicki. 2010. The effect of plant protein-based diet supplemented with dipeptide or free amino acids on digestive tract morphology and PepT1 and PepT2 expressions in common carp (*Cyprinus carpio* L.). Comp. Biochem. Physiol. Part A Mol. Integr. Physiol. 157: 158–169.

Paris, N.E., and E.A. Wong. 2013. Expression of digestive enzymes and nutrient transporters in the intestine of *Eimeria maxima* infected chickens. Poult. Sci. 92: 1331–1335.

Patten, B.M. 1971. Early Embryology of the Chick, 5th ed. McGraw Hill Book Co., New York.

Poole, C.A., E.A. Wong, A.P. McElroy, H.P. Veit, and K.E. Webb. 2003. Ontogenesis of peptide transport and morphological changes in the ovine gastrointestinal tract. Small Ruminant Res. 50: 163–176.

Pramod, A.B., J. Foster, L. Carvelli, and L.K. Henry. 2013. SLC6 transporters: Structure, function, regulation, disease association and therapeutics. Mol. Asp. Med. 34: 197–219.

Rimoldi, S., E. Bossi, S. Harpaz, A.G. Cattaneo, G. Bernardini, M. Saroglia, and G. Terova. 2015. Intestinal B(0)AT1 (SLC6A19) and PEPT1 (SLC15A1) mRNA levels in European sea bass (*Dicentrarchus labrax*) reared in fresh water and fed fish and plant protein sources. J. Nutr. Sci. 4: e21.

Romano, A., A. Barca, C. Storelli, and T. Verri. 2014. Teleost fish models in membrane transport research: The PEPT1(SLC15A1) H$^+$-oligopeptide transporter as a case study. J. Physiol. 592: 881–897.

Ronnestad, I., P.J. Gavaia, C.S. Viegas, T. Verri, A. Romano, T.O. Nilsen, A.E. Jordal, Y. Kamisaka, and M.L. Cancela. 2007. Oligopeptide transporter PepT1 in Atlantic cod (*Gadus morhua* L.): Cloning, tissue expression and comparative aspects. J. Exp. Biol. 210: 3883–3896.

Ronnestad, I., K. Murashita, G. Kottra, A.E. Jordal, S. Narawane, C. Jolly, H. Daniel, and T. Verri. 2010. Molecular cloning and functional expression of atlantic salmon peptide transporter 1 in *Xenopus* oocytes reveals efficient intestinal uptake of lysine-containing and other bioactive di- and tripeptides in teleost fish. J. Nutr. 140: 893–900.

Schioth, H.B., S. Roshanbin, M.G.A. Hagglund, and R. Frederiksson. 2013. Evolutionary origin of amino acid transporter families SLC32, SLC36 and SLC38 and physiological, pathological and therapeutic aspects. Mol. Asp. Med. 34: 571–585.

Seki, Y., K. Sato, T. Kono, H. Abe, and Y. Akiba. 2003. Broiler chickens (Ross strain) lack insulin responsive glucose transporter GLUT4 and have GLUT8 cDNA. Gen. Comp. Endocrinol. 133: 80–87.

Smith, D.E., B. Clemencon, and M.A. Hediger. 2013. Proton-coupled oligopeptide transporter family SLC15: Physiological, pharmacological and pathological implications. Mol. Asp. Med. 34: 323–336.

Speier, J.S., L. Yadgary, Z. Uni, and E.A. Wong. 2012. Gene expression of nutrient transporters and digestive enzymes in the yolk sac membrane and small intestine of the developing embryonic chick. Poult. Sci. 91: 1941–1949.

Su, S., K.B. Miska, R.H. Fetterer, M.C. Jenkins, and E.A. Wong. 2014. Expression of digestive enzymes and nutrient transporters in *Eimeria acervulina* challenged layers and broilers. Poult. Sci. 93: 1217–1226.

Su, S., K.B. Miska, R.H. Fetterer, M.C. Jenkins, and E.A. Wong. 2015. Expression of digestive enzymes and nutrient transporters in *Eimeria*-challenged broilers. Exp. Parasitol. 150: 13–21.

Sun, X., H. Zhang, A. Sheikhahmadi, Y. Wang, H. Jiao, H. Lin, and Z. Song. 2015. Effects of heat stress on the gene expression of nutrient transporters in the jejunum of broiler chickens (*Gallus gallus domesticus*). Int. J. Biometereol. 5: 127–135.

Terova, G., S. Cora, T. Verri, S. Rimoldi, G. Bernardini, and M. Saroglia. 2009. Impact of feed availability on PepT1 mRNA expression levels in sea bass (*Dicentrarchus labrax*). Aquaculture 294: 288–299.

Terova, G., L. Robaina, M. Izquierdo, A. Cattaneo, S. Molinari, G. Bernardini, and M. Saroglia. 2013. PepT1 mRNA expression levels in sea bream (*Sparus aurata*) fed different plant protein sources. SpringerPlus 2: 17.

Thamotharan, M., S.Z. Bawani, X. Zhou, and S.A. Adibi. 1999. Functional and molecular expression of intestinal oligopeptide transporter (PepT-1) after a brief fast. Metabolism 48: 681–684.

Thomas-Delloye, V., F. Marmonier, C. Duchamp, B. Pichon-Georges, J. Lachuer, H. Barre, and G. Crouzoulon. 1999. Biochemical and functional evidences for a GLUT-4 homologous protein in avian skeletal muscle. Am. J. Physiol. 277: R1733–R1740.

Tian, J., G. He, K. Mai, and C. Liu. 2015. Effects of postprandial starvation on mRNA expression of endocrine-, amino acid and peptide transporter-, and metabolic enzyme-related genes in zebrafish (*Danio rerio*). Fish Physiol. Biochem. 41: 773–787.

Verri, T., G. Kottra, A. Romano, N. Tiso, M. Peric, M. Maffia, M. Boll, F. Argenton, H. Daniel, and C. Storelli. 2003. Molecular and functional characterisation of the zebrafish (*Danio rerio*) PEPT1-type peptide transporter. FEBS Lett. 549: 115–122.

Wang, W., C. Shi, J. Zhang, W. Gu, T. Li, M. Gen, W. Chu, R. Huang, Y. Liu, Y. Hou, L. Peng, and Y. Yulong. 2009. Molecular cloning, distribution and ontogenetic expression of the oligopeptide transporter PepT1 mRNA in Tibetan suckling piglets. Amino Acids 37: 593–601.

Weintraut, M., S. Kim, R. Dalloul, and E.A. Wong. 2016. Expression of small intestinal nutrient transporters in embryonic and posthatch turkeys. Poult. Sci. 95: 90–98. doi: 10.3382/ps/pev310.

Wright, E.M. 2013. Glucose transport families SLC5 and SLC50. Mol. Asp. Med. 34: 183–196.

Yadgary, L., E.A. Wong, and Z. Uni. 2014. Temporal transcriptome analysis of the chicken embryo yolk sac. BMC Genomics 15: 690 doi:10.1186/1471-2164-15-690.

Yin, H., L.H. Sumners, R.A. Dalloul, K.B. Miska, R.H. Fetterer, M.C. Jenkins, Q. Zhu, and E.A. Wong. 2015. Expression of an antimicrobial peptide, digestive enzymes and nutrient transporters in the intestine of *E. praecox*-infected chickens. Poult. Sci. 94: 1521–1526.

Zeng, P., X. Li, X. Wang, D. Zhang, G. Shu, and Q. Luo. 2011. The relationship between gene expression of cationic and neutral amino acid transporters in the small intestine of chick embryos and chick breed, development, sex, and egg amino acid concentration. Poult. Sci. 90: 2548–2556.

Zwarycz, B., and E.A. Wong. 2013. Expression of the peptide transporters, PepT1, PepT2, and PHT1 in the embryonic and post-hatch chick. Poult. Sci. 92: 1314–1321.

CHAPTER-12

Novel Peptides in Poultry
A Case Study of the Expanding Glucagon Peptide Superfamily in Chickens (*Gallus gallus*)

Yajun Wang

||||||||||||||||||||||||||||||||||||||||||||||||||||||||||||||||||||||||||||||||||||||||||||||||||||||||||||||||||||||||||||||||||||||||||||||||||||||||||

INTRODUCTION

With the technological advancement which has allowed genome sequencing at lower costs, higher throughputs and accuracy, draft genomes of many invertebrate and vertebrate species including birds and mammals, have been sequenced (Lander et al., 2001; International Chicken Genome Sequencing, 2004). Based on these genome data, the ancestor of vertebrates is hypothesized to have experienced two rounds of genome duplication events (the "2R" hypothesis) ~520–550 million years ago (mya) which led to a higher number of genes in modern vertebrates including birds. Moreover, a third round of genome duplication event (the "3R" hypothesis) occurred in the ray-finned lineage ~350 mya which resulted in a further increase in number of genes in teleost fish (Van de Peer et al., 2009). Genome duplication, acting together with gene loss, gene innovation, and local gene duplication, have likely shaped the genomes of modern vertebrates and generated richer species diversity with divergent phenotypes, which may be beneficial for adaptation to the environment (Van de Peer et al., 2009). At present, the genomes of many birds have been sequenced, including chickens, turkeys (*Meleagris gallopavo*), and ducks (*Anas platyrhynchos*) (Huang et al., 2013). These species are of great importance to poultry

Key laboratory of Bio-resources and Eco-environment of Ministry of Education, College of Life Sciences, Sichuan University, Chengdu, 610064, PR China.
Email: cdwyjhk@gmail.com

industries and biomedical research. Their genomes greatly facilitate the identification and characterization of novel genes which likely affect important physiological processes and phenotypic traits of poultry species, such as growth, body composition, egg laying, and feather development.

Based on its draft genome, chickens were estimated to have more than 20,000 genes. Some of them encode bioactive peptides, which mainly act as hormones to regulate many vital physiological processes, such as growth, development, metabolism, energy balance, stress, reproduction and immunity. Since majority of these peptides have been documented in previous reviews (Scanes and Pierzchala-Koziec, 2014; Denbow and Cline, 2015; Scanes, 2015), therefore we will briefly introduce the peptides of glucagon superfamily in chickens in this chapter. Particular emphasis will be placed on glucagon-like peptide (GCGL) and secretin-like peptide (SCT-LP), which are two novel peptides recently identified and characterized in our laboratory (Wang et al., 2012a,b). The identification of novel members in the expanding glucagon superfamily caused by exon/genome duplication in chickens will hopefully facilitate the search for novel members in other gene families in birds and rethink on how they are related to the unique aspects of avian biology.

Glucagon Superfamily Members Co-exist in Chickens and Mammals

Glucagon (GCG) superfamily consists of 10 structurally-related peptides including GCG, glucagon-like peptide 1 (GLP1) and glucagon-like peptide 2 (GLP2), glucose-dependent insulinotropic polypeptide (GIP), secretin (SCT), pituitary adenylate cyclase-activating polypeptide (PACAP), PACAP-related peptide (PRP), vasoactive intestinal polypeptide (VIP), peptide histidine-isoleucine (PHI), growth hormone-releasing hormone (GHRH). They are encoded by six genes (Figure 1), namely *SCT, GHRH, GIP, GCG, PACAP* and *VIP* genes, which were likely generated by genome duplication during vertebrate evolution (Sherwood et al., 2000; Lee et al., 2007).

This repertoire of peptides is conserved between mammals and chickens (Figure 1), including the phenomenon of two (or three) structurally related peptides encoded by a single gene. For instance, *VIP* gene encodes bioactive VIP and PHI (PHM in humans), while *GCG* encodes three bioactive peptides, GCG, GLP1, and GLP2. This is hypothesized to have originated by exon duplication, although the functions these closely related peptides share may vary in tissues, probably due to gene innovation (Sherwood et al., 2000). Characterization of the avian GCG superfamily peptides by our laboratory and others have shown conservation over their structure and general actions between chicken and mammalian GCG superfamily peptides (Figure 1) (Irwin and Wong, 1995; Talbot et al., 1995; Irwin and

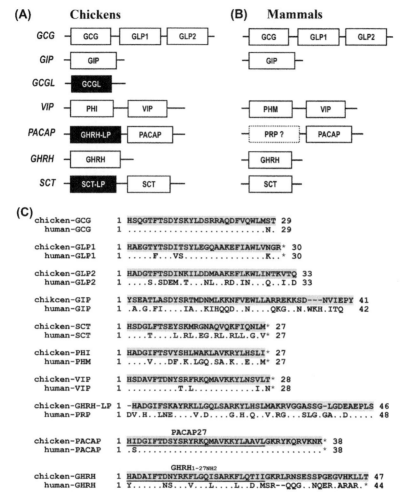

Figure 1: Glucagon superfamily genes and their encoded peptides in chickens and mammals. (A) In chickens, 7 genes (*GCG*, *GIP*, *GCGL*, *VIP*, *PACAP*, *GHRH* and *SCT*) of glucagon superfamily encode 12 bioactive peptides including three peptides unique to chickens, GCGL, GHRH-LP, and SCT-LP (shaded in black). (B) In mammals, 6 genes encode 9 bioactive peptides. PRP is inactive. (C) Sequence comparison of the 10 peptides identified in chickens and humans. PACAP has a short form of 27 amino acids (PACAP27, underlined); Chicken GHRH also has a short bioactive form of 27 amino acids with an amidated C-terminus (GHRH$_{1-27NH2}$, underlined); Asterisk (*) indicates peptide with an amidated C-terminus; Dot indicates the amino acid identical to chicken peptides.

Zhang, 2006; Wang et al., 2006; Wang et al., 2007), such as GHRH-induced GH secretion (Meng et al., 2014), however, actions of these peptides still await further characterization in birds. Of particular interest to note is that unlike PRP in mammals which has no biological activity, PRP (GHRH-LP) is bioactive in chickens, which has been shown to inhibit food intake after intracerebroventricular (icv) injection (Tachibana et al., 2015), possibly via activation of a functional GHRH-LP receptor (GHRH-LPR) expressed in the hypothalamus (Wang et al., 2010).

Novel Members of the Expanding Glucagon Superfamily in Chickens

Discovery of GCGL and SCT-LP in Chickens

Apart from the 10 classic peptides of GCG superfamily and their specific receptors in chickens identified by our and other teams (Table 1) (Wang et al., 2008; Huang et al., 2012; Mo et al., 2014), we recently identified two novel peptides belonging to the GCG superfamily, namely secretin-like peptide (SCT-LP) and glucagon-like peptide (GCGL) respectively, which are absent in mammals probably due to gene/exon loss during evolution (Wang et al., 2012a,b).

Table 1: The receptors for glucagon superfamily peptides identified in chickens.

Peptide	Peptide receptor(s)		Tissue expression of chicken receptors
	Chickens	Mammals	
GCG	GCGR	GCGR	Expressed in liver and other tissues
GLP1	GLP1R	GLP1R	Expressed in GI tract, CNS, and other tissues
GLP2	GLP2R	GLP2R	Expressed in GI tract and other tissues
PACAP	PAC1	PAC1	Widely expressed in CNS and other tissues
	VPAC1	VPAC1	Widely expressed in CNS and other tissues
	VPAC2	VPAC2	Widely expressed in CNS and other tissues
VIP	VPAC1	VPAC1	Widely expressed in CNS and other tissues
	VPAC2	VPAC2	Widely expressed in CNS and other tissues
SCT	SCTR	SCTR	Expressed in GI tract, liver, testis, and pancreas
GIP	GIPR	GIPR	Unknown
GHRH	GHRHR1	GHRHR	Predominantly expressed in pituitary
	GHRHR2	-	Expressed in CNS, pituitary and testis
PHI	VPAC1	?	Widely expressed in CNS and other tissues
	VPAC2	?	Widely expressed in CNS and other tissues
GHRH-LP (PRP)	GHRH-LPR	-	Expressed in the CNS, pituitary, and pancreas
SCT-LP	SCTR		Expressed in GI tract, liver, testis, and pancreas
GCGL	GCGLR	-	Mainly expressed in pituitary, CNS and testis

Notes: Three novel ligand-receptor pairs (GHRH-LP/GHRH-LPR; SCT-LP/SCT-LPR; GCGL/ GCGLR) and a novel GHRH receptor (GHRHR2) found only in chickens are underlined; The receptors for mammalian PHI (PHM) are unknown.

The 34-amino acid (aa) SCT-LP is encoded by chicken secretin (*SCT*) gene, which also encodes the 27-aa SCT (Figure 1). SCT-LP has also been identified in turkeys, ducks, geese and pigeons (*Columba livia*). Interestingly, SCT-LP can be found in *SCT* gene of western painted turtles (*Chrysemys picta bellii*) (XM_005290633) and alligators (*Alligator mississippiensis*) (XM_014604186.1) by searching their genomes. Since avian SCT-LP shares higher amino acid sequence identity (48–56%) to chicken and mammalian SCT than to other GCG superfamily members (Figure 2), it is highly probable that SCT-LP and SCT were generated by an exon duplication event in the avian/reptile lineage (Wang et al., 2012a).

Unlike SCT-LP, GCGL is 29 aa in length and is encoded by a glucagon-like gene (*GCGL*, EU718628), which is likely derived from a genome duplication event during vertebrate evolution (Figure 1). Interestingly, *GCGL* gene is lost in the mammalian lineage based on synteny analyses. Chicken GCGL shares high amino acid sequence identity (66%) with GCG of humans and chickens, but a comparatively low identity to other GCG superfamily members including GLP1 and GLP2. Supportive of our discovery, chicken *GCGL* was reported by others in aliases of exendin gene (Irwin and Prentice, 2011) or glucagon-related peptide gene (*GCRP*) (Park et al., 2013). GCGL has also been identified in other avian species (turkeys, ducks, geese), anole lizards (*Anolis carolinensis*), *Xenopus tropilcalis*, and teleosts (Figure 2).

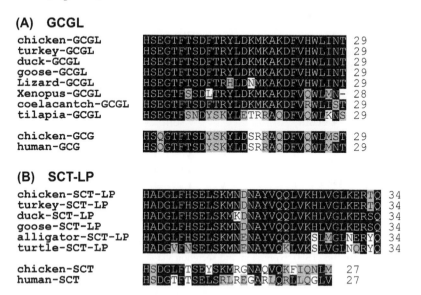

Figure 2: (A) Comparison of chicken GCGL with that of turkey, duck, goose, lizards, *Xenopus*, coelacantch and tilapia, or with chicken and human glucagon (GCG). (B) Comparison of chicken SCT-LP with that of turkey, duck, goose, alligators and turtles, or with that of chicken and human secretin (SCT).

Discovery of the Receptors for Chicken SCT-LP and GCGL

The presence of two novel peptides in chickens encourages us to search for their receptors through data mining and *in vitro* functional analyses. Consequently, we identified the receptors for SCT-LP and GCGL in chickens. SCT-LP potently activates (EC_{50}: 1.10 nM) secretin receptor (SCTR) expressed in Chinese hamster ovary (CHO) cells with high specificity and no cross-reactivity to receptors of other GCG superfamily members. This clearly indicates that SCTR, besides being a receptor for SCT, functions as a receptor specific to SCT-LP (Figure 3; Table 1) (Wang et al., 2012a).

Unlike SCT-LP, GCGL targets a novel member of GPCR B1-subfamily of 430aa, designated as GCGL receptor (GCGLR: EU718627), which shares high amino acid sequence identity (53–55%) to chicken and human glucagon receptor (GCGR). Like *GCGL, GCGLR* is lost in the mammalian lineage. Functional assay confirmed that chicken GCGL potently activates GCGLR (EC_{50}: 0.1 nM) expressed in CHO cells without cross-reactivity towards receptors for GCG, GHRH, VIP and SCT. Like other members of GPCR B1-subfamily, GCGLR activation leads to the activation of cAMP/protein kinase A and Ca^{2+} signaling pathways. All these findings firmly establish the new concept of the novel "GCGL-GCGLR" ligand-receptor pair functioning in chickens, which is absent in mammals (Wang et al., 2012b). This novel ligand-receptor pair has also been identified in genomes of other avian species, including ducks and turkeys (Figure 2).

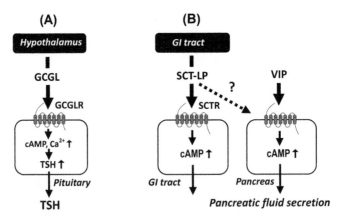

Figure 3: Potential roles of GCGL and SCT-LP in chickens. (A) Hypothalamic GCGL stimulates pituitary thyrotropin (TSH) expression and secretion via activation of GCGLR coupled to cAMP and [Ca^{2+}] pathways, suggesting that it may work with the other two TSH-releasing factors (TRH and CRH) to control TSH secretion in chickens. (B) Like SCT, SCT-LP produced by GI tract may also regulate the functions of GI tract and pancreas, via activation of SCTR in chickens. VIP can potently stimulate pancreatic fluid secretion, however, whether SCT-LP can also exert "secretin-like action" on avian pancreas remains unclear.

Potential Roles of SCT-LP and GCGL in Chickens

The presence of bioactive SCT-LP and GCGL in chickens and its absence in mammals pose a key question regarding their potential "unique" actions in birds. Since *SCT* mRNA transcripts are abundantly expressed in the duodenum, jejunum, ileum and caecum, it points out a possibility that both SCT-LP and SCT may be co-secreted from the GI tract (Wang et al., 2012a). In view of the fact that both SCT and SCT-LP activate SCTR, it is reasonable to speculate that intestine-derived SCT-LP is likely capable of regulating physiological processes analogous to SCT, such as modulation of GI tract activities, via activation of SCTR in the GI tract (Figure 3). Although *SCTR* mRNA is expressed in the pancreas and SCT has been shown to stimulate pancreatic bicarbonate secretion in chickens, chicken SCT is much less effective than VIP in stimulating pancreatic juice secretion in chickens and turkeys (Dimaline and Dockray, 1979; Duke et al., 1987). This indicates that VIP, instead of SCT or SCT-LP, may exert the "secretin-like action" named after the mammalian model in the avian pancreas (Figure 3). Thus, further studies are required to define the precise physiological roles of both SCT-LP/SCT in chicken tissues.

Although GCGL shows a remarkable degree of amino acid sequence identity (66%) with chicken and human GCG, the spatial expression pattern of *GCGL* differs from that of *GCG* significantly. Unlike *GCG* which is expressed in all chicken tissues examined with the highest levels detected in the pancreas and proventriculus (Richards and McMurtry, 2008), *GCGL* is mainly expressed in various brain regions with an abundant level noted in the hypothalamus. Meanwhile, GCGLR is abundantly expressed in the pituitary and brain regions, including the hypothalamus (Wang et al., 2012b). These tissue distribution profiles imply that GCGL mainly functions in the CNS and hypothalamus-pituitary axis. In agreement with these ideas, GCGL has been shown to stimulate TSH expression and secretion both potently and specifically via the activation of GCGLR expressed in the pituitary, suggesting that hypothalamic GCGL is a novel TSH-releasing factor (TRF) which may be involved in controlling growth, development and metabolism in chickens (Figure 3) (Huang et al., 2014). Given that the GCGL-GCGLR ligand-receptor pair only exists in birds, this finding also depicts fundamental differences in the hypothalamic control of the pituitary-thyroid axis between chickens and mammals. In mammals, TSH secretion is principally controlled by a thyrotropin-releasing hormone (TRH) (Morley, 1981), whereas in chickens, TSH secretion is likely controlled by GCGL and the other two TRFs (TRH and corticotrophin-releasing hormone) (De Groef et al., 2005; Huang et al., 2014). In addition to its action on the pituitary, GCGL has also been shown to suppress food intake in chicks after icv administration, suggesting that hypothalamic GCGL may also act as an anorexic peptide to control energy balance in chickens (Honda et al., 2014). However, the underlying mechanisms of GCGL actions on the chicken hypothalamus remain to be clarified.

Conclusion and Perspective

It is well documented that there are nine bioactive peptide hormones in the mammalian GCG superfamily, however, 12 bioactive peptide hormones of GCG superfamily including the two novel peptides, SCT-LP and GCGL, have been identified by us and others (Table 1). These two peptides have been lost in the mammalian lineage during evolution. Our functional studies further revealed that both SCT-LP and GCGL are bioactive and their actions are likely mediated by their specific receptors (SCTR and GCGLR respectively) expressed in chicken tissues, such as GCGL-induced TSH secretion. These preliminary findings support a concept that the novel GCG family members, SCT-LP and GCGL, may play unique roles in chickens. With a larger number of avian genome databases becoming available, it will favor the identification of novel peptides of other gene families, such as the RF-amide peptide family in chickens and other birds (Bechtold and Luckman, 2007). Concurrent with this notion, we recently identified a novel *Carassius* RF-amide peptide (C-RFa) and its two receptors (namely C-RFaR and PrRPR2), which are unique in birds (Wang et al., 2012c). Given the importance of peptides in avian physiology, further extensive studies are required to identify all these novel peptides from avian genomes and characterize their unique actions in birds. In addition, it should be bear in mind that despite the co-existence and sequence conservation of the 10 GCG superfamily peptides between chickens and mammals (Figure 1), some of them display similar-but-non-identical functions between them. For instance, SCT does not exert 'secretin-like action' (as the phenomenon well-characterized in mammals implied) in chickens. Therefore, tremendous efforts should also be paid by avian biologists to elucidate the roles of these seemingly well-studied peptides in birds in future. Undoubtedly, the systematic and comprehensive studies on each peptide family, including its novel member(s), will not only establish a clear molecular basis to address the unique aspects of avian biology and decipher the phenotypic traits of commercial interest in domestic birds such as growth, body composition, egg-laying and disease resistance, but also offer us a 'bird's-eye view' on the structural and functional changes of peptide families during evolution due to the unique evolutionary position of birds in vertebrates.

Acknowledgements

This work was supported by grants from the National Natural Science Foundation of China (30771171, 30871338, 31172202, 31271325). I would like give my special thanks to Dr. Juan Li (College of Life Sciences, Sichuan University) for her constructive suggestions to this manuscript.

Keywords: Chicken, peptide, receptor, Glucagon (GCG) superfamily, GCGL, Secretin, Secretin-like Peptide (SCT-LP), vasoactive intestinal polypeptide, gene duplication, exon duplication

References

Bechtold, D.A., and S.M. Luckman. 2007. The role of RFamide peptides in feeding. J. Endocrinol. 192: 3–15.

De Groef, B., K. Vandenborne, P. Van As, V.M. Darras, E.R. Kuhn, E. Decuypere et al. 2005. Hypothalamic control of the thyroidal axis in the chicken: Over the boundaries of the classical hormonal axes. Domest. Anim. Endocrinol. 29: 104–110.

Denbow, D., and M. Cline. 2015. Food intake regulation. pp. 469–485. *In*: C.G. Scanes (ed.). Sturkie's Avian Physiology, 6th edition. Academic Press, London, UK.

Dimaline, R., and G.J. Dockray. 1979. Potent stimulation of the avian exocrine pancreas by porcine and chicken vasoactive intestinal peptide. J. Physiol. 294: 153–163.

Duke, G.E., K. Larntz, and H. Hunt. 1987. The influence of cholecystokinin, vasoactive intestinal peptide and secretin on pancreatic and biliary secretion in laying hens. Comp. Biochem. Physiol. C 86: 97–102.

Honda, K., T. Saneyasu, T. Yamaguchi, T. Shimatani, K. Aoki, K. Nakanishi et al. 2014. Intracerebroventricular administration of novel glucagon-like peptide suppresses food intake in chicks. Peptides 52: 98–103.

Huang, G., J. Li, H. Fu, Z. Yan, G. Bu, X. He et al. 2012. Characterization of glucagon-like peptide 1 receptor (GLP1R) gene in chickens: functional analysis, tissue distribution, and identification of its transcript variants. Domest. Anim. Endocrinol. 43: 1–15.

Huang, G., C. He, F. Meng, J. Li, J. Zhang, and Y. Wang. 2014. Glucagon-like peptide (GCGL) is a novel potential TSH-releasing factor (TRF) in Chickens: I) Evidence for its potent and specific action on stimulating TSH mRNA expression and secretion in the pituitary. Endocrinology 155: 4568–4580.

Huang, Y., Y. Li, D.W. Burt, H. Chen, Y. Zhang, W. Qian et al. 2013. The duck genome and transcriptome provide insight into an avian influenza virus reservoir species. Nat. Genet. 45: 776–783.

International Chicken Genome Sequencing, C. 2004. Sequence and comparative analysis of the chicken genome provide unique perspectives on vertebrate evolution. Nature 432: 695–716.

Irwin, D.M., and J. Wong. 1995. Trout and chicken proglucagon: Alternative splicing generates mRNA transcripts encoding glucagon-like peptide 2. Mol. Endocrinol. 9: 267–277.

Irwin, D.M., and T. Zhang. 2006. Evolution of the vertebrate glucose-dependent insulinotropic polypeptide (GIP) gene. Comp. Biochem. Physiol. D. Genomics Proteomics 1: 385–395.

Irwin, D.M., and K.J. Prentice. 2011. Incretin hormones and the expanding families of glucagon-like sequences and their receptors. Diabetes Obes. Metab. 13 Suppl. 1: 69–81.

Lander, E.S., L.M. Linton, B. Birren, C. Nusbaum, M.C. Zody, J. Baldwin et al. 2001. Initial sequencing and analysis of the human genome. Nature 409: 860–921.

Lee, L.T., F.K. Siu, J.K. Tam, I.T. Lau, A.O. Wong, M.C. Lin et al. 2007. Discovery of growth hormone-releasing hormones and receptors in nonmammalian vertebrates. Proc. Natl. Acad. Sci. U.S.A. 104: 2133–2138.

Meng, F., G. Huang, S. Gao, J. Li, Z. Yan, and Y. Wang. 2014. Identification of the receptors for somatostatin (SST) and cortistatin (CST) in chickens and investigation of the roles of cSST28, cSST14, and cCST14 in inhibiting cGHRH1-27NH2-induced growth hormone secretion in cultured chicken pituitary cells. Mol. Cell. Endocrinol. 384: 83–95.

Mo, C., Y. Zhong, Y. Wang, Z. Yan, and J. Li. 2014. Characterization of glucagon-like peptide 2 receptor (GLP2R) gene in chickens: functional analysis, tissue distribution, and developmental expression profile of GLP2R in embryonic intestine. Domest. Anim. Endocrinol. 48: 1–6.

Morley, J.E. 1981. Neuroendocrine control of thyrotropin secretion. Endocr. Rev. 2: 396–436.

Park, C.R., M.J. Moon, S. Park, D.K. Kim, E.B. Cho, R.P. Millar et al. 2013. A novel glucagon-related peptide (GCRP) and its receptor GCRPR account for coevolution of their family members in vertebrates. PloS one 8(6): e65420.

Richards, M.P., and J.P. McMurtry. 2008. Expression of proglucagon and proglucagon-derived peptide hormone receptor genes in the chicken. Gen. Comp. Endocrinol. 156: 323–338.

Scanes, C.G., and K. Pierzchala-Koziec. 2014. Biology of the gastro-intestinal tract in poultry. Avian Biol. Res. 7: 193–222.

Scanes, C.G. 2015. Avian endocrine system. pp. 489–496. *In*: C.G. Scanes (ed.). Sturkie's Avian Physiology, 6th edition. Academic Press, London, UK.

Sherwood, N.M., S.L. Krueckl, and J.E. McRory. 2000. The origin and function of the pituitary adenylate cyclase-activating polypeptide (PACAP)/glucagon superfamily. Endocr. Rev. 21: 619–670.

Tachibana, T., I. Sugimoto, M. Ogino, M.S. Khan, K. Masuda, K. Ukena et al. 2015. Central administration of chicken growth hormone-releasing hormone decreases food intake in chicks. Physiol. Behav. 139: 195–201.

Talbot, R.T., I.C. Dunn, P.W. Wilson, H.M. Sang, and P.J. Sharp. 1995. Evidence for alternative splicing of the chicken vasoactive intestinal polypeptide gene transcript. J. Mol. Endocrinol. 15: 81–91.

Van de Peer, Y., S. Maere, and A. Meyer. 2009. The evolutionary significance of ancient genome duplications. Nat. Rev. Genet. 10: 725–732.

Wang, C.Y., Y. Wang, J. Li, and F.C. Leung. 2006. Expression profiles of growth hormone-releasing hormone and growth hormone-releasing hormone receptor during chicken embryonic pituitary development. Poult. Sci. 85: 569–576.

Wang, J., Y. Wang, X. Li, J. Li, and F.C. Leung. 2008. Cloning, tissue distribution, and functional characterization of chicken glucagon receptor. Poult. Sci. 87: 2678–2688.

Wang, Y., J. Li, C.Y. Wang, A.H. Kwok, and F.C. Leung. 2007. Identification of the endogenous ligands for chicken growth hormone-releasing hormone (GHRH) receptor: evidence for a separate gene encoding GHRH in submammalian vertebrates. Endocrinology 148: 2405–2416.

Wang, Y., J. Li, C.Y. Wang, A.Y. Kwok, X. Zhang, and F.C. Leung. 2010. Characterization of the receptors for chicken GHRH and GHRH-related peptides: Identification of a novel receptor for GHRH and the receptor for GHRH-LP (PRP). Domest. Anim. Endocrinol. 38: 13–31.

Wang, Y., G. Huang, J. Li, F. Meng, X. He, and F.C. Leung. 2012a. Characterization of chicken secretin (SCT) and secretin receptor (SCTR) genes: A novel secretin-like peptide (SCT-LP) and secretin encoded in a single gene. Mol. Cell. Endocrinol. 348: 270–280.

Wang, Y., F. Meng, Y. Zhong, G. Huang, and J. Li. 2012b. Discovery of a novel glucagon-like peptide (GCGL) and its receptor (GCGLR) in chickens: Evidence for the existence of GCGL and GCGLR genes in nonmammalian vertebrates. Endocrinology 153: 5247–5260.

Wang, Y., C.Y. Wang, Y. Wu, G. Huang, J. Li, and F.C. Leung. 2012c. Identification of the receptors for prolactin-releasing peptide (PrRP) and Carassius RFamide peptide (C-RFa) in chickens. Endocrinology 153: 1861–1874.

Index

〡〡

Printed and bound by CPI Group (UK) Ltd, Croydon, CR0 4YY

01/11/2024

01782624-0013